A FLORA OF
SAN NICOLAS ISLAND
CALIFORNIA

A Flora of
San Nicolas Island
California

Steve Junak

Illustrated by
Linda Ann Vorobik

Santa Barbara Botanic Garden
Santa Barbara, California
2008

© 2008 Santa Barbara Botanic Garden
1212 Mission Canyon Road
Santa Barbara, CA 93105

Graphic design, typesetting, and cover design by Katey O'Neill.

Front cover photographs (by Steve Junak):
Top: Daytona Beach, southeastern end of island, 18 April 1995
Center Left: Coastline near Rock Crusher, southwestern end of island, 17 May 1995
Center Right: Dudleya virens subsp. insularis, Live-forever Canyon, 13 June 1995
Bottom: Lower portion of Twin Rivers drainage, south side of island, 12 May 1992

Back cover photographs (by Steve Junak):
Top: Northeastern coastal area near Rock Jetty, 15 June 1995
Bottom: Northeastern coastal area near Rock Jetty, 19 April 1995

ISBN 978-0-916436-06-3

CONTENTS

A PRINTABLE MAP OF PLACE NAMES ON SAN NICOLAS ISLAND CAN BE DOWNLOADED AT
HTTP://WWW.SBBG.ORG/SANNICMAP

DEDICATED TO

Juana Maria and the Nicoleño people
Island explorer Blanche Trask
The sheep ranching families of San Nicolas Island

BENEFACTORS

This book would not have been possible without the generous financial support of

The Santa Cruz Island Foundation
The Legacy Program of the U.S. Navy

ACKNOWLEDGEMENTS

For allowing me to visit the study site and providing transportation to and from the mainland, I am most grateful to the Commanding Officers of the Naval Air Weapons Station, Point Mugu, and of Outlying Landing Field San Nicolas Island. I owe many thanks to Navy staff on San Nicolas Island and on the mainland for their help with transportation, lodging, and access.

The U.S. Navy's Environmental Division (Natural Resources Office) at Naval Base Ventura County provided invaluable logistical support. Special thanks are due to Ron Dow, Julie Vanderwier, Joe DiVittorio, and Matt Klope for many, many favors during my early visits to the island, and to Grace Smith, Steve Schwartz, Sandee Harvill, Tom Keeney, Ron Barrett, Lisa Thomas, and Gina Smith for their help and support during later visits.

I gratefully acknowledge the assistance of Michael Benedict, Fred Boutin, Ronilee Clark, Pat Corry, Marla Daily, Karen Danielsen, Stan Davis, Charles Drost, Jack Engle, Glenn Gorelick, Bill Halvorson, Tom Hesseldenz, Lyndal Laughrin, Keri Kirkland, Gail Meadows Milliken, Tom Murphey, Peter Raven, Peter Schuyler, Cathy Schwemm, Tim Thomas, Bob Thorne, Julie Vanderwier, Gary Wallace, Howie Wier, and especially Ron Foreman, Ralph Philbrick, and Grace Smith, who have shared collection information and observations on the island. David Hollombe kindly shared biographical information on several of the island's plant collectors. Judy Gibson provided information on botanical specimens collected by Norman Bilderback. Updated weather data for the island was provided by Charles Fisk.

For sharing their amazing historical photographs and their memories of the ranching era on San Nicolas Island, I am extremely grateful to Jeanne Lamberth Hess and Frances "Frankie" Agee Beard. Marla Daily of the Santa Cruz Island Foundation graciously allowed me to publish some of Agee Family photographs in this book, consult unpublished manuscripts written by Frankie Agee, and shared historical information from her unpublished encyclopedia and the Foundation's files. Brian Burd of the Santa Cruz Island Foundation scanned the Agee photographs and also shared his enhanced versions of the Blanche Trask photographs.

The hospitality and generous assistance provided by the curators and collections managers at the Field Museum, Harvard University Herbaria, Jepson Herbarium, Missouri Botanical Garden, New York Botanical Garden, Rancho Santa Ana Botanic Garden, San Diego Natural History Museum, Smithsonian Institution, and University of California at Berkeley made my trips to study their collections a real pleasure. I am also very grateful to the Arizona Historical Society, the Jepson Herbarium, and the Ventura County Museum of History and Art for sharing historical photographs and manuscripts in their collections.

Dieter Wilken and Bob Muller reviewed and proofread the manuscript and provided many helpful comments and corrections. Katey O'Neill also proofread the manuscript and provided valuable editorial comments and suggestions for improvement.

I sincerely thank Linda Ann Vorobik for her beautiful line drawings and Katey O'Neill for her skillful preparation of the photographs for publication and for designing the entire book from cover to cover. Don Matsumoto's help with the publication process is also much appreciated.

This project would not have been possible without the efforts of Bill Halvorson, Nancy Johnson, and Tom Keeney to obtain funding for this book and the financial support provided by the Santa Cruz Island Foundation, the U.S. Navy's Legacy Program, and the Santa Barbara Botanic Garden.

INTRODUCTION

THE CALIFORNIA CHANNEL ISLANDS

Off the coast of southern California are eight islands known for their natural beauty, windswept landscapes, rugged coastlines, and unspoiled beaches that are often teeming with marine mammals and birds. The islands' plant and animal life, rich in endemic species, have attracted the attention of biologists and horticulturists for more than a century.

The California Channel Islands, often shrouded in coastal fog but sometimes clearly visible from the mainland, have long captured the interest and imagination of southern Californians, beginning with the aboriginal inhabitants. The water barrier around the Channel Islands has discouraged casual human visitation, and several of the islands, relatively pristine and undeveloped, have preserved the ambiance of 19th century rural California.

The California Channel Islands are strikingly different from each other with respect to climate, geology, history, land forms, maximum elevation, size, and vegetation. The eight islands (Figure 1) have been divided into 2 major groups: 1) the Northern Channel Islands of San Miguel, Santa Rosa, Santa Cruz, and Anacapa, and 2) the Southern Channel Islands of Santa Barbara, San Nicolas, Santa Catalina, and San Clemente. The islands represent emergent portions of a complex system of submarine canyons and ridges in a geomorphic province referred to as the California Continental Borderland (Vedder 1976). They are situated between 32° 48' and 34° 05' N latitude and between 118° 21' and 120° 27' W longitude, and range in size from 1 to 96 square miles (2.6 to 249 square km) (Table 1). Maximum elevations vary between 635 feet (194 m) and 2,470 feet (753 m).

The Northern Channel Islands. Geographically, the Northern Channel Islands form a relatively cohesive group of "fringing islands" (Carlquist 1974) about 62 miles (100 km) long. They are a westward extension of the Santa Monica Mountains and are separated from the mainland by the Santa Barbara Channel. Distances from the mainland vary from 13 to 27 miles (20 to 44 km). Channels between individual islands range from 3 to 6 miles (5 to 9 km) in width. All four islands are oriented with their longest axes in an east-west direction. San Miguel, Santa Rosa, and Anacapa islands are part of Channel Islands National Park, although San Miguel is owned by the U.S. Navy. The western 75% of Santa Cruz Island is owned by The Nature Conservancy, an international nonprofit conservation organization, while the eastern 25% of the island is owned by the National Park Service. The Nature Conservancy and the National Park Service manage the island cooperatively. The University of California has included Santa Cruz Island in its Natural Reserve System and maintains a field station there, but is not a landowner.

The Southern Channel Islands. The Southern Channel Islands are more scattered geographically and more isolated from each other and from the mainland than the northern ones. Their longest axes are oriented in a north-south or northwest-southeast direction. They are separated from the mainland by the San Pedro Channel and are located 20 to 61 miles (32 to 98 km) from the mainland. Distances between individual islands range from 21 to 28 miles (34 to 45 km). Santa Barbara Island is part of Channel Islands National Park. San Nicolas and San Clemente islands are owned by the U.S. Navy and are closed to the public. Most of Santa Catalina Island is owned by the Santa Catalina Island Conservancy, a nonprofit conservation foundation, and the Santa Catalina Island Company. There also are private holdings at the city of Avalon.

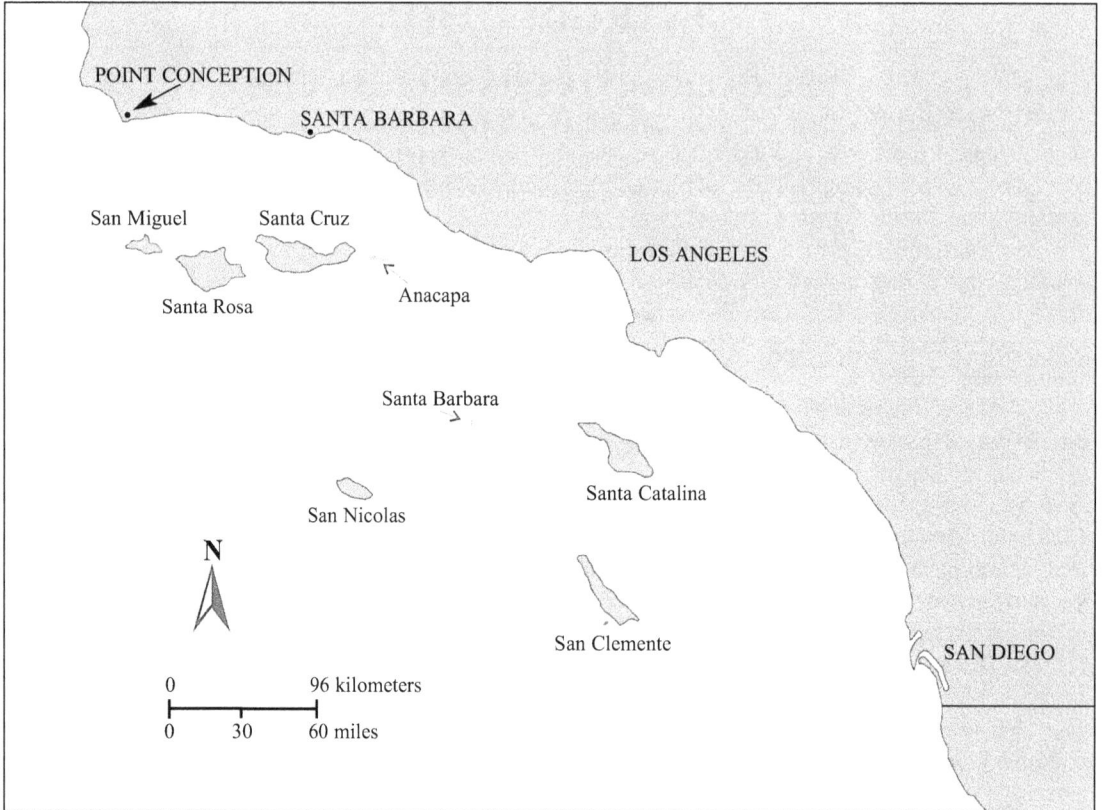

Figure 1. The California Channel Islands and coastline of southern California.

TABLE 1. Physical Characteristics of the California Channel Islands

	Area mi²	(km²)	Highest Elevation ft	(m)	Distance to Mainland mi	(km)
Northern Channel Islands						
San Miguel	14	(37)	830	(253)	26	(42)
Santa Rosa	84	(217)	1,589	(484)	27	(44)
Santa Cruz	96	(249)	2,470	(753)	19	(30)
Anacapa	1.1	(2.9)	930	(283)	13	(20)
Southern Channel Islands						
Santa Barbara	1	(2.6)	635	(194)	38	(61)
San Nicolas	22	(58)	910	(277)	61	(98)
Santa Catalina	75	(194)	2,125	(648)	20	(32)
San Clemente	56	(145)	1,965	(599)	49	(79)

REGIONAL GEOLOGIC AND TOPOGRAPHIC HISTORY

Vedder and Howell (1980) reviewed the geologic and topographic history of the southern California coastal region and the California Channel Islands. They reported that past sea-level fluctuations greatly affected size and shape of the Channel Islands. Large areas on Santa Rosa, Santa Cruz, Santa Catalina, and San Clemente islands apparently have been continuously above sea level for the last 500,000 years. In contrast, San Nicolas and Santa Barbara (and possibly San Miguel and Anacapa) islands were completely submerged during the maximum extent of the ocean, probably during one of the interglacial periods of the Pleistocene Epoch (which lasted from about 1.8 million to 10,000 years ago).

During the time of probable minimum sea levels about 17,000 to 18,000 years ago, the four present-day Northern Channel Islands apparently were all part of one large landmass, referred to as "Santarosae" (Orr 1968; Vedder and Howell 1980). This landmass has been estimated to have been approximately 724 square miles (1874 square km) in size. The four Southern Channel Islands were larger than they are today but probably were still separated from each other by ocean channels.

Although the four Northern Channel Islands are geographically and geologically aligned with the Santa Monica Mountains, there currently are no geologic or marine geophysical data that suggest the presence of a Pleistocene Epoch land bridge between the mainland and the islands. Scientists assumed that a land bridge connected "Santarosae" to the mainland, primarily because mammoth fossils were found on the islands and because it was believed that elephants did not or could not swim. Studies on the bathymetry of the channel floor have shown, however, that an ocean channel separated "Santarosae" from the mainland during the late Pleistocene Epoch (Junger and Johnson 1980, Vedder and Howell 1980). Furthermore, studies of modern elephants (Johnson 1980, Wenner and Johnson 1980) show that they are capable of swimming across channels as wide as the one that existed in the late Pleistocene.

SAN NICOLAS ISLAND

Over the years, San Nicolas Island (previously known as San Nicholas Island) has gained a reputation as one of the most desolate and mysterious places in California, a place where a Native American woman was abandoned and lived alone for 18 years. It's the famous Island of the Blue Dolphins. It is an island known for its wild and windswept landscape, an island where human skeletons used to cover parts of its expansive sand dunes. It's the island where Native Americans fashioned beautiful soapstone effigies and gathered interesting magic stones prized by their Shamans. It is the island where sea otters were exterminated for their fur, but were reintroduced there in the late 1900s. It's an island of contrasts—an island where an active military base is surrounded by the ever-changing ocean and beaches teeming with marine mammals.

San Nicolas Island has experienced three basic phases of human occupation. Native Americans known as the Nicoleño people were there for over 8,500 years until most were removed from the island in the early 1800s. After the removal of the Nicoleño people, only a few hardy fishermen, abalone divers, and sheepherders lived on this remote island from the mid-1800s until the mid-1900s, when the island became a military base and human occupation increased dramatically.

I think that Charles Frederick Holder, who visited San Nicolas Island in the late 1890s, aptly summarized the first impressions of many of the island's visitors when he wrote (Holder 1910):

> *I have heard of mild and beautiful days at San Nicolas, but the single herder, a Basque, informed me that it blew pretty much all the time. ... He told me that he had to pile big rocks on the roof of his house to keep it from blowing away into the sea. He said that the wind blew small stones into the air, and one could not face it. ... I have never felt a more irritating, searching, penetrating wind than this wraith of the spirits of San Nicolas, yet I found this island most attractive, from its very desolation. The desert is fascinating if one sees it at the right time, and is not hunting for water, or the right road, and it is not summer.*

San Nicolas Island definitely deserves its reputation as an exposed, windy area, but much has changed since the early 1900s. It is now an active military base with an airfield, modern facilities, and paved roads. It's also the home of unique plants and animals that are not found anywhere else in the world.

LOCATION AND TOPOGRAPHY

The most isolated of the eight California Channel Island (Figure 1), San Nicolas lies 61 miles (98 km) south southwest of the city of Ventura, with its center near 33°15' N latitude, 119°28' W longitude. The nearest neighboring island is Santa Barbara, 28 miles (45 km) to the northeast. San Nicolas is separated by deep ocean channels from Santa Barbara and the other islands in the chain, which are situated 50-60 miles (80-96 km) to the north and about 50 miles (80 km) to the east and southeast (Figure 1).

Approximately 22 square miles (58 square km) in size, San Nicolas is an oval-shaped island (see map at http://www.sbbg.org/SanNicMap) about 10 miles (15.5 km) long and 3 miles (5.4 km) wide. From the sea, San Nicolas presents a low, table-like profile. The island's topography is dominated by a broad central terrace or mesa with no distinctive peaks. The central mesa covers most of the surface area of the island, being roughly 6.2 miles (10.3 km) long and 2 miles (3.4 km) wide (USGS 1:24,000 Topographic Map, San Nicolas Island, CA). The main axis of the island runs from northwest to southeast and the central mesa slopes gently to the northeast from the highest points, which are near the south rim of the terrace. The maximum elevation on the island is 907 feet (277 m) at Jackson Hill, which is located near the southern rim of the mesa.

Along the island's coastline, there are sandstone ledges (many of which are usually occupied by roosting or nesting sea birds) and sandy beaches (some of which are often covered with marine mammals). Relatively flat coastal terraces encircle the perimeter of the island just inland from the coastline and are separated from the central mesa by rocky escarpments. The coastal terraces were apparently cut by the ocean's wave action when portions of the island were submerged in the past. Burnham et al. (1963) reported that as many as 15 to 20 wave-cut terrace levels can be discerned on the island.

The escarpment on the island's north side consists of slopes ranging from about 250 to 350 feet (76-107 m) high that are largely covered with vegetation. There are about 20 canyons on the north side of the island, several of which are about 1.5 miles (2.4 km) long and cut into the central mesa's surface. The escarpment slopes on the south side of the island range from about 400 to 800 feet (122-244 m) high. Most of these slopes are significantly eroded and vegetative cover is very sparse. A series of alternating sandstone and siltstone beds are dramatically exposed on the south escarpment and in the deep canyons that cut into it. There are about 20 canyons on the island's south side; these are mostly less than 1 mile (1.6 km) long and end at the central mesa's rim.

Aerial view of eastern portion of San Nicolas Island, looking west across central mesa, airfield, and coastal flats (S. Junak, March 1993).

Aerial view of coastal flats and northwestern shoreline of San Nicolas Island, looking west (S. Junak, March 1983).

Northern elephant seals on southeastern side of island, west of Daytona Beach (S. Junak, November 1989).

Daytona Beach, southeastern side of island, looking northwest (S. Junak, May 1983).

Celery Canyon, on north side of island (S. Junak, June 1995).

Towers Canyon, on southeast side of island (S. Junak, May 1992).

Aerial view of coastal terrace and eroded escarpment at northeastern end of island, looking southwest (S. Junak, December 1991).

Eroded escarpment on south side of island, west of Daytona Beach, looking northeast (S. Junak, May 1991).

Eroded sandstone and siltstone beds at mouth of Twin Rivers drainage, south side of island, looking northeast (S. Junak, May 1992).

CLIMATE AND HYDROLOGY

Climatic data, primarily of temperatures and precipitation, have been gathered on San Nicolas Island between September 1933 and August 1944 and continuously since 1947 (deViolini 1974). The island's climate is characterized by a Mediterranean pattern of relatively warm, wet winters and cool, dry summers. A dominant feature of the island's weather is the prevailing northwest wind, which averages 14 knots per hour (16.1 miles per hour) with an annual mean maximum of 52 knots per hour (59.8 mph).

Interactions between the strong winds, marine fogs, and the island's topography promote distinctive microclimatic zones which affect the distribution of individual species and of plant communities. Since the island presents an obstruction to the prevailing wind flow, the northern and western portions of the island may be covered by low stratus clouds if these low clouds are present over the adjacent waters, while the southern and eastern portions of the island may be clear (de Violini 1974). Mean monthly temperatures vary only slightly, from 54°F in January to 64°F in September (12.2°C-17.8°C), being moderated by the surrounding ocean. Freezing temperatures have not been recorded on the island; the minimum of 33°F (0.5°C) was recorded in January 1949 when a light snow fell. Temperatures above 100°F (37.8°C) are occasionally experienced on the island; a 25-year maximum temperature of 105°F (40.5°C) was recorded in September 1955 (de Violini 1974).

Average annual precipitation on the island between the 1948-1949 and 2004-2005 seasons was 8.21 inches (208.53 mm), with more than 70 percent of that amount typically falling between November and the end of February. During the 58-year period for which uninterrupted data are available (1949-2005), annual precipitation exceeded 10 inches (254 mm) in only 13 (22.4%) of those years (Appendix III).

The summer months average less than 0.1 inch (2.54 mm) of rainfall, and most of what falls is drizzle from stratus clouds. Annual precipitation extremes of 2.63 inches (66.80 mm) and 21.77 inches (552.95 mm) were recorded in the 1960-61 and 1997-98 seasons respectively (de Violini 1974; Charles Fisk, pers. comm. 2007; Appendix III).

Precipitation on the island is the only source of fresh ground water on San Nicolas Island, according to Burnham *et al.* (1963). During their investigations, Burnham *et al.* found no evidence of any geologic features that would allow fresh water to move between the California mainland and San Nicolas Island. They stated that the ground water on the island is recharged by deep penetration of rainfall and runoff into the highly absorptive fresh dune sand at the west end, which gravity moves downward to the main water table or to impermeable zones that divert the water laterally downslope. Ground water is discharged at the island's surface by a number of intermittent, as well as some perennial, springs and seeps. The perennial surface water is concentrated along the north side of the island between Red Eye Beach and Thousand Springs, where Burnham *et al.* (1963) mapped 12 small springs and seeps.

GEOLOGY AND SOILS

San Nicolas Island is a broad, complexly faulted anticline that parallels the long dimension of the island, with its crest near the southwestern shoreline (Vedder and Norris 1963). The exposed Tertiary Period section on the island consists of nearly 3,500 feet (1,067 m) of alternating marine sandstone and siltstone beds that contain minor amounts of interbedded conglomerate and pebbly mudstone. Vedder and Norris divided this sedimentary rock sequence into 35 mappable units, all of Eocene Epoch age (which extended from about 54.8 to 33.7 million years ago), and they reported that several small andesitic dikes of Miocene Epoch age (which extended from about 23.8 to 5.3 million years ago) intrude the sedimentary rocks near the southeastern end of the island. They found that dune sand and fossiliferous marine terrace deposits of Quaternary Period age (which extended from about 1.8 million years ago to present) cover most of the central and western parts of San Nicolas Island.

The U.S. Department of Agriculture (1985) mapped 27 soil units on San Nicolas Island, including beach and dune sand, Jehemy clay, several sandy loams and loamy sands, and rock outcrops, as well as eroded, channeled, and gullied complexes. Rock outcrops (3,700 acres/1,497 hectares), Vizcapoint severely eroded land complexes (1,270 acres/514 ha), dune land (1,160 acres/469 ha), and Vizcapoint sandy loam (1,080 acres/437 ha) were identified as the most common soil types on the island.

The distribution of at least two of the native plant taxa on the island appears to be correlated with the occurrence of particular siltstone layers on San Nicolas Island. Island poppy *(Eschscholzia ramosa)* and hydra stick-leaf *(Mentzelia affinis)* primarily occur along the portions of ridges on the south escarpment where some of the siltstone layers are exposed.

UNIQUE BIOLOGICAL RESOURCES

Besides several flowering plants that are known only from San Nicolas Island (discussed in the Floristic Analysis chapter below), there is an endemic fox, a mouse, a grasshopper, a beetle, a fly, a wasp, a bee, at least one spider, a centipede, and two terrestrial snails that are thought to occur nowhere else in the world (Table 2).

In addition to the organisms listed in Table 2 (facing page), there may be additional insects that are endemic to San Nicolas Island, such as a tortrix moth in the *Argyrotaenia franciscana* complex (Jerry Powell, personal communication 2007). The zodariid spider *Lutica nicolasia* may also be endemic to the island, but needs further study.

The island night lizard *(Xantusia riversiana)* is also worthy of special mention. This species, found only on Santa Barbara, San Nicolas, and San Clemente islands, is a remarkable lizard that lives in patches of *Opuntia* and *Lycium.* Lizards on San Nicolas Island may be endemic at the subspecies level.

Island night lizard, found only on the California Channel Islands (S. Junak, February 1990).

TABLE 2. Terrestrial Animals Restricted To San Nicolas Island

Common Name	Scientific Name
San Nicolas Island fox[1]	*Urocyon littoralis* subsp. *dickeyi*
San Nicolas Island deer mouse[1]	*Peromyscus maniculatus* subsp. *exterus*
Short-horned grasshopper[2]	*Microtes nicola*
Darkling beetle[2,3]	*Eleodes subvestitus*
San Nicolas Island robber fly[2]	*Cophura hennei*
San Nicolas Island sand wasp[2,4]	*Bembix americana* subsp. *nicolai*
San Nicolas Island solitary bee[4]	*Anthophora urbana* subsp. *nicolai*
San Nicolas Island funnel-weaver spider[2]	*Rualena alleni*
San Nicolas Island centipede[2]	*Geophilus nicolanus*
San Nicolas Island snail[5]	*Micrarionta feralis*
Prickly pear snail[5]	*Micrarionta opuntia*

Sources
[1]Paul Collins (personal communication 2007)
[2]Scott Miller (unpublished manuscript 1995)
[3]Mike Caterino (personal communication 2007)
[4]Robbin Thorpe (personal communication 2007)
[5]Barry Roth (personal communication 2007)

VEGETATION

Foreman (1967) described and mapped nine upland plant community types and two wetland community types for San Nicolas Island. His upland types included coastal strand and sand dunes, annual iceplant, six shrub communities *(Lycium-Lupinus, Lupinus,* mixed shrub, *Malacothrix, Coreopsis,* and *Baccharis),* and grassland. Foreman's two wetland types included coastal marsh and mesophytic communities (including moist areas in small isolated valleys near the water table, seeps, and canyon bottoms).

Philbrick and Haller (1977) described three upland plant communities for San Nicolas Island (southern coastal dune, coastal sage scrub, and valley and foothill grassland) and one wetland community (coastal marsh).

Halvorson *et al.* (1996) described and mapped nine upland plant communities and three wetland communities for the island. Their upland communities included coastal dune, inland dune, annual iceplant, caliche scrub, four coastal scrub communities *(Isocoma, Baccharis, Lupinus,* and *Coreopsis),* and grassland. Their wetland communities included coastal marsh, vernal pools, and riparian/deep drainages.

The following classification system for the island's plant communities is primarily based on Foreman (1967) and Halvorson *et al.* (1996), with additions from Junak *et al.* (1995a) and *Junak et al.* (2007). It includes eight upland communities, four terrestrial wetland communities, and one marine community. Upland communities include southern beach and coastal dune scrub, valley and foothill grassland, annual iceplant, caliche scrub, and four coastal scrub communities (goldenbush, coyote-brush, lupine, and *Coreopsis*).Wetland communities include coastal salt marsh, vernal pools, freshwater springs and seeps, and riparian vegetation. The marine community includes intertidal and subtidal areas around the perimeter of the island.

The areal extent of most of these plant communities (as mapped in 1992) are shown in Table 3. The four coastal scrub communities (along with caliche scrub and annual iceplant) covered more than 50% of the island at that time. Over 23% of the island's surface was barren or covered with very sparse vegetation, grassland covered 12.2%, and coastal dune scrub covered 6.5%. Developed areas and the other plant communities collectively covered about 5% of the island (Halvorson *et al.*1996).

TABLE 3. Land Cover on San Nicolas Island in 1992.
(Adapted from Halvorson *et al.* 1996)

Land Cover Class	Area (acres)	Area (ha)	% of island
Coastal scrub[1]	6,005	2,430	42.1
Barren or sparse	3,469	1,404	23.4
Grassland	1,740	704	12.2
Coreopsis scrub	1,349	546	9.5
Stabilized dune	783	317	5.5
Developed areas	324	131	2.3
Riparian	235	95	1.6
Unstabilized dune	138	56	1.0
Coastal marsh	10	4	0.1

[1]This map unit included the following communities: caliche scrub, goldenbush scrub, coyote-brush scrub, lupine scrub, and annual iceplant.

Southern Beach and Coastal Dune Scrub. Sandy beaches comprise much of the island's perimeter. The area above the mean high-tide line on these beaches is colonized by low-growing southern foredune dominants, including *Abronia maritima, Ambrosia chamissonis, Atriplex leucophylla, Cakile maritima,* and *Camissonia cheiranthifolia* subsp. *cheiranthifolia.* Unstabilized sand dunes, especially on the north side of the island, support patches of vegetation that are dominated by *Ambrosia chamissonis, Calystegia macrostegia* subsp. *amplissima, Mesembryanthemum crystallinum, Abronia maritima, Abronia umbellata* var. *umbellata, Malacothrix incana, Lupinus albifrons* var. *douglasii, Lotus argophyllus* var. *argenteus,* and *Camissonia cheiranthifolia* subsp. *cheiranthifolia.* Widespread associates in the unstabilized dunes include *Bromus rubens, Atriplex semibaccata, Sonchus oleraceus, Spergularia macrotheca* var. *macrotheca, Atriplex californica, Cakile maritima,* and *Melilotus indicus.*

On the west end of the island, some stabilized dune areas are dominated by large stands of *Astragalus traskiae,* which is endemic to San Nicolas and Santa Barbara islands. On Santa Barbara Island, *Astragalus traskiae* is quite rare and does not occur in large populations like those found on San Nicolas. The perennial non-native iceplant *Carpobrotus edulis* is common in stabilized or partially stabilized dunes. Widespread dominants in the stabilized dune areas, in addition to *Astragalus traskiae,* include *Abronia umbellata* var. *umbellata, Ambrosia chamissonis, Bromus rubens, Cakile maritima, Camissonia cheiranthifolia* subsp. *cheiranthifolia, Erodium cicutarium, Lotus argophyllus* var. *argenteus, Lupinus albifrons* var. *douglasii,* and *Malacothrix incana.* Widespread associates in stabilized dune areas include *Abronia maritima, Bromus diandrus, Medicago polymorpha, Mesembryanthemum crystallinum,* and *Sonchus oleraceus.*

Special status plants *Cryptantha traskiae* and *Dithyrea maritima* occur in isolated populations in unstabilized and partially stabilized coastal dunes, especially on the north side of the island. Other native taxa like *Amsinckia spectabilis* var. *spectabilis* and *Platystemon californicus* also occur on these same general areas.

Several non-native plant taxa threaten native plants in both the stabilized and unstabilized dunes on the island. Highly invasive *Ammophila arenaria,* which was planted for erosion control, is spreading. *Carpobrotus edulis, Cakile maritima,* and *Lobularia maritima* are widespread, and *Brassica tournefortii,* which has been rapidly spreading in recent years, represents a special threat to the endemic *Cryptantha traskiae.*

Coastal dunes on north side of island, just west of Thousand Springs (S. Junak, March 1991).

Stabilized dunes and coastal flats on southwest side of island, dominated by Astragalus traskiae *(S. Junak, March 1993).*

Dithyrea maritima in dunes on Vizcaino Point peninsula, northwest end of island (S. Junak, April 1992).

Valley and Foothill Grassland. Large areas on the island's mesa are dominated by non-native grasses, especially *Avena barbata, Bromus diandrus, Bromus hordeaceus,* and *Hordeum murinum.* Widespread associates, especially in disturbed areas, include *Atriplex semibaccata, Erodium cicutarium, Erodium moschatum, Medicago polymorpha,* and *Sonchus oleraceus.* Other associates in the grasslands include *Centaurea melitensis, Vulpia myuros, Spergularia macrotheca* var. *macrotheca, Lamarckia aurea, Galium aparine, Mesembryanthemum nodiflorum* (in open sites), *Calystegia macrostegia* subsp. *amplissima, Coreopsis gigantea, Amblyopappus pusillus, Deinandra clementina, Avena fatua, Achillea millefolium, Dichelostemma capitatum, Lactuca serriola, Lotus argophyllus* var. *argenteus,* and *Melilotus indicus. Lolium multiflorum* and *Lolium perenne* dominate some grasslands, especially near the airfield. *Frankenia salina,* which is usually found along the immediate coast or near salt marshes in southern California, is common in some grassy areas, even on the central mesa.

There are a few small patches of native grassland dominated by *Nassella pulchra,* especially near Jackson Hill and near the southeastern end of the central mesa. On open flats and in shallow depressions in the eastern portion of the central mesa, the native annual *Hordeum intercedens* can be very common in wet years and can dominate small areas.

Small, but very interesting, populations of native annuals occur in scattered areas, especially in grassy openings between shrubs on the northeastern coastal flats and in the northeastern portion of the central mesa. *Lasthenia gracilis, Castilleja densiflora, Daucus pusillus, Trifolium* spp., *Claytonia perfoliata* subsp. *mexicana, Crassula connata* (open sites), and *Malacothrix foliosa* subsp. *polycephala* can be locally common in these locations.

Disturbed grassland dominated by non-native species on island's central mesa, looking toward Nicktown from airfield area (S. Junak, March 1991).

Lasthenia gracilis *in grassy opening between shrubs, northeastern coastal flats east of West Mesa Canyon (S. Junak, March 1993).*

Annual Iceplant. At the west end of the island, on exposed, disturbed flats and slopes that are used seasonally by nesting Western gulls, the annual non-native iceplant *Mesembryanthemum crystallinum* is dominant. Associates include *Parapholis incurva, Atriplex semibaccata, Mesembryanthemum nodiflorum, Atriplex watsonii, Ambrosia chamissonis, Abronia umbellata* var. *umbellata, Cakile maritima, Spergularia macrotheca* var. *macrotheca, Lepidium lasiocarpum* var. *lasiocarpum,* and *Frankenia salina.*

Caliche Scrub. On the island's central mesa (especially on its southern edge), there are some flats that have been completely denuded of top soil. At these sites, a calcium carbonate layer called caliche is exposed on the landscape. The caliche layer, which can form a sub-surface hardpan layer in arid-land soils, is typically covered by top soil. On San Nicolas Island, the caliche layer has presumably been exposed by soil erosion caused by wind and water after the vegetation was removed by overgrazing. Similar habitats can be seen in many areas on San Miguel Island, at the western end of Santa Rosa Island, and at scattered locations on San Clemente Island. Vegetation in these caliche areas is sparse because there are few pockets of soil where plants can grow. Eighteen of the 28 plant taxa typically found in this community are non-native, presumably due to the disturbed nature of the sites (Halvorson *et al.* 1996).

On San Nicolas Island, the most important native dominants in this community are *Isocoma menziesii* and *Achillea millefolium. Bromus rubens, Parapholis incurva, Mesembryanthemum nodiflorum, Daucus pusillus, Erodium cicutarium,* and *Medicago polymorpha* are the most common associates. Other associates include *Spergularia macrotheca* var. *macrotheca, Lamarckia aurea, Sonchus oleraceus, Astragalus traskiae, Avena barbata, Mesembryanthemum crystallinum, Vulpia myuros, Bromus hordeaceus, Calystegia macrostegia* subsp. *amplissima, Crassula connata, Hordeum murinum, Plantago ovata, Senecio vulgaris, Amblyopappus pusillus, Atriplex semibaccata, Centaurea melitensis, Erodium moschatum, Lepidium lasiocarpum* var. *lasiocarpum,* and *Oligomeris linifolia* (Halvorson *et al.* 1996).

Disturbed flats dominated by Mesembryanthemum crystallinum *in gull colony at west end of island (S. Junak, June 1987).*

Caliche scrub vegetation on southern edge of island's central mesa (S. Junak, April 1992).

Mixed goldenbush scrub, dominated here by Isocoma, Lotus, *and* Lupinus, *on southwestern side of island (S. Junak, May 1993).*

Mixed Goldenbush Scrub. *Isocoma menziesii,* one of the most common and widespread shrubs on San Nicolas Island, characterizes this low-growing scrub community. This community is the most diverse association on the island and no single plant taxon dominates over large areas. Rather, the community is composed of vegetation patches of varying sizes and species assemblages. Numerous patches of *Opuntia* occur in this community and can be locally dominant (Halvorson *et al.* 1996, Junak 2003a). *Artemisia nesiotica* and *Artemisia californica* occur in small patches.

The most widespread taxa in this community are *Isocoma menziesii, Lotus argophyllus* var. *argenteus, Daucus pusillus, Sonchus oleraceus, Bromus rubens,* and *Mesembryanthemum crystallinum.* Other common associates are *Achillea millefolium, Amblyopappus pusillus, Erodium cicutarium, Bromus diandrus, Atriplex semibaccata, Lomatium insulare, Avena barbata, Crassula connata, Plantago ovata, Spergularia macrotheca* var. *macrotheca, Eriogonum grande* var. *timorum, Bromus hordeaceus, Parapholis incurva,* and *Lamarckia aurea* (Halvorson *et al.* 1996).

Coyote-Brush Scrub. This scrub community, dominated by *Baccharis pilularis,* is scattered on flats of the central mesa, especially where there are shallow depressions in the landscape or where the topography provides some protection from the prevailing winds. Coyote-brush scrub is also found in some of the canyons and drainages on the north side of the island (Halvorson *et al.* 1996).

The most widespread taxa in this community, besides *Baccharis pilularis,* are *Avena barbata, Bromus diandrus, Bromus rubens, Atriplex semibaccata, Lupinus albifrons* var. *douglasii,* and *Malacothrix incana.* Other common associates include *Isocoma menziesii, Melilotus indicus, Sonchus oleraceus, Erodium cicutarium, Erodium moschatum, Medicago polymorpha, Calystegia macrostegia* subsp. *amplissima, Achillea millefolium, Ambrosia chamissonis, Camissonia cheiranthifolia* subsp. *cheiranthifolia,* and *Malacothrix foliosa* subsp. *polycephala.*

Lupine Scrub. This scrub community, dominated by *Lupinus albifrons* var. *douglasii,* occurs primarily on stabilized dunes on the north side of the island. Large populations of this lupine can be found in

Lupine scrub on coastal flats east of Corral Harbor, on north side of island (S. Junak, May 1984).

similar habitats on San Miguel and Santa Rosa islands.

Common associates in this community include *Abronia umbellata, Ambrosia chamissonis, Bromus rubens, Camissonia cheiranthifolia* subsp. *cheiranthifolia, Coreopsis gigantea, Lotus argophyllus* var. *argenteus,* and *Malacothrix incana.*

Coreopsis Scrub. Dominated by *Coreopsis gigantea,* this is the second most diverse community on San Nicolas Island (Halvorson *et al.* 1996) and is very conspicuous because of its dense cover and tall stature. Most of the scrub associations on the island are composed of low-growing perennials and annuals that are usually not much more than a meter tall, while *Coreopsis* scrub can form nearly continuous stands that are more than 2 meters tall. Dense stands consist of an overstory of *Coreopsis gigantea* and an understory of vines, grasses, and herbs. The appearance of *Coreopsis* scrub changes dramatically as the dry summer season approaches. *Coreopsis gigantea* typically drops its green leaves during the early summer.

The most widespread associates in this community typically include *Avena barbata, Bromus diandrus, Bromus rubens, Erodium cicutarium, Melilotus indicus, Crassula connata, Daucus pusillus, Lupinus albifrons* var. *douglasii, Isocoma menziesii, Malacothrix foliosa* subsp. *polycephala, Sonchus oleraceus, Atriplex semibaccata, Ambrosia chamissonis, Medicago polymorpha, Bromus hordeaceus, Lotus argophyllus* var. *argenteus,* and *Malacothrix incana.* Other common associates include *Amblyopappus pusillus, Mesembryanthemum crystallinum, Vulpia myuros, Camissonia cheiranthifolia* subsp. *cheiranthifolia, Lomatium insulare, Claytonia perfoliata* subsp. *mexicana, Cryptantha traskiae, Mesembryanthemum nodiflorum,* and *Erodium moschatum.*

Coastal Salt Marsh. A very small marsh area occurs on flats near the base of the Sandspit at the southeastern end of San Nicolas Island. *Salicornia virginica* and *Frankenia salina* are the dominants in this community. Associates include *Distichlis spicata, Mesembryanthemum nodiflorum, Atriplex californica, Calystegia macrostegia* subsp. *amplissima, Parapholis incurva, Abronia maritima, Ambrosia chamissonis, Bromus hordeaceus, Spergularia macrotheca* var. *macrotheca,* and *Mesembryanthemum crystallinum.*

Coreopsis *scrub on coastal flats east of Corral Harbor, on north side of island (S. Junak, March 1992).*

Coreopsis *scrub on coastal flats near the Rock Jetty, on northeastern side of island (S. Junak, March 1993).*

Vernal pool on northeastern end of island's central mesa, near airfield (S. Junak, March 1991).

Vernal Pools. Several depauperate vernal pools occur on the central mesa, especially in its northeastern portion. Most of these seasonally-wet depressions have been artificially created by Navy operations, but a least two sizeable natural depressions apparently occurred on the mesa before its surface was disturbed (Forney 1879a). The two dry lake beds shown on surveyor Stehman Forney's map were in the area where the island's airfield was subsequently built.

Blanche Trask, a botanist who first visited the island in 1897, saw "a tiny lake fringed with *Eleocharis*" and found "*Lupinus micranthus* [*L. bicolor*], several clovers, *Pectocarya,* and *Orthocarpus*" nearby (Eastwood 1898). During a second visit in 1901, Trask also found *Gnaphalium palustre,* which typically occurs in vernal depressions. All of these taxa can still be found in small, widely scattered populations near the airfield, even though their original habitat has been fragmented.

Currently, the dominant species around the margin of the vernal pools on the central mesa is *Eleocharis macrostachya.* Associates include *Juncus bufonius, Parapholis incurva,* and *Polygonum argyrocoleon* at some locations.

Freshwater Springs and Seeps. Freshwater springs and seeps, often with water that is slightly brackish, occur in the western and northwestern portions of the island (e.g., Red Eye Beach, Thousand Springs, NavFac Beach, and Army Springs). These springs and seeps have been important water sources for people and animals living on the island.

Anemopsis californica occurs at a seep at Red Eye Beach. *Soleirolia soleirolii* is found on moist, shaded bluffs at Thousand Springs. Other plants occurring at seeps include *Apium graveolens, Rumex salicifolius* var. *salicifolius, Salix lasiolepis, Hordeum murinum,* and *Polypogon monspeliensis.*

Riparian Vegetation. This community occurs in some of the island's canyon bottoms with intermittent streams and in low spots where soil moisture accumulates (e.g., the borrow pit along Monroe Road). This vegetation type is very depauperate on San Nicolas Island and most canyon bottoms are dominated by non-native plants.

The most significant riparian area on the island apparently occurred at Thousand Springs, but the area has been greatly disturbed by human activities. *Berula erecta* and *Ruppia maritima* previously occurred there in a small estuary, but have not been seen on the island since the early 1980s.

Plant taxa now found in canyon bottoms and other moist areas include *Typha domingensis, Typha latifolia, Cotula coronopifolia, Polypogon monspeliensis, Distichlis spicata, Atriplex prostrata, Festuca arundinacea, Rumex salicifolius* var. *salicifolius, Rumex crispus* subsp. *crispus, Baccharis salicifolia, Salix lasiolepis, Myoporum laetum, Marrubium vulgare, Apium graveolens,* and *Lythrum hyssopifolium.*

Salix exigua is currently known only from the lower portion of Grand Canyon. *Nasturtium officinale* and *Schoenoplectus americanus* are locally common in Tule Creek. *Rumex obtusifolius* is localized in a ditch near the airfield.

The non-native tree *Tamarix ramosissima* is a threat to the island's riparian areas. It is currently known from several sites on the central mesa and in one of the canyons on the north side of the island.

Intertidal and Subtidal Marine Community. Several species of marine angiosperms occur in intertidal and subtidal zones around the perimeter of the island. *Phyllospadix scouleri* and *Phyllospadix torreyi* are widespread on shallow reefs and rocky shorelines in the intertidal zone and are often exposed in tidepools at low tides. A significant population of *Zostera pacifica* occurs in water 45 to 55 feet (13.7-16.8 m) deep off Coast Guard Beach, about 0.5 miles (0.8 km) northwest of the Rock Jetty (Jack Engle, personal communication 2002).

Phyllospadix *washed up on beach after storm, just east of NavFac Grade on north side of island (S. Junak, July 1989).*

HISTORY OF LAND USE AND VEGETATION CHANGES

Native American Period, 8500 Years BP to the 1850s. The earliest human occupation of San Nicolas Island apparently began more than 8,500 years ago (Schwartz and Martz 1992). The Native Americans who lived on San Nicolas relied primarily on the rich marine resources surrounding the island (e.g., marine mammals, fish, and shellfish) for their food (Martz 2005). By the middle Holocene (about 6,000 to 3,000 years ago), the Nicoleño people occupied at least 22 village sites on the island, primarily on the west end (Martz 2005). By the early 1800s, occupation of the Native American villages on the island had dwindled and almost all of the island residents were moved to the mainland in 1835 (Ellison 1937). Juana Maria, a Native American woman, was left behind on San Nicolas Island in 1835. She lived alone on the island for 18 years until she was taken to the mainland in 1853.

Ranching Period, 1850s to the 1930s. Like several other islands off the coast of southern California (e.g., San Miguel, Anacapa, and Santa Barbara islands), San Nicolas was not granted to an individual during the Mexican Period in California. San Nicolas Island thus became the property of the U.S. government after the Mexican-American War (under the terms of the Treaty of Guadalupe Hidalgo in 1848). Two 20-acre parcels, located on the southwestern and southeastern portions of the island, were reserved for lighthouse purposes by order of the President of the United States in January 1867. The U.S. government, however, essentially neglected San Nicolas Island until it was reserved in its entirety for lighthouse uses by order of President Roosevelt in November 1901. Because the island is so remote, it was apparently easy for unregulated ranching to begin there shortly after Juana Maria left in 1853. Unregulated ranching activities continued on the island until the early 1900s.

The most significant factors affecting native plants on San Nicolas Island during the ranching period were the introduction of domestic grazing animals (especially sheep) and the arrival and spread of aggressive non-native plants. Because of the relatively gentle topography, almost all of the island was accessible to grazing animals and very few areas have escaped decades of disturbance. The most extreme impacts apparently occurred in the 1860s and 1870s, when sheep numbers were highest and a series of dry years drove the animals to strip the island of vegetation. Evidence for vegetation changes during the ranching period on the island comes from a variety of sources, including historical descriptions and maps, examination of historical photographs, and herbarium records of non-native plant introductions.

Very detailed accounts of sheep ranching activities on the island can be found in Swanson (1993) and McCawley (1997). Sheep numbers on the island fluctuated over time as individual operations came and went, but they were apparently present for almost 100 years. During the sheep ranching period, the comments of visitors indicated that woody plants were nearly eliminated from the island a number of times.

Unfortunately, there are no detailed descriptions of vegetation on San Nicolas Island before sheep were introduced there. There is, however, an early reference to woody vegetation. This observation was made by Captain George Nidever, a sea otter hunter who visited San Nicolas Island in the fall of 1852. Nidever reportedly found "some high bushes, called by the natives malva real ..." (Ellison 1937).

Sheep were first brought to the island by Captain Martin M. Kimberly in the mid-1850s and multiplied for several years. According to his wife Jane Merritt Kimberly (1988):

> *Captain Kimberly stocked San Nicolas with sheep, and they increased very rapidly. The ewes had young twice a year and two were almost always born each time. The flocks increased until they numbered 15,000 ... Then came the dry year of 1864, which dealt Captain Kimberly a very hard blow. There was no rain at all, and many of the sheep died. Another dry year, in 1869 or '70, turned San Nicolas into a desert and drove my husband out of the sheep business with heavy loss. In those days, San Nicolas was luxuriantly covered with vegetation, but the sheep, in their frantic efforts to get water, clipped off all that survived the dry, hot winds. The wild carrot [giant coreopsis (Coreopsis gigantea)?], with long, strong roots which went far down into the soil, had moisture at the bottom of them and the sheep dug two or three feet into the ground to get at the bottom of them. The winds blew sand completely over the island, burying the roots and the seeds that remained so deep that they were smothered and have made the island simply a waste of yellow sand ... Captain Kimberly saw that he would lose all his sheep unless he could get them off, so he chartered a large vessel and took the sheep, 1,000 at a time, to San Francisco, where he sold them to the butchers. The last 4,000 he could not get off and they remained on the island when he sold it to Mr. Hamilton, a San Francisco banker, in 1870 or '71.*

William E. Greenwell, who established the first triangulation stations on the island for the U.S. Coast Survey, visited the island in 1858. Greenwell found woody vegetation in the immediate vicinity of most of his stations. He also reported succulent plants that may have been *Dudleya* and *Mesembryanthemum*. He described the vegetation that he found in the vicinity of eight of his ten stations as follows (Greenwell 1858):

> Δ *"Kelp" (located at extreme se end of central mesa): The ground is covered with low sage bushes at this end of the island but in the immediate vicinity of the signal it is more or less free of these.*
> Δ *"Cliff"(located at s edge of central mesa e of Building 186): Some small green bushes are within a few meters of it to the north.*
> Δ *"N Base" and* Δ *"S Base" (located at airfield on central mesa): It [The base line] is on a level piece of ground covered with low sage bushes and a species of cabbage plant, known here as 'siempre vive'.*
> Δ *"Bluff"(located at nw end of airfield on central mesa): A knoll or hill covered with the*

same low sage bushes as seen about the base.

Δ *"Port" (located near n edge of central mesa just ne of Nicktown): The ground about the signal is covered with sage bushes which are higher here than about the stations above described.*

Δ *"Slope" (located in central portion of mesa just e of Shannon Road): The soil at this station is loose and sandy and the ground thickly covered by a sort of succulent plant, called in these parts 'soldier tree'.*

Δ *"North Head" (located on nw side of island, ene of Benchmark 616 and w of Tule Creek): To the eastward of it is a fine spring of water and a few trees growing around it.*

Martin Kimberly formally claimed 160 acres on the island in 1858 (Kimberly 1858), although he had apparently been occupying the island since 1856 (Kimberly 1870). Once sheep were taken to San Nicolas Island, a man named Lennie Crabb and his wife reportedly lived on the island and took care of them (Kimberly 1961). According to the Agricultural Census Schedule for Santa Barbara County in June 1860, Kimberly had 800 sheep and 10 horses on San Nicolas Island in June 1860. Annual wool production was listed as 2,000 pounds.

Kimberly sold his interest in San Nicolas Island (as well as another 160 acres that had been claimed by a James Crabb and previously quitclaimed to Kimberly) to William Hamilton and Abraham Halsey in September 1870 (Kimberly 1870, Halsey 1872). Kimberly's sale also included the sheep, cattle, and horses that were on the island at that time (Kimberly 1870). Kimberly reportedly had 3,400 sheep and 50 horses on San Nicolas Island in 1870, according to that year's Agricultural Census Schedule for Santa Barbara County.

After William Hamilton died in the early 1870s, Abraham Halsey sold his interest in San Nicolas Island to Hamilton's heirs (Halsey 1872), who in turn sold their interests to the Pacific Wool Growing Company in September 1872 (Hamilton 1872).

Paul Schumacher (1877), an archaeologist working for the Smithsonian Institution, visited San Nicolas Island for ten days in June 1875 and observed that:

> *The vegetation is like that of San Miguel [Island], and also ruined by overstocking it with sheep, which are here found in a like starving condition. Near the house on the northeast side, we found some malva bushes [malva rosa (Lavatera assurgentiflora)?] cleared of their foliage to the reach of sheep, which gave them the appearance of scrub oak trees when seen from a distance.*

Schumacher stated that the vegetation on San Miguel Island (which he compared to San Nicolas Island in the quote above) consisted of "low bushes, cactus, and grass, but no trees". Maps of San Nicolas Island in Schumacher's report show four sheds and a sheep corral at Corral Harbor, as well as an adobe house located about 0.75 miles southwest of the harbor.

By 1876, the Pacific Wool Growing Company had 3,000 sheep and 400 lambs on San Nicolas and Anacapa islands (Ventura Signal 1876). The animals were presumably split between the two islands, but since San Nicolas is over 20 times larger than Anacapa, the majority were probably on San Nicolas. The Pacific Wool Growing Company was based in San Francisco and owned by the Mills brothers, Hiram and Warren (Roberts 1991). The Mills brothers also raised sheep on San Miguel Island from 1869 until at least 1887 (Roberts 1991).

In 1879, Stehman Forney thoroughly explored San Nicolas while recovering old triangulation stations and installing new ones for the U.S. Coast Survey. He spent over two months there (from July 7 to September 16, 1879). At that time, the island was reportedly occupied by H. D. Mills, who had about 1,000-2,000 head of sheep (Forney 1879a, 1879b). Forney's map of the island (Forney 1879a) shows extensive areas covered with sand at the west end and grassland covering most of the central mesa. His map also shows two dry lake beds in the eastern portion of the central mesa. Forney (1879b) commented that:

About two-thirds of the island is covered with sand, and the remainder with the wild grasses found on the mainland. Small patches of scrub oak [malva rosa (Lavatera assurgentiflora)?] are found in a few remote parts of the island.

Hiram W. Mills and E.L. Tuttle sold their interest in San Nicolas Island to Ezekiel Elliott and Joseph V. Elliott in the summer of 1882 (Mills and Tuttle 1882). Mills and Tuttle also sold their interest in Anacapa Island at the same time.

Stephen Bowers, self-taught archaeologist and geologist, camped on the island for 19 days from mid-October until early November 1889. He was there at the request of the California State Mineralogist, who was interested in a geologic reconnaissance of the island. Bowers (1890) reported that:

San Nicolas is entirely destitute of timber ... At the present time there is not even a bush growing on it except a stunted kind of thorn, scarcely two feet high [boxthorn (Lycium californicum)?], and a few species of the tree cactus.

The surface is comparatively level, sufficiently so to till with little trouble. This cultivable land embraces about two thirds of the island's area, and much of it is apparently rich and fertile.

However, Bowers (1889) also noted that:

Owing to the flocks of sheep which have been kept here for the last twenty-five years, the island is washing into deep gullies, thus injuring it for tillable land or for grazing purposes. The island is covered with untold millions of dead land shells which are evidence of moisture and much vegetation in past time.

There were reportedly about 2,000 sheep on San Nicolas Island in the fall of 1891 and "from a sort of rough grass they seem to keep fat ..." (Anonymous 1891). By the late 1890s, the sheepherders' headquarters had moved from Corral Harbor on the island's north shore to the east end of the island near the Sand Spit. This move was apparently made because loose sand was moving into the area around Corral Harbor.

Blanche Trask, botanist and intrepid island explorer, was the first known person to document the island's plant life with scientific specimens. Unfortunately, her first trip to San Nicolas Island wasn't until April 1897, 40 or more years after sheep had been introduced there. The vegetation had already been severely disturbed and a number of non-native plants had already been introduced to the island by the time of Trask's visit. Over 23% (19 taxa) of the nearly 80 plant taxa that she found are not native to California.

Trask did not "wholly agree with Dr. Bower's account" of the island's cultivable landscape and (in Eastwood 1898) commented that:

There is no soil on the broad level top; but tons of pebbles, round as shot and of a like size. Even here the ice-plants flourish and an occasional gay patch of Hordeum *or foxtail is seen. Everywhere the rocks are visible and the soil thin.*

Trask (Eastwood 1898, Trask 1900) saw significant erosion and sand movement on the island during her 1897 visit. She noted that:

Even on the comparatively "level top" of the island one must pass through gorge after gorge fantastically wind- and sand-carved. It is not unusual to be stopped by an erosion from 10 to 100 feet in depth, when following the main ridge, and to have to go far out of your way to reach its head.

Trask commented that "great cañons hundreds of feet deep are really 'snowed in' by the sand", observed "an old house built of stones yet standing, half 'snowed in' by sand, at Corral Harbor", and described the difficulties of tent camping in the strong winds and blowing sands.

Trask commented further (Eastwood 1898, Trask 1900) that:

The cañons are not what we usually call cañons; arroya is a fitter term. In them we hear no sound of bird, no whirr of wing; we see no bright flowers, only the ice plants. There is no ripple of stream, only the briny tidal waters which glide but do not flow ...

In March, after the rains, here and there about the central summit are gay sparkles of little flowers. ... San Nicolas is indeed a dying land. In all his length was found but one shrub seven feet high [Baccharis pilularis], *and in three or four localities Leptosynes* [Coreopsis gigantea] *grow from four to six feet high ... One tiny lake, too, was found begemmed with bullrush* [Eleocharis]. . . .

At the "east end" there are a cabin, a barn, shearing sheds, a cistern and a platform which drains its rain water into a reservoir. All these improvements are due to the once ambitious ranchmen who seem now to have abandoned the sheep; about 500 are occasionally seen, with long and beautiful white wool ... They doubtless persist largely upon the ice-plant which here, owing to the briny streamlets, thrives upon the summits, growing to a height of three and four feet, and wetting through both boot and leggin [sic] as the plant is crushed; becoming slippery on the broken gorges; indeed in this land of erosions the ice-plant is to be avoided as extremely dangerous. The cactus which on Santa Catalina or San Clemente one goes many a mile to avoid in the course of a day, in San Nicolas is met but rarely.

Trask's observations indicate that woody vegetation was scarce on the island in 1897. She did not see any trees and her comments indicate that most of the shrubs that she saw on the island were already quite rare (Eastwood 1898). Trask found only a dozen different shrubs (Table 4) and noted only one plant of the native vine *Calystegia macrostegia* subsp. *amplissima* (comments on herbarium specimen label).

TABLE 4. Native Shrubs Seen on San Nicolas Island in the Spring of 1897, with Comments on their Distribution or Abundance.

Shrub	Comments on Distribution or Abundance
Ambrosia chamissonis	"common on the sand hills"
Baccharis pilularis	"only one individual"
Coreopsis gigantea	"four or five localities"
Deinandra clementina	"found on sea cliffs in only one place"
Isocoma menziesii	"the only plant seen was about two feet in height"
Lotus argophyllus var. *argenteus*	"rare, collected in one moist flat"
Lupinus albifrons	"infrequent, growing on bare stretches of sand"
Lycium californicum	"in sheltered, moist nooks with Opuntia"
Lycium verrucosum	"several localities on arroya cliffs"
Malacothrix saxatilis var. *implicata*	
Opuntia littoralis	"rarely met"
Opuntia prolifera	

The *Opuntia* cacti that Trask saw, although scarce, apparently protected smaller plants from the sheep. Trask noted (Eastwood 1898) that "wherever … cactus grew, it protected a small colony of grasses and tender herbs". Six of the eleven grass taxa that Trask collected were found only in cactus patches, as were other herbaceous plants (e.g., *Parietaria hespera* and *Chenopodium californicum*).

On the other hand, with the sheep numbers down, at least two annual plant species (i.e., *Lasthenia gracilis* and *Malacothrix foliosa* subsp. *polycephala*) were widespread and very conspicuous on San Nicolas Island in the spring of 1897 (comments on Trask's herbarium specimen labels). For *Lasthenia*, Trask commented that "from [the] ship, all the uplands are golden with this plant". *Malacothrix* was reportedly "covering large areas on the ridge".

John L. Kelley spent a few days exploring the island in the fall of 1897. He and his friend Will Squires camped in the old stone house that Trask had seen earlier in the year. Kelley (1923) commented that:

There were a few sheep still on the island. Some years previous there had been a large herd there but the owners had for some reason taken them away, all but a few, which it seems had escaped them in some way, and these had multiplied until now there were probably in the neighborhood of a hundred or more.

There was at that time, absolutely no trees, or even brush wood growing there. I did not see a single shrub with a trunk as thick as my wrist, growing in any place upon the island. In early days there must have been brush wood of some sort growing there, for the old camp fires of the Indians, or rather ash heaps where such fires had been were very much in evidence along the north end of the island. They surely had some other source than the small amount of drift wood to depend on.

Charles F. Holder visited San Nicolas Island in September 1898 (Anonymous 1898, Holder 1899). He described the island's sole inhabitant, a Basque sheepherder with two dogs, and a barren mesa covered with small polished pebbles. He noted (Holder 1910):

As I walked up the island I did not see a green thing; it was summer when all vegetation was dead, but at least two-thirds of the island is sand-dune, except where the sand has been blown away. Parts are covered with coarse grass, and there are a few scrub oaks.

Barren as it appears, I fancy a botanist or even a layman would find a remarkably large list of plants here in the spring after heavy rains as there were few sheep.

In a newspaper article published in early 1899, S.J. Mathis (1899) described the island's lone sheepherder, noted that the only buildings on the island were the sheepherder's shack and the old stone hut that was mostly buried in the sand, and commented that:

A few sheep are kept on the island, though in the absence of bushes or grass one wonders how they exist. Their principal food is a thick, waxy "ice" plant, which grows out of the sand at the eastern end of the island.

In the spring of 1900, it was reported that (Anonymous 1900):

The sheep herder has not been off the island for more than two years, and much of the time he is the only human being on that desert spot, which does not now produce a stick or a shrub on its entire area.

Philip Mills Jones, a physician associated with the Anthropology Department at the University of California at Berkeley, spent five days exploring San Nicolas Island in early February 1901. Jones landed near the Sand Spit and reported a "hut for the sheep herder, and wool packing and shearing sheds" near the landing.

Charles W. Merritt of Santa Barbara, brother-in-law of Martin Kimberly and an experienced rancher, told Jones (Jones 1969) that Kimberly "put the first sheep on San Nicolas Island in 1852 or 1853, and this lot consisted of 500 Mexican ewes and a few rams".

Jones (1969) reported that:

Recent destructive erosions by wind, rain, and sand undoubtedly date their commencement from the time when these sheep were first placed on the island by Mr. Kimberley [sic]. Indeed the next succeeding twenty-five years must have produced great changes in the surface topography of the island, for the destructive effect of the sheep was made evident almost at once. At that time the island was well covered with grass, weeds, and in places, with a considerable growth of brush, six to eight feet high. The first dry season, however, saw the end of the brush; the sheep ate it off as high as they could reach, and their sharp hoofs cut up the ground about its roots. At the western end of the island, where the sand dunes now stretch inland as far as the second cañon from that extremity of the island, the surface was well covered with vegetation to within a quarter mile from the coast line; now there are about three miles of dunes, and the house and buildings erected by Mr. Kimberley [sic] are covered many feet deep with drifting sand which has almost obliterated the cañon just west of Corral Harbor. ...

As nearly as I could judge from the information gathered from Mr. Merritt, the formation and migration of the extensive sand dunes now so noticeable on San Nicolas Island commenced about the year 1866. Up to that time he had not noticed them, but shortly afterward he observed that large tracts of land, formerly covered with vegetation and furnishing good pasture, were now bare and had become simply shifting sand dunes.

Jones (1969) also speculated that:

... it is probable that during the time of its aboriginal occupancy, some parts of the island were covered with brush and scrubby trees, the charred roots of which may still be found, here and there.

U. Sebree and E. Davis, inspector and engineer with the 12th Lighthouse District, visited the island on 10 September 1901. They reported that there was a man named Fred Julian staying at the east end of the island, where there were two huts and a sheep corral. Julian was caring for the 1,200 sheep that were reportedly on the island. One of the owners of the sheep, George Le Mesnager of Los Angeles, told Sebree and Davis that he and a partner, Peter Cazes, had purchased the sheep of Anacapa and San Nicolas islands about five years earlier from E. Elliott (Sebree and Davis 1901). Official records in Ventura County indicate that J.V. Elliott sold his interest in San Nicolas Island to Peter Cazes in November 1897 (Elliott 1897).

After the inspection by lighthouse personnel, the entire island was reserved by order of President Roosevelt in November 1901. Once the U.S. Lighthouse Bureau was given control of San Nicolas Island, they quickly moved to end unauthorized grazing on the island and started to lease the grazing and other agricultural rights to the highest bidder in 5-year increments.

Inspectors with the 12th Lighthouse District visited San Nicolas Island again on 13 June 1902. They reported (Handbury 1902) that:

> We landed upon the east end of the island and there found two cheap wooden structures with some fencing, forming a sheep corral. One of the houses was occupied by the two men in charge of sheep, the other had been used for storing wool and supplies for the herders.

> There were several hundred sheep in the corral at that time, which one of the herders informed us had been collected at that point with view of their being taken away in response to the notice that had been given by me to their owner, Mr. Mesnager. They were expecting a schooner to come for them the following Wednesday (June 13). There was every evidence that Mr. Mesnager was taking his property off the island. ...

A few days after the inspectors visited San Nicolas Island, it was reported (Anonymous 1902d) that:

> A report is going the rounds of seafaring men and others that all the buildings and shacks on the island were seen burning the latter part of last week. The island was visited Tuesday by the United States lighthouse tender Madrona, when notices were posted ordering a French sheepherder to vacate, for the reason that the island property had been leased by the government. A few days later the scow Brothers, Capt. Winters, conveyed the sheepherder, with 3000 sheep, to San Pedro.

Successful bidders for the leases granted by the Lighthouse Bureau included W.J. McGimpsey (1902-1907), D.R. Weller (1907-1909), Joseph G. Howland (1909-1919), and Edward N. Vail (1919-1934). It appears that the first two lessees (McGimpsey and Weller) did not actively manage the sheep on San Nicolas Island after most of the animals were removed in 1902, but at least a few sheep apparently remained on the island. McGimpsey was reportedly most interested in prospecting for oil on the island and also ran a fishing company (Anonymous 1902e, 1902f). The captain of a lobster schooner reported that the island was essentially uninhabited in the fall of 1904, "save for the fisherman that go there during the lobster season". He speculated that a lone "wild man" that he saw on San Nicolas Island "must have been living on the sheep that still remained there after the herds were removed two years ago" (Anonymous 1904).

North side of San Nicolas Island in the early 1900s, looking east. This view shows sheep trails on the coastal flats and adjacent slopes. A few annual plants were protected from sheep by the cactus clump in the foreground (Ventura County Museum of History and Art).

Archaeologist De Moss Bowers on San Nicolas Island in May 1907, in one of the sandy areas where little vegetation survived overgrazing by sheep in the late 1800s. His father, Stephen Bowers, explored San Nicolas Island in 1889 (Gladys Bibb Collection, Ventura County Museum of History and Art).

De Moss Bowers at one of the huge shell mound areas left behind by the Nicoleño people (Gladys Bibb Collection, Ventura County Museum of History and Art).

Corral Harbor area in May 1907, the site of the first sheep ranch operation on San Nicolas Island. This view shows the old stone house mentioned by many early visitors to the island (Gladys Bibb Collection, Ventura County Museum of History and Art).

Entrance to Corral Harbor in May 1907 (Gladys Bibb Collection, Ventura County Museum of History and Art).

Coastal dunes at Corral Harbor, on north side of island, in May 1907 (Gladys Bibb Collection, Ventura County Museum of History and Art).

Coastal dunes at Corral Harbor, on north side of island, in May 1907 (Gladys Bibb Collection, Ventura County Museum of History and Art).

Old stone house at Corral Harbor, already partially buried by sand in May 1907 (Gladys Bibb Collection, Ventura County Museum of History and Art).

Old stone house at Corral Harbor, nearly buried by drifting sand in 1929. Coastal vegetation visible in 1907 had also been buried by sand (Agee Family Collection, Santa Cruz Island Foundation).

During Joseph G. Howland's lease on the island, which began in 1909, sheep multiplied. William McCoy (1917) reported that there was a lone, very aloof "Basque shepherd-hermit" living near Corral Harbor and caring for a "band of a few hundred sheep" when he visited San Nicolas Island. Besides the shepherd's hut, McCoy saw " a couple of sheds and a few sheep corrals on the sand mesa" above Corral Harbor. By the spring of 1917, Howland was reportedly living on the island alone and had 1,400 sheep there (Anonymous 1917).

Edward N. Vail apparently bid for the grazing lease on San Nicolas Island with the hopes of putting cattle there. When he and others inspected the island in April 1919, however, Vail found that conditions were not suitable for cattle (Visel 1923):

> *About April 1919, the writer [C.P. Visel], in company with E.N. Vail, made a thorough examination of San Nicholas Island. We found the island to be in a deplorable condition having been overstocked with sheep, apparently for a number of years, and at the time of our examination was practically barren of any feed. The seed condition on the ground was very sparse. However, it was Mr. Vail's opinion that with several years of rest, the island might be brought back in point of feed to a condition where it would be practical to put on a few hundred head of cattle.*

Vail did not immediately introduce cattle after his lease began, but chose to let the island recover from the effects of past overgrazing. Conditions on the island, however, apparently did not improve much in the following years (Visel 1923):

> *Upon examination by either Mr. Vail or Mr. C.W. Smith, Mr. Vail's foreman on Santa Rosa Island, in the years 1920, 1921 and 1922, it was found that San Nicholas Island was very slow in recovering; 1920 developing hardly more than a seed crop and 1921 and 1922 did not make sufficient gains in point of feed to support cattle.*

In 1923, conditions improved somewhat but Vail abandoned the idea of putting cattle on San Nicolas Island and decided that it would be more viable to reintroduce sheep there (Visel 1923):

> *Within the past ninety days, after a recent examination, he [Mr. Vail] has developed the fact that it is doubtful whether San Nicholas can be brought back to the point of supporting cattle. However, the island has made rapid strides in the point of feed propagation and Mr. Vail decided that it would be commercially possible to put on sheep this year. Before putting on sheep, it will be absolutely necessary to build a certain amount of fence and equip the island with facilities for loading and unloading.*

Vail, in partnership with Robert L. Brooks, brought thousands of sheep from San Miguel Island to San Nicolas Island in 1924. Brooks had a sheep ranching operation on San Miguel Island at that time (Roberts 1991). Newspaper accounts in October 1924 (Anonymous 1924a, 1924b) reported that 2,500 sheep had been transferred from San Miguel Island to prevent their starvation:

> *When another dry year became a certainty this spring, the lessees of San Miguel Island [Robert L. Brooks and his partners] had a contract with a big feed corporation to feed the San Miguel sheep for the summer and everything seemed fine for the sheep owners. Then the foot-and-mouth disease broke out on the mainland and the feed corporation by the terms of its contract was released from its unprofitable bargain. It was then up to the San Miguel lessees to sell the sheep at any price and get them off the island before they died of starvation. The sheep could not be brought to shore because of the foot-and-mouth disease quarantine and the only place to take them was San Nicholas Island, owned by Vail and Vickers.*

The sheep were taken from San Miguel aboard the old schooner Vacquero *and were float-ed from the* Vacquero *to the shore of San Nicholas aboard a raft made of barrels and boards. Horsemen ashore with ropes attached to the raft pulled it from the ship to the beach and kept it there until the sheep were unloaded and by many such trips all but about 500 of the several thousand sheep that had been on San Miguel were transported about 30 miles to San Nicholas.*

There were reportedly 2,500 sheep on San Nicolas Island in the fall of 1924 (Rhodes 1924). In the early winter of 1925, it was reported that a man named McArthur and his wife were taking care of about 2,000 sheep on the island (Anonymous 1925).

Shortly after his grazing lease was renewed in 1924, Vail reportedly built a dock, shearing pens, a three-room house, three cisterns for domestic water, seven miles of hog wire fencing for pastures and drift fences (Visel 1928). By the spring of 1926, however, Vail was planning to remove most of the livestock from San Nicolas Island after three consecutive years of drought (Visel 1926, 1928).

Bruce Bryan, archaeologist with the Southwest Museum, explored San Nicolas Island in 1926 and met a Captain Nelson, "the 76-year-old sheepherder (an ex-sea captain), who lived on the island 'tend-ing' about 2,500 head of half-wild sheep" (Bryan 1970).

During his lease, Edward N.Vail reportedly introduced seeds of a number of non-native plants to the island for sheep forage, including *Atriplex semibaccata, Erodium, Medicago polymorpha,* and some un-named grasses (Visel 1928).

In March 1929, Lyman P. Elliott and his wife B. Edna Elliott accepted a job as island caretakers for Robert Brooks. They moved to San Nicolas Island with Milton Prentice, Edna's teenaged son from another marriage. Unfortunately, Milton was arrested, convicted, and imprisoned for awhile as a result of a shooting incident with several sheep thieves in the winter of 1929. As a result of this family dis-aster, the Elliotts convinced their daughter Margaret and her husband Roy E. Agee to help with the sheep operation. The Agees soon moved to San Nicolas Island with their daughter Frances (Swanson 1993, Beard 1994).

Sheep numbers had apparently been reduced again by the early 1930s. Luis E. Kemnitzer, gradu-ate student in geology at the California Institute of Technology in Pasadena, explored San Nicolas Island in the early 1930s. Over the span of three separate trips to the island, he spent about a month there. Kemnitzer (1933) reported that

At the present time the island is inhabited by one family of sheep ranchers who graze some twelve hundred head on the rather meager growth of grass. ... Plant life is sparse, being limited to a few scattered shrubs and cactus, ice plant, and some rather large tracts of grass used for grazing, principally on the upper slopes on the eastern half of the island.

John Thomas Howell, botanist at the California Academy of Sciences, visited the island on two days in March 1932. Howell (1935) reported that he saw numerous sheep grazing on the mesa. *Lycium* and *Opuntia* were apparently still acting as refuges for smaller plants (protecting them from sheep), as Howell found some species growing "among protecting thickets ...". Howell saw "badland" areas dur-ing his visit and reported areas where there was "little or no vegetation to cover the eroding slopes".

Transition to Military Control, 1930s-1940s. In January 1933, the island was transferred to the U.S. Navy by Executive Order of President Hoover, but sheep ranching continued on the island until the 1940s.

Robert L. Brooks sold his interest in San Nicolas Island to Roy Agee, Margaret Agee, Lyman Elliott, and Edna Elliott in September 1933. The sale included about 1,000 sheep, horses, cows, equipment, sheds, fencing, and houses located on the island (Brooks 1933). The U.S. Navy granted a revocable permit for sheep grazing on the island to the Agees and Elliotts in June 1934 (U.S. Navy 1934). A couple of years later, the Elliotts moved to Buellton on the mainland and sold their interest in the island to the Agees (Swanson 1993, Beard 1994).

The Agee family bought a ranch on the mainland in 1937 and began to hire other families to manage the sheep ranch on San Nicolas Island. The first managers that they hired were Roy McWaters and his family. The McWaters worked on San Nicolas Island from February 1937 until June 1938 (Beard 1994).

Reginald "Reggie" Lamberth, his wife Evelyn, their children Jeanne and Dennis, and Evelyn's two brothers (Lawrence and Raymond Foster) came out to San Nicolas island in June 1938. The Lamberths were hired by Roy Agee to replace the McWaters family as ranch managers. The Lamberths and Fosters lived there until the fall of 1939 and had some truly amazing adventures (Lamberth and Lamberth 1939, Woodward 1939, Foster 1992).

Loye Miller, avian paleontologist with the University of California at Los Angeles, spent about 10 days on the island in July 1938. The only woody plants that he reported from the mesa were a few *Baccharis pilularis* shrubs. He commented that: "How the Indians made fires enough to blacken the soil of their enormous mounds is a mystery to me" [an observation remarkably similar to that made by John Kelley in 1897]. He described the island's mesa as a "great rolling plain that might be a Kansas landscape for looks" (Miller 1938).

The Los Angeles County Natural History Museum's Channel Islands Biological Survey team of nine people spent a week on the island in July 1939. Donald Meadows (1939), one of the entomologists on the expedition, noted that:

> Great sand dunes cover the northern end, and the upper parts of the island are almost des-
> titute of vegetation. ... Sheep have been run on the island for many years and as a conse-
> quence the island is badly eroded and vegetation is sparse.

Meryl B. Dunkle (1939), the botanist on the expedition, thoroughly explored the north side of the island and observed that:

> Due to apparent overgrazing during the past, most of the shrubs and summer perennials
> have been destroyed, except for a few in large but scattered clumps of Opuntia littoralis and
> on the faces of cliffs. Destructive competition with the indigenous flora has been applied
> through the broadcast seeding of Atriplex semibaccata and Medicago hispida.

Open area near west end of San Nicolas Island in 1929 (Agee Family Collection, Santa Cruz Island Foundation).

View of ranch area, coastal flats, and north escarpment of San Nicolas Island in the early 1930s, looking to the south from a boat at sea (Agee Family Collection, Santa Cruz Island Foundation).

Edna Elliott and her son Milton at the older ranch house on San Nicolas Island in 1929. The three-room house was built by Ed Vail and Robert Brooks in the mid-1920s. Rainwater was collected from the metal roof and saved (Agee Family Collection, Santa Cruz Island Foundation).

The ranch area on San Nicolas Island in 1933 or 1934, looking to the north. The barn and sheep shearing shed is shown in the middle of this view (Agee Family Collection, Santa Cruz Island Foundation).

San Nicolas Island ranching families in front of the Agee ranch house (building on left) and the island's one-room schoolhouse (building on right) in 1933 (Standing left to right: the Agees, the Elliotts, Agnes Sanger Mundon, Ed Rucker, a friend of Agnes', Waif McAllister. In front with dogs: Shorty Daily and Frances Agee). Wynona "Waif" McAllister, a retired teacher from Idaho, came out to the island to teach Frances Agee (Agee Family Collection, Santa Cruz Island Foundation).

The island's north escarpment and southern portion of ranch area on San Nicolas Island in the mid-1930s, looking northwest. This view shows the storage shed for dry goods on the left and the Agee ranch house area just to the right of center. The barn and sheep corrals, which are out of view to the right, are shown on the facing page (Jeanne Lamberth Hess collection).

The northern portion of ranch area on San Nicolas Island in the mid-1930s, looking to the north. The blacksmith shop, barn, and sheep corrals in the foreground. The old ranch house area and Navy radio shack are shown in the background (Jeanne Lamberth Hess collection).

The ranch area on San Nicolas Island in the mid-1930s, looking to the south-southeast. The sheep corrals, barn, and blacksmith shop are visible on the left. The fuel storage building is just to the right of the tall pole. The dry goods storage shed and edge of the Agee ranch house are on the extreme right (Jeanne Lamberth Hess collection).

View of Agee ranch house and surroundings in 1940, looking to the south. The water tank was built by the Navy in 1939. The schoolhouse building, to the right of the water tank, had been enlarged and converted into a bunkhouse by this time (Agee Family Collection, Santa Cruz Island Foundation).

In their early days on San Nicolas Island, there was only one harness, so Roy Agee tied a rope to the saddle on Big Ben to help Monty haul this load of hay in 1930 (Agee Family Collection, Santa Cruz Island Foundation).

Sled used for hauling materials on San Nicolas Island. Roy Agee driving and Frances Agee riding in July, 1931 (Agee Family Collection, Santa Cruz Island Foundation).

In 1935 or 1936, an attempt was made to cultivate this piece of ground on the north side of San Nicolas Island, but it was a dry year and the winds reportedly blew away all but the hardiest, deeply-rooted plants (Agee Family Collection, Santa Cruz Island Foundation).

Using a truck frame, Roy Agee built a wheeled wagon that made it much easier to carry heavy loads on the island (Left to right: Roy McWaters, Roy Agee, Jimmy McWaters, Navy man and wife, Mrs. McWaters, Waif McAllister, and Frances Agee). The McWaters family came out to the island in 1937 to manage the sheep ranch (Agee Family Collection, Santa Cruz Island Foundation).

Roy Agee and crew on a fence repair expedition in 1934 (Agee Family Collection, Santa Cruz Island Foundation).

Roy Agee with team and wagon loaded with supplies brought to ranch from beach in 1935 (Agee Family Collection, Santa Cruz Island Foundation).

Moving supplies from beach to ranch on San Nicolas Island in the late 1930s. Imported materials such as the hay bales shown here probably contained seeds of non-native plants (Jeanne Lamberth Hess collection).

Cattle and fenced garden area at Thousand Springs in the late 1930s (Jeanne Lamberth Hess collection).

Fenced garden area at Thousand Springs in 1937. A variety of vegetables were grown here. Shorty Daily, one of the ranch hands, cut tin cans in half lengthwise and soldered them together to bring water from the spring to the garden area (Agee Family Collection, Santa Cruz Island Foundation).

Roy Agee with a prized watermelon grown on San Nicolas Island in 1932 (Agee Family Collection, Santa Cruz Island Foundation).

Sheep in corrals at ranch on San Nicolas Island, mid-1930s (Jeanne Lamberth Hess collection).

Sheep in corrals at ranch on San Nicolas Island, mid-1930s (Agee Family Collection, Santa Cruz Island Foundation).

Sheep on San Nicolas Island were rounded up for shearing once a year in the 1930s. This photograph shows the roundup and shearing crew at work in March 1930 (Agee Family Collection, Santa Cruz Island Foundation).

Sheep corrals at the ranch on San Nicolas Island. Newly-shorn sheep are in holding pen on the right (Agee Family Collection, Santa Cruz Island Foundation).

Sheep roundup and shearing crew on San Nicolas Island in March 1930 (Left to right in front: Roy Agee, Frances Agee, Edna Elliott. Left to right standing: Lyman Elliott, "Tin Can" Arnold, unknown, sheep shearers Frank and Fred.) (Agee Family Collection, Santa Cruz Island Foundation).

Sheep roundup and shearing crew on San Nicolas Island in February 1937 (Left to right in front: Lyman Elliott, "Tin Can" Arnold, Jimmy McWaters holding lariat, rider named Bill. Left to right in back: rider named Juan, Al Jensen, Roy Agee, Roy McWaters) (Agee Family Collection, Santa Cruz Island Foundation).

Big Ben and Roy Agee on the annual sheep roundup, ca. 1937 (Jeanne Lamberth Hess collection).

Roy and Frances Agee with other riders assembled for last sheep roundup on San Nicolas Island, June 1943 (Agee Family Collection, Santa Cruz Island Foundation).

Ramp leading down to ranch pier on San Nicolas Island in the early 1930s. The narrow chute prevented sheep from turning around when they were driven down to the pier. Full wool sacks awaiting shipment off the island are visible at the top of the chute (Agee Family Collection, Santa Cruz Island Foundation).

Loading sheep onto Alvin Hyder's 70-foot boat Nora II *in the mid-1930s. On this occasion, 309 sheep were loaded in 9 minutes and 3 seconds. Alvin Hyder hauled sheep, wool, and supplies on and off the Channel Islands from the 1920s until 1938, when ocean waves capsized the* Nora II *and took his life near San Nicolas Island (Agee Family Collection, Santa Cruz Island Foundation).*

After Alvin Hyder's boat was lost in 1938, island ranchers had to use tug boats and barges to send their sheep to the mainland (Agee Family Collection, Santa Cruz Island Foundation).

Loading sheep on barge for shipment to the mainland in the late 1930s or early 1940s (Jeanne Lamberth Hess collection).

There were still thousands of sheep on the island in 1938 and 1939, along with chickens, horses, mules, and cattle (Lamberth and Lamberth 1939, Foster 1992). Nearly 3,000 sheep were rounded up and sheared in July 1938 (Foster 1992). Arthur Woodward (1939), archaeologist on the Los Angeles County Museum's expedition, wrote that there were about 1,650 sheep on the island in July 1939 and that about 1,000 sheep had just been shipped off the island by barge (in mid-June 1939).

The McWaters family came back to San Nicolas Island in February 1940 and worked there again until March 1941 (Beard 1994).

Military Control, 1940s to the Present. In November 1942, the U.S. Army was given temporary jurisdiction over the island, and the western end reportedly became a range for artillery and bombing. The Army built an airstrip near the current airfield on the island and built facilities at the site of the sheep ranch. An Army Air Surveillance Squadron was stationed on San Nicolas Island during World War II (Schwartz 1994).

Most of the sheep were removed from San Nicolas Island by 1943 (Swanson 1993). A few sheep, however, remained on the island as late as the spring of 1949 (Anonymous 1949).

Wheeler (1944) reported that "in past years some 2,500 head of sheep have been supported by the grass that grows on the high central plateau [of San Nicolas Island]. The sheep have been removed for the duration of the war [World War II]".

Phil Orr and Egmont Rett, anthropologist and zoologist with the Santa Barbara Museum of Natural History, spent several weeks on the island during the spring and fall of 1945. Rett (1947) reported that:

> *Vegetation is sparse. The greater portion of the western half of San Nicolas is barren rock and shifting sand. The rather flat top of the eastern half and some of the less steep slopes are covered for the most part with foxtail grass* (Hordeum murinum), *bur clover* (Medicago hisp- ida), *and filaree* (Erodium cicutarium). *There are scattered patches of bush lupine* (Lupinus albifrons) *and in the canyons, washes, and other places, more or less protected from the wind, are found prickly pear cactus* (Opuntia) *and coyote brush* (Baccharis pilularis).

Orr (Irwin 1945, Orr 1945) commented that:

> *In late February ... the island plateaus were carpeted with fields of cream cups and gold- fields. ... there were wild flowers and a velvet green carpet of grasses ... covered parts of the island. The tallest plants on the island are the tree lupine and coyote brush growing in the shelter and moisture of the canyons where they reach a height of ten feet.*

> *The western end is lower than the plateau as a whole ... Vegetation is nonexistent except for a few small patches around springs. The remainder is a waste of sand and rock ...*

San Nicolas Island was returned to the Navy following the end of World War II. In 1947, the island came under the control of the Naval Air Station at Point Mugu as part of the Naval Air Missile Test Center (de Violini 1974, Schwartz 1994).

Since the ranching era ended, the impacts on the island's flora and plant communities have been from Naval operations of building and road construction, accidental fires, excavation of borrow pits, erosion from roads and pipelines, and attempts at revegetation with introduced species. Unfortunately, none of these activities are well documented (Halvorson *et al.* 1996).

The numerous wells drilled on the island by the Navy have probably affected the natural springs and seeps (Schwartz 1994). The fresh water pools just downstream from the seeps at Thousand Springs, which were definitely affected by the gardening activities of the sheep ranchers, suffered even more damage from the construction of a cement water catchment system by the Navy.

In the mid 1960s, efforts to promote revegetation on the island were begun. During this era, the Navy apparently fertilized portions of the island (especially around the airfield and on eroded slopes) to "encourage

Reggie Lamberth and his family came out to San Nicolas Island to manage the ranch for the Agees in 1938 (Left to right: Reggie and Evelyn Lamberth, Dennis Lamberth, Alfred Perry, and Evelyn's father Theodore Foster). A fresh water seep on the north side of the island is visible in the background (Jeanne Lamberth Hess collection).

Hundreds of sheep on the landscape of San Nicolas Island, mid-1930s. The ranching era on the island came to an end in the early 1940s, when sheep were removed (Jeanne Lamberth Hess collection).

Coastal lookout tower on mesa of San Nicolas Island in 1942 (photograph by S. Wheeler, Woodward Collection, Arizona Historical Society).

Ranch on north shore of San Nicolas Island in 1942, looking north-northeast. The Agee ranch house is visible on the left. The blacksmith shop and barn are shown on the right (photograph by S. Wheeler, Woodward Collection, Arizona Historical Society).

growth of vegetation and promote erosion control" (Anonymous 1965). About 116 tons of urea fertilizer were reportedly spread over 2,330 acres by four planes in 1965. The Navy barber on the island, Felix Fernandez, started to plant trees next to the Navy Exchange Building in 1965. By 1971, the trees covered nearly two acres (Anonymous 1971).

A test plot for fire-resistant plants was apparently established at Nicktown sometime in the early- to mid-1960s. The plot on San Nicolas Island was part of a program established by the Los Angeles County Foresty Department and the Los Angeles State and County Arboretum. Plantings on the island included *Atriplex* spp. and *Cistus* spp. (Fred Boutin, personal communication 2007). *Atriplex canescens* was one of the species included in the study (Gonderman 1966) and may have been introduced to the island as part of this test planting.

In the spring of 1970, volunteers with the Ventura County Nurserymen's Association reportedly took 200 specimens of trees and shrubs to the island. Austen Perley of the Ventura Park Department, four nurserymen, and a college student spent a weekend on the island installing the plants. The nurserymen "spent the day planting on the leeward side of hills. They placed trees near buildings in some cases but mainly they searched for and found isolated areas where plant life was desperately needed. One such location was One Thousand Springs where they planted material indigenous to California's arid areas". The list of trees that were planted included "junipers, pines, eucalyptus, and Washington palms". For ground cover, they planted "rosemary prostrata, myoporum, and osteospermum", an un-named "ground cover not unlike North African ice plant" and also spread seeds of "salt bush". They also "brought out plant material native to Australia and New Zealand". Navy personnel also reportedly made plantings of their own (Anonymous 1970). Many of the plants that the nurserymen and Navy personnel brought to the island have apparently persisted and at least one taxon *(Myoporum laetum)* continues to spread.

San Nicolas Island is now used primarily as a range instrumentation test site by the U.S. Navy. The island is equipped with facilities supporting metric radar, telemetry, Extended Area Test System (E.A.T.S.), optics, communications, microwave, missile launching, drone launching, surveillance radar, and target control (Dulka *et al.* 1993). The main support facilities include a 10,000 foot runway, an air terminal, housing, a power plant, a fuel farm, a reverse osmosis water system, and a public works and transportation detachment.

The introduction of non-native plants on San Nicolas Island has increased dramatically since the Navy took over the island in the 1940s (Table 5). Even though sheep were apparently moved on and off the island many times during the ranching era and ranchers deliberately introduced some non-native plants as forage and for erosion control, only 12 new non-native plant taxa were collected on San Nicolas Island between the 1890s and the 1940s. Perhaps sheep or horses were eating new plant arrivals before they had a chance to get established on the island. Since the early 1960s, 110 new records of non-native plant taxa have been documented. Many of these new introductions have been associated with gravel and other construction materials that have been brought to the island. The Navy's environmental staff has been making valiant efforts to control and/or eliminate new introductions of non-native plants and animals to the island, but staff and funding levels are extremely limited.

Some components of the island's native vegetation have recovered dramatically since the sheep ranching era. As mentioned above, woody plants were extremely rare on the island for many years. Nearly all of the native shrubs noted as rare by Blanche Trask in 1897 (Table 4) are now common on the island. The one exception to that general trend is *Lycium verrucosum,* which is now presumed to be extinct.

There have been significant changes on the island's landscape since the 1800s. Fragile lichen crusts have still not recovered in many areas on the island and sheep trails made more than 60 years ago are still visible. There has been a tremendous amount of erosion, as evidenced by isolated pedestals of original soil that can be seen near the northern and southern escarpments. The recovery process will be extremely slow in many parts of the island, but will hopefully be assisted by the efforts of current and future caretakers of San Nicolas Island.

TABLE 5. First Known Records of Non-Native Plant Taxa on San Nicolas Island, by Decade

Decade	New Records Found During Each Decade	Total Records to Date
1890-1899	19	19
1900-1909	0	19
1910-1919[1]	0	19
1920-1929[1]	0	19
1930-1939	7	26
1940-1949	5	31
1950-1959[1]	0	31
1960-1969	32	63
1970-1979	15	78
1980-1989	39	117
1990-1999	19	136
2000-2005	5	141

[1] There were no botanical collectors on San Nicolas Island during this decade.

HISTORY OF BOTANICAL EXPLORATION AND DISCOVERY

The 1860s to the 1880s. One of the first scientists to visit San Nicolas Island was James G. Cooper (1830-1902), a physician and field naturalist who visited on June 29, 1863. He also explored Santa Barbara, Santa Catalina, and San Clemente islands between May and July 1863 (Coan 1982).

Cooper may have collected plants on San Nicolas Island but, unfortunately, no specimens have been found. It is known that he collected at least one botanical specimen on the California Channel Islands in 1863 (Junak *et al.* 1993). In a letter to botanist Blanche Trask, Alice Eastwood of the California Academy of Sciences mentioned that another collector had visited San Nicolas Island before 1897, but those collections were lost before they could be studied. It is possible that Eastwood was referring to specimens collected by J.G. Cooper.

The 1890s to the 1900s. In the spring of 1897, the first documented specimens of vascular plants on San Nicolas Island were collected. Island explorer and field botanist Blanche Trask (Luella Blanche Engles, 1865-1916) visited the island in April 1897. On that trip, she traveled by schooner and landed at Corral Harbor. She apparently covered much of the island and collected 80 plant taxa. Her specimens were sent to Alice Eastwood, who identified them and published the results of the trip (Eastwood 1898).

Trask made additional plant collections on San Nicolas Island in April and early May 1901. On that trip, she was accompanied by friends, including Mrs. E.A. Ledbrook and Gus Knowles. She and her friends traveled to San Nicolas Island from Santa Catalina Island aboard the yacht *Avalon* and returned to Catalina on May 14, 1901 (Anonymous 1901). Trask returned to San Nicolas Island in late April 1902, accompanied by Mrs. E.A. Ledbrook, Gus Knowles, and Harry Doss. Trask had chartered the *Avalon* again and had a difficult time landing on San Nicolas Island because of rough seas; it was almost a week before she and her friends were able to get ashore. They spent a month on San Nicolas Island before returning to Avalon (Anonymous 1902a, 1902b, 1902c). It is not known whether Trask collected plant specimens on her trip in 1902.

Blanche Trask lived on Santa Catalina Island from the early 1890s until about 1915 (Jepson 1908, Windle 1940, Cantelow & Cantelow 1957). She lived much of the year in the town of Avalon and also had a house at Fisherman's Cove near the Isthmus (Jepson 1908, Windle 1940). In 1909, it was reported that "Sixteen years ago Mrs. Trask came to the island an invalid, today she can outwalk nearly every one here.

Blanche Trask, island explorer and field botanist who collected plants on San Nicolas Island in April 1897. This photograph, taken ca. 1900, shows her dressed in field clothes (Jepson Herbarium, University of California at Berkeley).

Often during the winter she walks from the Isthmus, taking to the trails, and covers the journey of fifteen miles in a little over three hours" (Anonymous 1909). Her husband, Walter J. Trask, was a prominent attorney in Los Angeles and divorced her in December 1895, after Blanche "deserted" him in October 1894 (public records for Los Angeles County). Willis Linn Jepson, professor of botany at the University of California at Berkeley, explored Catalina Island with Trask in 1908 and they became friends. Jepson (1908) wrote:

No one knows so much about Catalina Island as Mrs. Blanche Trask, who has been here about 17 years. ... For the island as a whole, its rocks, cliffs and cañons, as well as its plants, trees, and shrubs, this woman has a most remarkable love. ... I have never known anyone anywhere who knows the plants individually over such a large area as she does. She seems to know the individual trees and shrubs like old friends and knows whether they have changed in the last ten years and how much. If a Dendromecon *shrub has disappeared from the flood bed of Swain's Cañon, she misses it and finally locates the old stump. ... Mrs. Trask has lived so long in the open (she has a camp on the south side) that in appearance she does not suggest the woman she is, being bronzed by the desert sun. A heavy head of brown, slightly gray above the temples, good features, a happy smile when she is making some quaint joke, brown eyes (I think) – she is in her curious shepherd-like costume, with the blanket and staff or stock for climbing which she always carries, her short skirt scarcely reaching the knees, the lower part of the leg from the knee down encased in leggings of leather, a tall peaked straw hat with broad brim. This costume would be the wonderment of the Avalonians but that she departs on her trips early in the morning when only a few people are astir and returns in the watches of the night.*

In November 1916, Blanche Trask died in northern California. Jepson attended her funeral in San Francisco and wrote (Jepson 1916):

... Mrs. Trask was, as Miss Eastwood expressed it, "a wild woman". She had given up all that wealth could afford and the pleasures of a social career to live her life on Catalina! If she had died on Catalina it would have seemed fitting. But she was buried in a great city with only two or three persons present who had known her and no relatives! It seemed tragic, and as the words of the service went on, my mind left the confines of the undertaking chapel and I saw Mrs. Trask, once again, standing on a high ridge beyond Avalon in the moonlit shadows far in the night in silent worship of the sea and air, completely controlled by love of strange beauty and mysticism. Mrs. Trask botanized ardently on her island. She took long journeys on foot, with a shepherd's staff and a bit of food. She discovered several new species and collected many rarities. ...

Blanche Trask collected extensively on the Southern Channel Islands during the late 1890s and early 1900s. She collected vascular plants, lichens, Native American artifacts, minerals, and other natural history specimens. She published notes on the floras of Santa Catalina and San Clemente islands (Trask 1899, 1904) and also collected scientific specimens on Santa Rosa, Santa Cruz, and Santa Barbara islands.

Unfortunately, Trask's prime botanical specimens were deposited at the California Academy of Sciences and were destroyed during San Francisco's 1906 earthquake and fire (Millspaugh & Nuttall 1923, Cantelow & Cantelow 1957). Her personal herbarium collection was also destroyed in a large fire at Avalon in November 1915 (Millspaugh & Nuttall 1923), so only duplicate specimens distributed to other institutions can be examined today.

Seven plants restricted to one or more of the Channel Islands have been named in honor of Blanche Trask (i.e, *Astragalus traskiae* of Santa Barbara and San Nicolas islands, *Cercocarpus traskiae* of Santa Catalina Island, *Cryptantha traskiae* of San Nicolas and San Clemente islands, *Dudleya traskiae* of Santa Barbara Island, *Eriodictyon traskiae* subsp. *traskiae* of Santa Catalina Island, *Lotus dendroideus* var. *traskiae* of San Clemente Island, and *Mimulus traskiae* of Santa Catalina Island). A distinctive variety of mountain mahogany *(Cercocarpus betuloides* var. *blancheae)* found on Santa Rosa and Santa Cruz islands and in the Santa Monica Mountains also was named in her honor.

The 1910s to the 1920s. Between 1902 and 1932, there was a significant hiatus in the botanical exploration of San Nicolas Island. Apparently, no plant specimens were collected on the island during that time period. Luckily for those interested in early observations and documentation of the island's flora, Blanche Trask's irreplaceable contributions have survived.

The 1930s. John Thomas Howell (1903-1994), botanist at the California Academy of Sciences, visited the eastern end of San Nicolas Island on the afternoon of March 12, 1932 and climbed up to the island's mesa on the morning of March 13 (Howell 1932, 1935). He visited the island aboard the yacht *Zaca* as the first stop of the Templeton Crocker Expedition of 1932, which also explored several islands off the west coast of Mexico and the Galapagos Islands of Ecuador (Crocker 1933). Scientists on the expedition had intended to land again on San Nicolas Island when their boat returned to the area in late August 1935, but heavy surf prevented them from doing so. Howell (1935) published an update to Eastwood's list for the island, adding seven plant taxa that had not been reported earlier. Howell also collected plants on Santa Cruz, Anacapa, and on several of the Baja California islands. He published a number of papers on the plant life of the California Islands (e.g., Howell 1933, 1941, 1942).

Norman E. Bilderback collected botanical specimens on Santa Nicolas Island on April 21, 1938. He also collected on San Miguel, Santa Cruz, Santa Barbara, and Santa Catalina islands between April 14 and May 5, 1938. Bilderback and four other scientists associated with the San Diego Natural History Museum and the San Diego Museum of Man traveled to the islands aboard the yacht *Novia del Mar* on an expedition sponsored by the Scripps family (Anonymous 1938, MacMullen 1938).

Blanche Trask and her friend Mrs. Kinney in field clothes on Santa Catalina Island. This photograph was labelled "two of a kind" and may have been taken at Middle Ranch (Jepson Herbarium, University of California at Berkeley).

Loye H. Miller (1874-1970), zoologist and ornithologist with the University of California at Los Angeles, collected *Astragalus traskiae* on San Nicolas Island on July 15, 1938. Miller, who specialized in avian paleontology, was the first scientist to study the fossils at the famous La Brea tar pits in Los Angeles. He reached San Nicolas Island aboard the yacht *E.W. Scripps* and was there from July 7-17 with T.D.A. Cockerell (1866-1948), who was a biologist with Colorado State University, and geologist Harvey Allen (Miller 1938). Cockerell collected at least one plant *(Amsinckia spectabilis)* on San Nicolas Island and also collected botanical specimens on San Miguel and Santa Catalina islands. Cockerell focused his island studies on the native bees and also published papers on the plant life of the Channel Islands (e.g., Cockerell 1937, 1939).

Meryl B. Dunkle (1888-1969) collected extensively on the California Channel Islands, beginning with botanical specimens taken on Santa Catalina Island between February 1928 and May 1932. Dunkle was the principal of the Santa Catalina Island Schools in Avalon from 1923 until 1932 and head of the Science Department at Wilson High School in Long Beach from 1933 until 1954 (Windle 1940; David Hollombe, personal communication 1998). As field botanist for the Los Angeles County Natural History Museum's Channel Islands Biological Survey, he visited the other seven California Channel Islands between April 1939 and September 1941. Dunkle was on San Nicolas Island from July 22-28, 1939 and collected 391 sheets of 55 plant taxa (Dunkle 1939).

The 1940s. George P. Kanakoff (1897-1973), invertebrate zoologist and later curator of invertebrate paleontology at Los Angeles County Natural History Museum, collected botanical specimens from San Nicolas Island on April 12-24, 1940. He also made collections on San Miguel Island in April 1940 and on Santa Barbara and Middle Anacapa islands in August 1940.

Channel Islands Biological Survey team from Los Angeles County Natural History Museum on deck of California Fish and Game vessel Bluefin before their trip to San Nicolas Island (Left to right: George Kanakoff, Jewel Lewis, Don Meadows, Capt. W. Engelke, Capt. A. Groat, Russ Sprong, Art Woodward, Meryl Dunkle, Lloyd Martin, Jack von Bloeker, Jr.). At Pier 1, Long Beach, CA (Jeanne Lamberth Hess collection, 21 July 1939).

Members of the Channel Islands Biological Survey team, island foxes, and ranchers' children (Left to right: Dennis Lamberth, Jack von Bloeker, Jr., Jeanne Lamberth, and Russ Sprong). At camp on San Nicolas Island (Jeanne Lamberth Hess collection, July 1939).

Meryl B. Dunkle and Norman C. Cooper collected plants in the vicinity of Dutch Harbor on San Nicolas Island on November 24, 1940. These were apparently the last botanical specimens collected on the island before World War II.

Another significant hiatus in the botanical exploration of San Nicolas Island occurred during most of World War II. Under war-time regulations, the Channel Islands were closed to private yachts, and travel in the waters off the mainland coast was rigidly controlled by defense authorities (Wheeler 1944, Gherini 1966).

Phil C. Orr (1903-1991), curator of anthropology and paleontology at the Santa Barbara Museum of Natural History, spent about a month on San Nicolas Island between February and April 1945 (Irwin 1945, Orr 1945). He was accompanied by the Museum's curator of ornithology and mammalogy, Egmont Z. Rett (1897-1963). Orr and Rett collected a number of plants in mid-March, in addition to archaeological and zoological specimens. Orr explored many of the Channel Islands and wrote a book on prehistoric human use of Santa Rosa Island (Orr 1950). Rett (1947) published a report of the birds they recorded on San Nicolas.

Reid V. Moran (b.1916), herbarium curator and botanist at the San Diego Natural History Museum for many years, collected 14 numbers during a trip to San Nicolas Island on February 11-12 , 1949, on a cruise which also included Santa Barbara and Santa Catalina islands. He collected the type specimen of *Eriogonum grande* var. *timorum* on that trip. He visited the other Channel Islands on trips between February 1941 and August 1975, acting as botanist for the Los Angeles County Natural History Museum's Channel Islands Biological Survey on many of his early trips. Moran has published a number of papers on endemic plants of the California Channel Islands and on the flora of the Baja California islands (e.g., Moran 1995, 1996).

The 1950s to the 1960s. Another significant hiatus in botanical exploration on San Nicolas Island occurred during the 1950s. Although the staff at the Santa Barbara Botanic Garden began a study of the flora of the California Channel Islands in 1958, no trips were made to San Nicolas Island until 1961. E.R. "Jim" Blakley (b.1924), the Garden's grounds superintendent, collected on all eight Channel Islands between October 1958 and September 1964 as part of this program. Blakley visited San Nicolas Island on April 21-22, 1961 and collected 125 numbers.

Ralph N. Philbrick (b.1934), taxonomist at the Santa Barbara Botanic Garden and director from 1973 until 1987, continued and expanded the Garden's floristic projects. He has collected extensively on all of the California Islands and has published a number of papers on their flora (e.g., Junak *et al.* 1993, Philbrick 1964, 1972, 1980, Philbrick & Haller 1977). Philbrick began his trips to San Nicolas Island with a brief visit on February 9, 1964, when he collected cacti. He made general plant collections on the island on June 10-12, 1969, accompanied by Michael R. Benedict (b.1940), Garden research associate. Benedict, who has collected on all of the California Islands, made additional collections on San Nicolas Island on August 7, 1969 and on November 16-17, 1971. Philbrick returned to San Nicolas Island on June 22-23, 1977 with George Kritzman, curator of archaeology at the Southwest Museum in Los Angeles.

Ronald E. Foreman (b.1938), now emeritus professor of botany at the University of British Columbia, made three collecting trips to San Nicolas and prepared a report on the island's vegetation and flora for the U.S. Navy (Foreman 1967). When he visited the island, he was a civilian employee of the Navy and a graduate student at the University of California at Berkeley. On July 27-30, 1965, Foreman was accompanied by Evan C. Evans III and Sam C. Rainey. He returned to the island on December 17-22, 1965 with Robert M. Lloyd, and again on April 7-11, 1966 with Donald R. Smith. Foreman's thorough checklist documented 147 plant taxa for the island (excluding 35 previously reported plant taxa for which he did not find voucher specimens during herbarium research).

Peter H. Raven (b.1936), now director of the Missouri Botanical Garden, and Henry J. Thompson (b.1921), emeritus professor of botany at the University of California at Los Angeles, collected 116 numbers on San Nicolas Island on April 23-24, 1966. Raven has also collected botanical specimens on Santa Rosa, Santa Cruz, Santa Catalina, and San Clemente islands. He has published a flora of San Clemente Island (Raven 1963) and other papers on the flora of the California Channel Islands (e.g., Raven 1967).

Fredrick C. Boutin (b.1940) and Robert L. Gonderman (1917-1991), research associate and plant physiologist at Los Angeles State and County Arboretum, collected over 50 numbers at San Nicolas Island on May 17-18, 1967. Boutin also collected botanical specimens on San Clemente Island in April 1967. Boutin and Gonderman went to San Nicolas Island to examine a test plot for fire-resistant plants which had previously been established at Nicktown.

The 1970s. David B. Weissman (b.1947), entomologist associated with the California Academy of Sciences, collected botanical specimens on San Nicolas Island on June 11-12, 1971. He has published papers on the Orthoptera of the Channel Islands (e.g., Weissman and Rentz 1976).

Tom Hesseldenz (b.1953), who worked for many years with The Nature Conservancy and is currently a landscape architect in northern California, collected botanical specimens on San Nicolas Island on June 6, 1977. When he visited the island, he was a student at the University of California at Santa Barbara and helping with shorebird surveys. Hesseldenz worked for many years on Santa Cruz Island and has collected plants there.

Marla D. Daily (b.1950), now president of the Santa Cruz Island Foundation, collected botanical specimens on San Nicolas Island on April 28-29, 1978. At the time of her visit, she was working for the University of California Field Station on Santa Cruz Island. Daily returned to the island and made additional collections with Ann Bromfield (b.1949) on July 27-30, 1979. Daily has collected botanical specimens on Santa Rosa, Santa Cruz, Santa Catalina, and San Clemente islands, as well as on several of the Baja California islands. She has written a general book (Daily 1987) and coauthored a number of papers and books on the history of the Channel Islands (e.g., Daily & Stanton 1983). Bromfield has also collected plants on Santa Rosa Island and has coauthored papers on the vegetation of the Channel Islands (Carroll et al.1993, Laughrin et al.1994).

R. Mitchel "Mitch" Beauchamp and Harold A. "Howie" Wier (1952-2001), botanists employed by WesTec Services, surveyed the island between June 28 and July 5, 1978 and collected a number of botanical voucher specimens (WesTec Services 1978). Beauchamp has collected plants on Santa Catalina and San Clemente islands, as well as on several of the islands off Baja California, and has published papers on the flora of the Southern Channel Islands (e.g., Beauchamp 1987, 1997).

Glenn A. Gorelick, entomologist and biology teacher at Citrus College, collected botanical specimens on San Nicolas Island on July 29-30, 1978 as part of his studies of island butterflies. He has also collected on San Miguel, Santa Cruz, and Santa Catalina islands.

Robert F. Thorne (b.1920), emeritus taxonomist and herbarium curator at Rancho Santa Ana Botanic Garden, collected 116 numbers on San Nicolas Island on April 3-4, 1979. Thorne has collected extensively on nearly all of the California Islands and has published a flora of Santa Catalina Island (Thorne 1967) and other papers on the California Islands (e.g., Thorne 1969a, 1969b). He was accompanied on his trip to San Nicolas Island by Mark L. Hoefs (b.1942), former director of the Wrigley Botanical Garden on Santa Catalina Island, Clarence W. "Dick" Tilforth (1916-1994), former horticulturist at Rancho Santa Ana Botanic Garden, and Robert R. Given (b.1932), former director of the Catalina Marine Science Center. Hoefs has collected extensively on Santa Catalina Island, and has also collected botanical specimens on Santa Rosa, Santa Cruz, and San Clemente islands. Hoefs has also collected plant specimens on several of the Baja California islands.

Gary D. Wallace (b.1946), then botanist with the Los Angeles County Museum of Natural History, was also on the April 3-4, 1979 trip. Wallace collected plants under his own numbers, which he incorporated into his very detailed vascular plant checklist for the Channel Islands (Wallace 1985). He reported 114 native and 66 non-native plant taxa for San Nicolas Island and cited voucher specimens, confirming many of the taxa that Foreman (1967) did not locate during his herbarium work. Wallace has also collected botanical specimens on Santa Rosa, Santa Cruz, Santa Catalina, and San Clemente islands.

The 1980s to the Present. Judith Newman, Julie M.Vanderwier (b.1953), and Thomas G. Murphey (b.1957), biologists with the Environmental Division at Point Mugu Naval Air Station, collected botanical specimens on the island in the early to mid-1980s. Newman collected plant specimens on the island every month between March and August 1980, November 19-20, 1980, and July 18-20, 1981 (with Tom Murphey). Murphey collected additional plant specimens in July 1982 and April 1983. Vanderwier made additional collections between April 1982 and May 1985, primarily during the springtime. Both Vanderwier and Murphey have also collected plants on Santa Cruz and Santa Barbara islands, and Murphey has collected on San Miguel Island as well. Vanderwier coauthored an annotated checklist for San Nicolas Island (Junak & Vanderwier 1990) which documented 133 native and 116 non-native plant taxa.

My own work on the island began in March 1983, during a wet El Niño year. Steven L. Timbrook (b.1938) and Caroline Kuizenga (b.1940) of the Santa Barbara Botanic Garden accompanied me on that first, very memorable trip. Timbrook collected *Toxicodendron* on our trip to San Nicolas Island and has also collected botanical specimens on San Miguel, Santa Cruz, Anacapa, and Santa Barbara islands. Kuizenga has collected plants on Anacapa Island as well.

I made 96 additional trips to San Nicolas Island between May 1983 and March 2003, reaching the island by fixed-wing plane or helicopter each time (much easier than going by boat!). I have collected a total of 1,744 numbers on San Nicolas Island and, after repeated searches, have been able to relocate populations of almost every plant taxon reported by Blanche Trask in 1897. On most of my trips to San Nicolas Island, other botanists and I were mapping rare plants, invasive plants, or plant communities for the U.S. Navy (Junak *et al.* 1995b, 1996; Halvorson *et al.* 1996; Junak 2003a, 2003b, 2003c). I have been fortunate enough to explore and collect botanical specimens on all of the California Islands.

William S. "Stan" Davis (b.1930), emeritus professor of botany at the University of Louisville, collected voucher specimens for his systematic studies of *Malacothrix* nearly every year between May 1985 and May 1990. Davis has published several papers on the genus (e.g., Davis & Junak 1993, Davis 1997). He has also collected voucher specimens on San Miguel, Santa Cruz, Anacapa, Santa Barbara, and San Clemente islands.

Other plant collectors during the mid- to late-1980s included Bill Wright on February 18, 1985 and L. Gallop on March 26, 1988. Joe Divittorio (b.1952), biologist with the Environmental Division at Point Mugu Naval Air Station from 1986-1988, collected plants on the island during this time period. Grace Smith (b.1953), biologist with the Environmental Division at Point Mugu Naval Air Station since 1989, has also been collecting botanical specimens on San Nicolas Island since the early 1990s.

Botanists with the National Park Service and U.S. Geological Survey have collected additional plant specimens on San Nicolas Island since the early 1990s as part of contract studies for the U.S. Navy. Ronilee A. Clark (b.1954), Karen C. Danielsen (b.1958), William L. Halvorson (b.1943), Katherine A. "Katie" Chess (b.1967), and others have conducted botanical monitoring programs, rare plant surveys, or revegetation activities in Channel Islands National Park and on San Nicolas Island (e.g., Chess *et al.*1996, Clark & Halvorson 1989, Clark *et al.*1990, D'Antonio *et al.*1992, Halvorson *et al.*1992). Most of these botanists have also collected plants on the Northern Channel Islands and on Santa Barbara Island.

Tim Thomas, now retired from the U.S. Fish and Wildlife Service, collected plants on San Nicolas Island on April 8-9, 1992, while helping with the U.S. Geological Survey projects on San Nicolas Island. He has also collected botanical specimens on Santa Rosa, Santa Cruz, and San Clemente islands.

Charles A. Drost (b.1957), biologist with the U.S. Geological Survey, has collected plants on San Nicolas Island since the 1990s, while studying the island night lizard. He coauthored a paper on the flora of Santa Barbara Island (Junak *et al.* 1993) and has published papers on the island night lizard and island deer mice (e.g., Drost and Fellers 1991, Fellers and Drost 1991). He has collected botanical specimens on the Northern Channel Islands and on Santa Barbara Island as well.

Marla Daily of the Santa Cruz Island Foundation made a few more botanical collections on the island on May 9,1994. Travis Olson collected a few plants on August 19,1994. Travis Columbus of Rancho Santa Ana Botanic Garden collected grasses on May 21-23, 2001.

In addition to herbarium specimens, botanists and horticulturists have also collected seeds and cuttings from plants on San Nicolas Island since at least the 1960s. Living collections can be seen and studied at several California institutions, including Rancho Santa Ana Botanic Garden, Santa Barbara Botanic Garden, and the University of California (Berkeley) Botanical Garden.

FLORISTIC ANALYSIS

The vascular flora of San Nicolas Island includes 52 families of plants, 172 genera, 273 species, and 1 common hybrid *(Pelargonium* x *hortorum)*. Only four species are represented by more than one variety or subspecies; there are eight such taxa. Consequently, 278 taxa at or below the species level occur on San Nicolas Island. The three largest families are Asteraceae, Poaceae, and Fabaceae; the 3 largest genera are *Atriplex, Bromus,* and *Trifolium*.

San Nicolas Island has limited species diversity when compared to the other California Channel Islands (Table 6), presumably due in large part to its isolated location and relative lack of habitat diversity compared to the other islands in the group. It is also possible that some native plants were eliminated by intense grazing pressures on the vegetation and thus were not found even by the earliest botanical explorers (Raven 1967).

TABLE 6. Numbers of Vascular Plant Taxa on the California Channel Islands.

Island	Plant Taxa			
	Native[1]	Endemic[2]	Non-Native[3]	Total
Northern Channel Islands				
San Miguel	209	18 (9)	82 (28)	291
Santa Rosa	403	42 (10)	120 (23)	523
Santa Cruz	488	46 (9)	191 (28)	679
Anacapa	190	22 (12)	80 (30)	270
Southern Channel Islands				
Santa Barbara	88	14 (16)	45 (34)	133
San Nicolas	137	15 (11)	141 (51)	278
Santa Catalina	431	37 (9)	203 (32)	634
San Clemente	297	47 (16)	135 (31)	433

[1]Totals shown for native taxa include those endemic to the California Islands.
[2]Endemic taxa are those found only on one or more of the California Islands. Number in ( ) represents percentage of total native taxa in island flora.
[3]Number in ( ) represents percentage of total taxa in island flora.

Sources: Raven (1963), Thorne (1967, 1969), Philbrick (1972), Junak *et al.* (1993), Junak *et al.* (1995), Ross *et al.* (1997), Junak *et al.* (1997), and unpublished updates.

Non-Native Species. Eleven of the families and 78 of the genera on San Nicolas Island are represented exclusively by non-native taxa. At least 141 taxa (including 140 species) are considered to be introduced to San Nicolas Island and represent almost 51% of the island's total flora. Percentages of non-native taxa on the other California Channel Islands range from 23% to 34% (Table 6). Most of the non-native taxa are native to the Mediterranean region (Europe, n Africa, and w Asia) or Australia. Sixteen of the introduced taxa are native to the California mainland *(Amsinckia menziesii* var. *intermedia, Atriplex argentea* var. *mohavensis, Atriplex canescens* var. *canescens, Atriplex lentiformis, Chenopodium berlandieri, Encelia californica, Eriogonum cinereum, Eriogonum fasciculatum* var. *polifolium, Eschscholzia californica* subsp. *californica, Grindelia hirsutula, Heterotheca grandiflora, Lotus salsuginosus* var. *salsuginosus, Lupinus succulentus, Spergularia salina, Verbena lasiostachys* var. *lasiostachys,* and *Xanthium strumarium)* but were either intentionally or unintentionally introduced to San Nicolas Island during the last 100 years. One taxon *(Lavatera assurgentiflora* subsp. *assurgentiflora)* is native to other California Channel islands but was apparently planted on San Nicolas Island and has subsequently escaped from cultivation. At least ten non-native plant taxa, which have become naturalized on San Nicolas Island, have not yet been reported on the other California Channel Islands *(Ammophila arenaria, Calendula officinalis, Calystegia malacophylla* subsp. *pedicellata, Daucus carota, Delosperma litorale, Pelargonium peltatum, Rumex obtusifolius, Salsola kali* subsp. *pontica, Soleirolia soleirolii,* and *Stenotaphrum secundatum).*

Hybridization. The flora of San Nicolas Island includes a few natural hybrids and at least one horticultural introduction of hybrid origin *(Pelargonium* x *hortorum).* The most conspicuous examples of hybridization occur in the genera *Abronia, Malacothrix* (Davis and Junak 1993), and *Opuntia* (Philbrick 1963, Junak 2003a). Putative natural hybrids between species has also been observed in the genus *Cakile.*

Endemism. The California Channel Islands are known for high levels of endemism (Raven 1967). Some of the characteristics of plant taxa endemic to the California Channel Islands have been discussed by Philbrick (1980) and Thorne (1969). At least 12 of the island's native taxa are island endemics, not found on the mainland but occurring on at least two of the California Channel Islands *(Artemisia nesiotica, Astragalus traskiae, Calystegia macrostegia* var. *amplissima, Cryptantha traskiae, Deinandra clementina, Eschscholzia ramosa, Gilia nevinii, Jepsonia malvifolia, Lomatium insulare, Lotus argophyllus* var. *argenteus, Malacothrix saxatilis* var. *implicata,* and *Trifolium palmeri).* Unfortunately, *Gilia nevinii* has not been seen for many years on San Nicolas Island. One additional taxon occuring on San Nicolas Island *(Marah macrocarpus* var. *major)* is probably endemic to the California Islands but needs further study. Three taxa are endemic to San Nicolas Island *(Eriogonum grande* var. *timorum, Lycium verrucosum,* and *Malacothrix foliosa* subsp. *polycephala).* Regrettably, *Lycium verrucosum* is now thought to be extinct.

Relationships Among Islands. Floristic relationships among the California Channel Islands have been examined and discussed by Raven (1967), Thorne (1969), and Wallace (1985), Moody (2000), and Oberbauer (2002). San Nicolas Island shares from 13 to 113 of its native taxa (9.5% to 82.5%, respectively) with other California and Baja California islands (Table 7). Floristic affinities are related to distance and size of island, with the largest islands (Santa Rosa, Santa Cruz, Santa Catalina, and San Clemente) having the highest numbers of shared native taxa with San Nicolas. Among the Baja California islands, Guadalupe shares the greatest number of native taxa with San Nicolas. With respect to insular endemics, San Nicolas Island has its closest affinities with the other Southern Channel Islands, especially Santa Barbara and San Clemente (Table 7).

TABLE 7. **Floristic Relationships between San Nicolas and Other Islands of Southern and Baja California.**

Number of Native Plant Taxa Shared with San Nicolas Island

	Native[1]	Endemic
California Channel Islands		
San Miguel	95 (69)	1
Santa Rosa	103 (75)	4
Santa Cruz	113 (82)	5
Anacapa	90 (66)	4
Santa Barbara	59 (43)	8
Santa Catalina	112 (82)	6
San Clemente	105 (77)	10
Baja California Islands		
Los Coronados	38 (28)	1
Todos Santos	36 (26)	1
San Martin	34 (25)	1
Guadalupe	47 (34)	5
San Benito	13 (9)	1
Cedros	42 (31)	1
Natividad	13 (9)	1

[1]Totals for native taxa include those endemic to the California Islands. Number in ( ) represents percentage of total native taxa known from San Nicolas Island.

GUIDE FOR THE READER

Included Taxa. All known native and introduced (non-native) vascular plant taxa that are growing and reproducing without cultivation on San Nicolas Island are included. Introduced or non-native plants include those originally planted (e.g., *Opuntia ficus-indica*) and now naturalized. Taxa that were apparently planted but have not subsequently spread to surrounding areas or established new populations (e.g., *Aeonium* spp., *Cotyledon orbiculata*, *Eucalyptus globulus*, *Heteromeles salicifolia*, *Phoenix canariensis*, *Pinus* spp., and *Washingtonia* spp.) have been excluded, even if individual plants are persisting without cultivation. Taxa planted near buildings or along main roads are not included unless they appear to be spreading. Representative herbarium specimens are cited in Appendix I. Doubtful or unsubstantiated reports for San Nicolas Island are listed in Appendix II.

Arrangement of Taxa. Plant taxa are arranged alphabetically by family within related plant groups (i.e., ferns and fern allies, dicotyledonous flowering plants, and monocotyledonous flowering plants).

Identification Keys. Dichotomous identification keys are provided for all taxa (families, genera, species, subspecies, and varieties) known to occur on San Nicolas Island. For convenience, a few large keys (families of flowering plants, Asteraceae, and Poaceae) are separated into two or more "Groups", each of which shares a pattern of morphological variation. Dichotomous keys are composed of a series of consecutively numbered, comparative statements or couplets. Couplets within each key are assigned a unique number. The first phrase of a couplet begins with a number followed by a period; the alternate second phrase begins with the same number followed by an apostrophe.

Names of Taxa. Scientific names used in this book are generally those considered to be the most correct, based on published studies of taxonomic relationships and conventions established by the International Code of Botanical Nomenclature. Most of the names used are consistent with Hickman (1993) and Junak *et al.* (1995a). Use of other names for some taxa at the specific, subspecific, or varietal levels are based on more recent literature. A limited number of synonyms are included for selected taxa, especially those that have been cited in earlier floras for the California Channel Islands. Citation and abbreviation of authors follow Brummitt and Powell (1992). Common names are mostly according to Smith (1976) and Junak *et al.* (1995a).

Plant Families and Genera. Statements regarding life cycle or habit of families and genera are based on the range of variation found on San Nicolas Island. This is followed by information regarding approximate number of included genera or species, their worldwide distribution, and comments regarding toxicity and ornamental, economic, or medicinal uses. These data are mostly according to Mabberley (1997) or completed volumes of the Flora of North America series (e.g., Flora of North America Editorial Committee 1993). Lastly, information on etymology of generic names is given (enclosed in parentheses).

Entries for species, subspecies, and varieties known from San Nicolas Island include information (if pertinent) that is organized as follows:

Native versus Introduced Status. All taxa are presumed to be native to the island unless their scientific name is preceded by an asterisk. Such non-native species are considered introduced to California by human intervention beginning with Spanish colonization in the late 1700s. An estimate of their earliest known occurrence on the island, based on herbarium specimens and publications, is provided for most non-native taxa by a note that begins "First collected ..." The geographic origin of alien taxa is given as part of the distribution statement (see below). If native or non-native status is in doubt (e.g., *Galium aparine*), appropriate comments are appended. Several taxa (e.g., *Atriplex lentiformis* subsp. *breweri, Conyza canadensis, Encelia california, Grindelia camporum* var. *bracteosum, Heterotheca grandiflora, Lotus salsuginosus,* and *Lupinus succulentus*) are considered to be introduced on San Nicolas Island, even though they are native to the California mainland. Some of these native California taxa appear to have been intentionally planted on the island, while others have been found on imported gravel piles or in severely disturbed areas.

Habit and Size. Habit and size information is based on morphological variation of plants found on San Nicolas Island, as represented by plants in the field and herbarium specimens deposited at the Santa Barbara Botanic Garden and other institutions.

Time of Flowering. The time during which plants are producing spores or flowers is given as a range from earliest to latest month. Months are abbreviated to the first three letters. Exceptional flowering times are enclosed in parentheses.

Frequency of Distribution. Frequency of distribution was divided into five categories, each of which provides an estimate of relative abundance. These categories are based on extensive field observations and by the known number of localities as determined from herbarium specimens. Localities in this system are considered to be populations separated by distances of at least 0.25 mile (ca. 0.4 kilometer).

Taxa that are exceptionally widespread or conspicuous in many habitat types and known from more than 30 localities are treated as "**Abundant.**" Taxa that are widespread or conspicuous within one to several habitat types, regularly encountered throughout the island, and known from 21-30 localities are treated as "**Common.**" Taxa that are fairly widespread within one or a few habitat types, have a scattered distribution, and known from 11-20 localities are treated as "**Occasional.**" Taxa that are somewhat widespread, often locally common, but known from 4-10 localities are treated as "**Scarce.**" Taxa that are known from fewer than 3, often very restricted, localities are considered to be "**Rare.**" However, in some cases (e.g., *Orobanche fasciculata*) such taxa may be common elsewhere and not considered rare for the purposes of legal protection.

Habitat Types. Brief descriptive phrases (e.g., disturbed flats, sandy flats, grassy sites, coastal slopes) are given to describe the general physiographic conditions in which each taxon is likely to occur.

Geographic Distribution. Geographic distribution, with the exception of San Nicolas Island endemics, is composed of two separate statements: distribution on the island and distribution elsewhere.

The distribution on San Nicolas Island is divided into three major areas corresponding to major physiographic features. General or specific localities, if pertinent, are listed for each of the three major areas: the north side of the island, the mesa, and the south side of the island. Localities are listed in a roughly west-to-east sequence in each of the three major areas. Placenames on the island are shown on the map included in this publication. When two or three major areas are represented, statements for each are separated by semicolons. The "north side" of San Nicolas Island extends from Vizcaino Point on the west to the Sand Spit on the east end; the northern coastline, northern coastal flats, and northern escarpment (the steep slopes and associated canyons between the northern coastal flats and the northern edge of the mesa) are included in this area. The elevated tableland portion of the island, including the upper portions of Tule, Mineral, Celery, West Mesa, and East Mesa canyons, is considered to be the "mesa." The "south side" of the island extends from Vizcaino Point to the Sand Spit; it includes Dizon's Ravine, Army Springs, the southern coastline, southern coastal flats, and southern escarpment (the steep slopes and associated canyons between the southern edge of the mesa and the southern coastal flats).

The geographic distribution of taxa occurring outside of San Nicolas Island is given (if pertinent) for other California Channel Islands, Baja California Islands (on the west side of the peninsula), the general mainland distribution, and, in the case of non-native taxa, other areas of the world where the taxa are considered to be invasive, and the general region of origin. Mainland distributions for the United States are given as a range from north to south, supplemented by eastward distribution beyond California (if appropriate). The sequence for California islands includes San Miguel, Santa Rosa, Santa Cruz, Anacapa, Santa Barbara, Santa Catalina, and San Clemente. The sequence for Baja California includes Los Coronados, Todos Santos, San Martin, San Geronimo, Guadalupe, San Benito, Cedros, and Natividad islands. United States Postal Service abbreviations are used for the continental states (e.g., AZ for Arizona, CA for California, WA for Washington). The geographic origin of "Introduced" taxa is given as "native of ..."

Endemic Taxa. Taxa believed to occur only on the island are noted as "Endemic to San Nicolas Island." Taxa known to occur on other islands, but not on the mainland, are noted as "Endemic to ... (followed by a list of islands)."

Other Comments. Other comments include notes pertaining to economic use (especially by Native Americans), toxicity, taxonomic relationships, and first records of occurrence (for introduced taxa). Native American tribes are not identified in most cases and exact medicinal uses are not given. Before attempting to use plants for food or medicine, the reader should consult appropriate references. Primary sources of information on edible and medicinal plants included Bean and Saubel (1972), Clark (1977), Strike (1994), Tanaka (1976), Timbrook (1984), and Uphof (1968). Information on toxic plants was obtained mostly from Fuller and McClintock (1986).

Illustrations. With a few exceptions, every species, subspecies, and variety known to occur without cultivation on San Nicolas Island is illustrated. In most cases, at least one habit drawing is included for each genus; at least one diagnostic character is illustrated for each species, subspecies, or variety. Taxa not illustrated include those either reported for San Nicolas Island (but not documented with herbarium specimens) or those observed and documented after final illustration plates were prepared. Illustrations are accompanied by vertical or horizontal lines that are drawn to scale and measured in cm or mm. When appropriate, labels are used to identify individual structures. The pertinent scientific name appears immediately below the habit and/or structure(s).

KEY TO MAJOR TAXONOMIC GROUPS

1. Plants reproducing by means of spores; spores produced in sporangia on surfaces of leaf blades; flowers and seeds not produced ....**Pteridophytes**
1' Plants reproducing by seeds and pollen; seeds enclosed in a dry or fleshy fruit; ovary with style(s) and stigma(s) ....**Angiosperms**

PTERIDOPHYTES
FERNS AND FERN ALLIES

Key to Families
1. Sori borne along midvein or away from blade margin; indusia absent ....**Polypodiaceae**
1' Sori borne along blade margin; indusia absent or sori partly covered by reflexed or inrolled leaf blade (false indusium) ....**Pteridaceae**

POLYPODIACEAE Bercht. & J.Presl
POLYPODY FAMILY
Perennials with rhizomes. Ca. 40 genera and 500 species; worldwide, especially in tropical and subtropical regions.

POLYPODIUM L. POLYPODY
Ca. 100 species; worldwide. Some species are cultivated as ornamentals. (Latin: many feet, referring to the branched rhizome)

P. californicum Kaulf. CALIFORNIA POLYPODY Plants 15-25 cm tall. Dec-Feb. Rare; n facing slopes. North side in Live-forever and Celery canyons, and in canyon just e of Airfield Grade on Beach Road. All CA Channel Islands except San Miguel; Los Coronados, Todos Santos, Guadalupe, and Cedros islands; n CA (Mendocino Co.) to Baja CA. Rhizomes were eaten by Native Americans.

PTERIDACEAE Rchb.
BRAKE OR MAIDENHAIR FAMILY
Perennials with rhizomes. Ca. 40 genera and 1000 species; temperate and tropical regions worldwide.

1. Lower surfaces of leaf blades covered with a waxy yellow or silvery powder (farina) which ages blackish ....**Pentagramma**
1' Lower surfaces of leaf blades (except for sori) ± glabrous
  2. Ultimate leaf segments thin, membranous; margins lobed or toothed ....**Adiantum**
  2' Ultimate leaf segments thick, leathery; margins entire ....**Pellaea**

ADIANTUM L. MAIDENHAIR FERN
Ca. 150-200 species; many in tropical America but found nearly worldwide. Some species are cultivated as ornamentals. (Greek: unwettable, referring to the water-repellent leaves)

A. jordanii Müller Halle CALIFORNIA MAIDENHAIR [*A. emarginatum* D.C.Eaton] Plants 45-60 cm tall. Feb-Jun. Rare; n-facing slopes and flats near canyon bottoms. North side in second canyon w of "L" Canyon and in Jetty Canyon. All CA Channel Islands except Santa Barbara; OR to Baja CA. Plants were used medicinally by Native Americans.

POLYPODIACEAE

PTERIDACEAE

2 cm

2 cm

5 mm

sorus

ultimate segment

Polypodium californicum

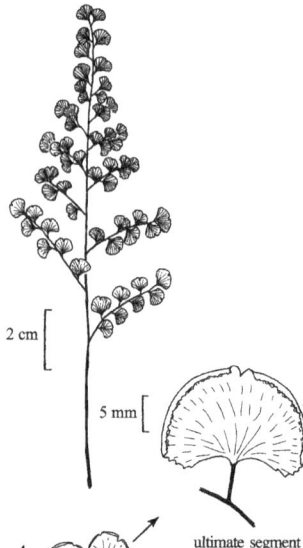

pinna

Adiantum jordanii

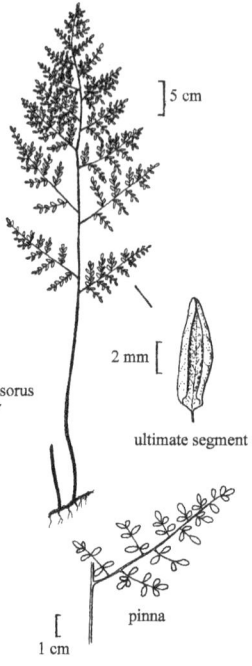

5 cm

2 mm

ultimate segment

1 cm

pinna

Pellaea andromedifolia

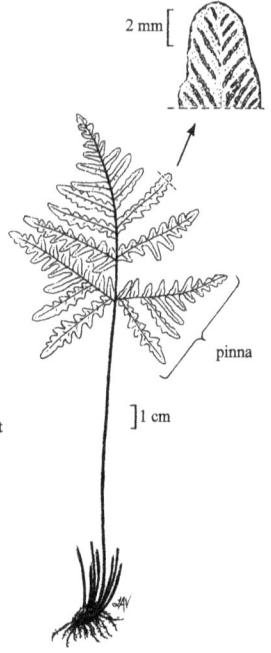

2 mm

1 cm

pinna

Pentagramma triangularis
subsp. triangularis

AIZOACEAE

1 cm

5 mm

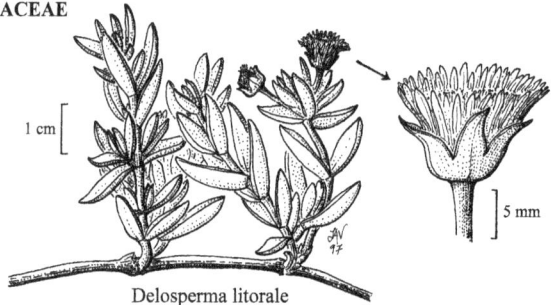

Delosperma litorale

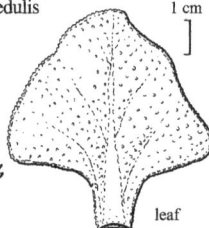

1 cm

C. edulis

Carpobrotus chilensis

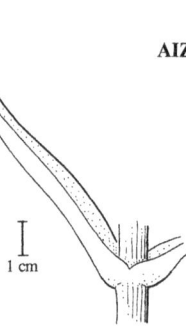

1 cm

leaf

M. crystallinum

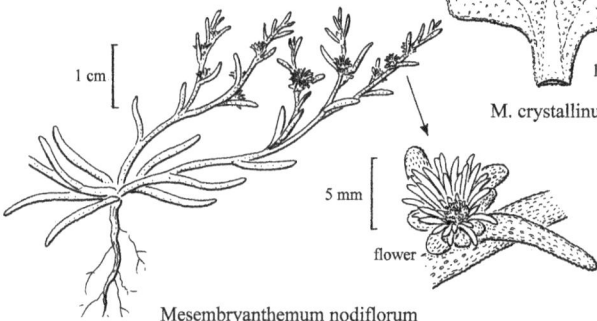

1 cm

1 cm

5 mm

flower

Mesembryanthemum nodiflorum

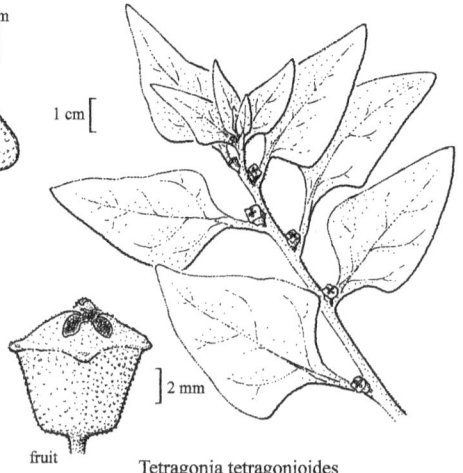

2 mm

fruit

Tetragonia tetragonioides

PELLAEA Link CLIFF-BRAKE

Ca. 40 species; worldwide but mostly in W Hemisphere. (Greek: dusky or dark, referring to the dark brown or blackish petioles)

P. andromedifolia (Kaulf.) Fée COFFEE FERN Plants 6-9 dm tall. Feb-Aug. Scarce; n-facing canyon walls, slopes, and flats. North side on escarpment and in unnamed canyons on both sides of Airfield Grade on Beach Road, in first and second canyons w of "L" Canyon, in "L" Canyon, and in S Spur Canyon. All CA Channel Islands except Santa Barbara; Los Coronados and Cedros islands; OR to Baja CA. Plants were used by Native Americans in basketry and to make medicinal tea.

PENTAGRAMMA Yatsk., Windham, & E.Wollenw.
GOLDBACK or SILVERBACK FERN

Two species; w N America. California species were previously included in *Pityrogramma* Link, a tropical genus. (Greek: five lines, referring to the pentagonal leaf blades)

P. triangularis (Kaulf.) Yatsk., Windham, & E.Wollenw. subsp. **triangularis** GOLDBACK FERN [*Pityrogramma t.* (Kaulf.) Maxon] Plants 30-45 cm tall. Mar-Jun. Occasional; n-facing slopes, canyon walls, and flats. North side in scattered populations on coastal flats and escarpment between W Mesa and S Spur canyons; on mesa just e of Nicktown and in E Mesa Canyon. All CA Channel Islands except Santa Barbara; Cedros Island; British Columbia to Baja CA, e to ID. Petioles were used by Native Americans in basketry and for medicinal purposes.

ANGIOSPERMS
FLOWERING PLANTS

Key to Families and Groups

1. Leaf venation parallel, simple; perianth parts, if present, in multiples of 3; vascular bundles in stem cross-section usually scattered ....**GROUP 1 (MONOCOTS)**
1' Leaf venation pinnate or palmate; perianth parts, if present, in multiples of 4 or 5; vascular bundles in stem cross-section usually organized in ring ....**(DICOTS)**
  2. Shrubs or trees ....**GROUP 2**
  2' Perennial herbs or annuals
    3. Perianth parts apparently absent or only in 1 whorl....................**GROUP 3**
    3' Perianth parts in 2 or more whorls
      4. Petals free from each other ....**GROUP 4**
      4' Petals fused together ....**GROUP 5**

GROUP 1. MONOCOTS

1. Plants aquatic and often completely submerged
  2. Plants of salt water at or below sea level; inflorescences one-sided, ± flattened . .**Zosteraceae**
  2' Plants of fresh or brackish water at or above sea level; inflorescences neither flattened nor one-sided ....**Potamogetonaceae**
1' Plants terrestrial; if aquatic, then emergent, with roots and bases of stems rooted in mud and most leaves and flowers above water
  3. Perianth present; inner parts purplish to whitish, outer ones also brightly colored . .**Alliaceae**
  3' Perianth either absent or composed of greenish to brown, scale-like to bristly parts
    4. Perianth parts 6, scale-like; fruit a capsule with many seeds ....**Juncaceae**
    4' Perianth parts absent or, if present, bristle-like; fruit indehiscent, usually with 1 seed
      5. Inflorescence a dense cylindrical spike 1.5-3.5 cm wide ....**Typhaceae**

     5' Inflorescences of various types (racemes, spikes, or panicles) but not dense and cylindrical
          6. Stems usually sharply or obtusely 3-angled, usually with solid internodes; leaves in 3 ranks; stamens and pistil usually subtended by 1 bract; fruit lenticular or 3-angled . . . .**Cyperaceae**
          6' Stems circular to terete, usually with hollow internodes; leaves in 2 ranks; stamens and pistil usually subtended by 2 bracts; fruit usually elliptic or oblong in outline, not compressed or angled . . . .**Poaceae**

GROUP 2. DICOT SHRUBS or TREES

1. Perianth parts apparently absent or only in 1 whorl
    2. Leaves opposite at most nodes; leaves scale-like; stems ± succulent, with short, joint-like internodes . . . . **Chenopodiaceae *(Salicornia)***
    2' Leaves alternate at most nodes; leaves not scale-like; stems not succulent; internodes not joint-like
        3. Inflorescence a panicle, terminal; leaves palmately lobed; stipules persistent . . . .
        . . . .**Euphorbiaceae *(Ricinus)***
        3' Inflorescence a catkin, axillary; leaf margins entire or nearly so; stipules deciduous . . . .
        . . . .**Salicaceae**
1' Perianth parts in 2 or more whorls
    4. Petals fused, forming a ring or tube, falling from flower as 1 unit
        5. Ovary inferior; calyx modified into a pappus or absent; fruit an achene . . . . .**Asteraceae**
        5' Ovary superior; calyx composed of 5 green sepals or lobes; fruit not an achene
            6. Leaves opposite at most nodes; foliage usually aromatic when crushed; corolla bilateral . . . .**Lamiaceae**
            6' Leaves alternate at most nodes; foliage not aromatic when crushed; corolla radial
                7. Stamens many, fused into a tube surrounding style; leaves palmately veined . . .
                . . . .**Malvaceae**
                7' Stamens 5, not fused into a tube; leaves pinnately veined . . . .**Solanaceae**
    4' Petals free from each other, usually falling separately
        8. Ovary inferior; stems succulent, spiny; leaves apparently absent; sepals and petals many . . . .**Cactaceae**
        8' Ovary superior; stems not succulent and spiny; leaves usually present; sepals and petals less than 10
            9. Perianth with 6 lobes, in 2 whorls of 3; stamens 9 . . . . .**Polygonaceae *(Eriogonum)***
            9' Perianth composed of 4-5 petals and 4-5 sepals; stamens 5-10
                10. Leaves opposite at most nodes . . . .**Frankeniaceae**
                10' Leaves alternate at most nodes
                    11. Leaves simple, scale-like . . . .**Tamaricaceae**
                  11' Leaves pinnately or palmately compound, conspicuous; leaflets 3-many
                      12. Flowers bilateral (papilionaceous) . . . .**Fabaceae**
                      12' Flowers radial . . . .**Anacardiaceae**

GROUP 3. DICOT HERBS; flowers with perianth apparently absent or only in 1 whorl

1. Most cauline leaves alternate
    2. Ultimate inflorescence (sometimes all flowers) dense, subtended by an involucre of free to fused bracts
        3. Ovary completely inferior or deeply embedded in inflorescence rachis

4. Style branches 2; calyx (if present) modified into a pappus; bracts subtending each flower (if present) not white ....**Asteraceae**

4' Style branches 3-4; perianth absent; each flower subtended by 1 white bract ....**Saururaceae**

    3' Ovary superior

5. Inflorescence ± terminal; flowers bisexual ....**Polygonaceae**

5' Inflorescence axillary; flowers unisexual ....**Urticaceae**

2' Flowers in racemes, in panicles, or solitary; ultimate clusters (if any) not subtended by an involucre

    6. Ovary inferior

7. Plants vine-like or twining; stems with tendrils; fruit bristly ....**Cucurbitaceae** *(Marah)*

7' Plants neither vine-like nor twining; tendrils absent; fruit smooth, except for persistent, horn-like perianth parts ....**Aizoaceae** *(Tetragonia)*

    6' Ovary superior

8. Stipules present, fused and partly to completely sheathing stem ......**Polygonaceae**

8' Stipules apparently absent ....**Chenopodiaceae**

1' Most cauline leaves opposite or whorled, leaves sometimes all basal

  9. Inflorescence an umbel

10. Perianth parts free; leaves deeply lobed to dissected or compound ....**Apiaceae**

10' Perianth parts fused, tubular; leaves simple, entire ....**Nyctaginaceae** *(Abronia)*

  9' Inflorescence various, but not an umbel

11. Ovary inferior; perianth parts fused; fruit densely hispid ....**Rubiaceae** *(Galium)*

11' Ovary superior; perianth parts (if present) free from each other; fruit not hispid

    12. Styles 4-5; ovules 3-many per ovary; fruit a capsule ....**Caryophyllaceae**

    12' Styles 2-3; ovules only 1 per ovary; fruit an achene or utricle

13. Stems and leaves either succulent or scurfy or both; fruit ± circular in cross-section; sepals usually 3-5 ....**Chenopodiaceae**

13' Stems and leaves neither succulent nor scurfy; fruit triangular in cross-section; sepals usually 6 ....**Polygonaceae**

GROUP 4. DICOT HERBS; flowers with perianth parts in 2 or more whorls; petals free

1. Flowers bilateral; some petals dissimilar in size or shape, or petals displaced to one side of floral axis

  2. Leaves compound, leaflets 3-many; petals 5 ....**Fabaceae**

  2' Leaves simple; petals 2 ....**Resedaceae** *(Oligomeris)*

1' Flowers radial, ± equal in size and shape

  3. Ovary partly or completely inferior

4. Perianth parts many; stems and leaves (if present) succulent

    5. Stems spiny; leaves apparently absent; ovary with 1 locule ....**Cactaceae**

    5' Stems not spiny; leaves present; ovary with 5-20 locules ....**Aizoaceae**

4' Sepals 2-5; petals 3-5; stems and leaves usually not succulent (except some Portulacaceae)

    6. Stamens many ....**Loasaceae**

    6' Stamens 1-10

7. Sepals 2 ....**Portulacaceae**

7' Sepals 5 ....**Apiaceae**

  3' Overy completely superior

8. Stamens many, at least more than twice the number of petals

    9. Leaves palmately veined and lobed; trichomes, if present, stellate; sepals 5 ....**Malvaceae**

        9' Leaves pinnately veined, entire to dissected; trichomes, if present, not stellate; sepals 2-3 . . . .**Papaveraceae**

    8' Stamens 1-12, fewer than to no more than twice the number of petals

        10. Most leaves compound or nearly so

            11. Leaves palmately compound, leaflets 3 . . . .**Oxalidaceae**

            11' Leaves pinnately compound or deeply lobed, leaflets or leaf segments 5 or more

                12. Sepals 4; petals 4; stamens 6 . . . .**Brassicaceae**

                12' Sepals 5; petals 5; stamens 5 . . . .**Geraniaceae** *(Erodium)*

        10' Most leaves simple, entire to deeply lobed

            13. Most leaves alternate and cauline

                14. Hypanthium present, cylindrical . . . .**Lythraceae**

                14' Hypanthium absent

                    15. Sepals 2; ovary with 1 locule, seeds 2 or more per fruit . . . .**Portulacaceae** *(Calandrinia)*

                  15' Sepals 3-5; ovary with 2-5 locules, if with 1 locule, then with 1 seed and triangular in cross-section

                      16. Ovary with 1 locule and 1 ovule; fruit triangular in cross-section . . . .**Polygonaceae**

                      16' Ovary with 2 locules; fruit ± circular or flattened in cross-section . . . .**Brassicaceae**

        13' Most leaves opposite, basal, or both

            17. Pistils 3-5; stems and leaves succulent . . . .**Crassulaceae**

            17' Pistil usually 1; stems and leaves not succulent (except in some Portulacaceae)

                18. Flower with a hypanthium

                    19. Lower leaves opposite, upper ones alternate; flowers usually axillary at leafy nodes . . . .**Lythraceae**

                    19' Leaves basal; flowers on scapes . . . . . .**Saxifragaceae** *(Jepsonia)*

                18' Flower without a hypanthium, calyx free from corolla

                  20. Sepals fused into a tube

                    21. Leaves not decussate; stamens in 1 whorl, ± equal . . . .**Caryophyllaceae**

                    21' Leaves decussate; stamens in 2 unequal whorls, with outer ones shorter than inner ones . . . .**Frankeniaceae**

                20' Sepals free from each other

                  22. Sepals 2; stipules absent . . . .**Portulacaceae** *(Claytonia)*

                  22' Sepals 5; stipules usually present

                    23. Leaf margins usually entire; ovary not lobed, with 1 locule . . . .**Caryophyllaceae**

                    23' Leaves lobed to dissected; ovary with 5 lobes and 5 locules . . . .**Geraniaceae**

GROUP 5. DICOT HERBS; flowers with perianth parts in 2 or more whorls; petals connate

1. Plants without green leaves; stems light yellow or yellowish-brown . . . .**Orobanchaceae**

1' Plants with green leaves and often with green stems

    2. Flowers in dense heads subtended by an involucre of bracts; calyx absent or modified into a pappus . . . .**Asteraceae**

    2' Flowers not in dense heads, if bracteate then bracts not arranged in an involucre; calyx usually herbaceous

        3. Corolla bilateral

            4. Ovary deeply 4-lobed; style arising from notch between lobes . . . .**Lamiaceae**

        4' Ovary not clearly lobed; style at tip of ovary . . . .**Scrophulariaceae**

3' Corolla radial

    5. Ovary partly to completely inferior

        6. Cauline leaves opposite . . . .**Nyctaginaceae**

        6' Cauline leaves alternate . . . .**Cucurbitaceae**

    5' Ovary completely superior

        7. Leaves opposite at most nodes

           8. Stamens 4 . . . .**Verbenaceae**

           8' Stamens 5

              9. Style unbranched, stigma capitate, oblong, or fan-shaped

                  10. Stigma oblong or fan-shaped; flowers clustered, terminal, erect in

                      fruit . . . .**Gentianaceae**

                  10' Stigma capitate; flowers solitary, axillary, recurved in fruit . . . . .

                . . . .**Primulaceae** *(Anagallis)*

             9' Style with 2 branches or lobes

                  11. Leaves entire; ovary with 4 lobes; style arising from notch between

                      lobes . . . .**Boraginaceae**

                  11' Leaves deeply lobed or dissected; ovary globose to ellipsoid; style

                      arising from tip of ovary . . . .**Hydrophyllaceae**

        7' Leaves alternate at most nodes or leaves all basal (plant scapose)

           12. Sepals 4; petals 4; stamens 4 . . . .**Plantaginaceae**

           12' Sepals 5; petals 5; stamens 5

              13. Inflorescence umbellate; stamens opposite corolla lobes . . . .

              . . . .**Primulaceae** *(Primula)*

             13' Inflorescence a raceme, panicle, or flowers solitary, but not umbellate;

                stamens alternate with corolla lobes

                  14. Ovary deeply 2- or 4-lobed; fruit dehiscing into 2 or 4 segments . .

             . . . .**Boraginaceae**

                  14' Ovary globose to ellipsoid, not obviously lobed; fruit a capsule

                      15. Style branches 3; ovary with 3 locules . . . . . .**Polemoniaceae**

                    15' Style either unbranched or with 2 branches; ovary with 1-2 locules

                        16. Style 1, stigma capitate or minutely 2-lobed

                          17. Stigma capitate, not evidently lobed . . . .**Solanaceae**

                        17' Stigma with 2 lobes

                            18. Flowers axillary; leaves entire; ovules usually 4 per

                              ovary . . . .**Convolvulaceae**

                          18' Flowers usually in a terminal raceme; leaves deeply

                              lobed or dissected; ovules 6 or more per ovary . . .

                          . . . .**Hydrophyllaceae**

                    16' Style evidently 2-branched from near base to above middle

                      19. Leaves deeply lobed or dissected . . . .

                  . . . .**Hydrophyllaceae** *(Phacelia)*

                    19' Leaves entire . . . .**Convolvulaceae** *(Cressa)*

DICOTYLEDONOUS ANGIOSPERMS (DICOTS)

AIZOACEAE Martinov
FIG-MARIGOLD FAMILY

Annuals, perennial herbs, or subshrubs. Ca. 130 genera and 2500 species; tropical and subtropical regions, mostly S Hemisphere and especially S Africa. Many species are cultivated as ornamentals.

1. Petals absent; fruit nut-like, sepals adnate to fruit and becoming horn-like . . . .**Tetragonia**
1' Petals present, many; fruit a berry or ± dehiscent capsule, sepals not becoming horn-like in fruit
    2. Annuals; leaves cylindrical to flat, with conspicuous, shining, watery papillae . . . .
. . . .**Mesembryanthemum**
    2' Perennials; leaves cylindrical to triangular in cross-section, glabrous
        3. Fruit fleshy, indehiscent; stigmas and ovary chambers 8-12 . . . .**Carpobrotus**
        3' Fruit a persistent capsule, valves finally separating; stigmas and ovary chambers 4-11
            4. Ovary chambers 4-6; petals whitish . . . .**Delosperma**
            4' Ovary chambers 8-11; upper surface of petals orangish, lower surface purple . . . .
. . . .**Malephora**

CARPOBROTUS N.E.Br. FIG-MARIGOLD

Subshrubs. Ca. 13 species; S Africa; introduced in Mexico, S America, Europe, and Australia. Many species are used ornamentally. (Greek: fruit edible)

1. Petals rose-magenta; flower 3-5 cm wide, sessile; leaves 3-5 cm long, with smooth outer angle . . .
C. chilensis
1' Petals pink to yellow and aging pink; flower 8-10 cm wide, on a peduncle; leaves 7-10 cm long, with
    outer angle serrulate near apex . . . .**C. edulis**

*C. chilensis** (Molina) N.E.Br. [The name *C. aequilaterus* (Haw.) N.E.Br. has been misapplied to this species] SEA-FIG Stems to 2.5 m long. Apr-Sep. Scarce; sandy slopes and flats. North side near Red Eye Beach and at w end of Tender Beach; on mesa near wells area on Tufts Road; s side on coastal flats and dunes at sw end and on s escarpment. All CA Channel Islands except Santa Barbara and Santa Catalina; Los Coronados and Todos Santos islands; OR to Baja CA; S America; native of S Africa. First collected on s side of island, s of Jackson Hill, in June 1969. Leaves and fruits are edible.
*C. edulis** (L.) N.E.Br. HOTTENTOT-FIG Stems to 3 m long. Apr-Oct. Common; sandy slopes and flats. Widespread in many locations throughout much of island, especially on n side. San Miguel, Santa Rosa, Santa Cruz, Anacapa, and San Clemente islands; OR to Baja CA; FL; S America (Chile); Australia; New Zealand; Europe; native of S Africa. First collected in Nicktown in April 1961, where it reportedly had been "planted near buildings in the living quarter complex and subsequently escaped into nearby areas" (Foreman 1967). During the last decade, some of the populations on San Nicolas Island have been attacked by scale insects. Populations that were attacked declined dramatically, suggesting that scale insects may prove to be an effective bio-control agent for *Carpobrotus*. Leaves and fruits are edible; leaves have been used medicinally in S Africa.

DELOSPERMA N.E.Br.

Subshrubs. Ca. 160 species; S Africa; introduced in Asia. (Greek: visible seed, referring to the seeds exposed in the open chambers of the capsule)

*D. litorale** (Kensit) L.Bolus SEASIDE DELOSPERMA Stems to 3 m long. Apr-Jun. Rare; sandy slopes and flats. North side in lower Tule Creek and on adjacent sand dunes. CA (Los Angeles Co.); native of S Africa. First collected in Tule Creek area in April 1979, but already established at that time.

MALEPHORA N.E.Br.

Subshrubs. Ca. 15 species; S Africa. (Greek: bearing arm-holes, referring to seed pockets in the fruit)

*M. crocea (Jacq.) Schwantes var. **crocea** Stems to 2 m long. Apr-Jun. Rare; gully bottoms. Known only from s edge of mesa, in gully just s of Building 112. Anacapa and San Clemente islands; Cedros Island (cultivated); central CA (Santa Barbara Co.) to Baja CA; native of S Africa. First noted at s edge of mesa in May 1993, but already established at that time.

MESEMBRYANTHEMUM L. ICEPLANT

Annuals or short-lived perennials. Ca. 74 species; s and w Africa; introduced in Mexico, Australia, Atlantic Islands, Mediterranean Europe, and w Asia. Leaves of some species are cooked and eaten like spinach. (Greek: flowering at mid-day, alluding to the fact that the only known species at that time bloomed at noon)

1. Leaf flat, 2-20 cm long; flower 7-10 mm wide . . . .**M. crystallinum**
1' Leaf cylindrical, 1-2 cm long; flower 4-5 mm wide . . . .**M. nodiflorum**

*M. crystallinum L. [*Gasoul* c. (L.) Rothm.] CRYSTALLINE ICEPLANT Plants prostrate; branches 2-6 dm long. Mar-Oct. Common; eroded slopes and flats. Throughout much of island, especially in Western Gull colony at w end, where it is dominant, and on s escarpment. All CA Channel Islands; all Baja CA islands; central CA (Monterey Co.) to Baja CA, e to AZ and a disjunct occurrence in PA; S America; Australia; Atlantic Islands; Europe; Asia; native of s and w Africa. Reported as "abundant" and "growing on cliffs 1000 feet high" by Eastwood (1898). Stems, leaves, and fruits are edible; plants and leaves have been used medicinally.

*M. nodiflorum L. [*Gasoul* n. (L.) Rothm.] SMALL-FLOWERED ICEPLANT Plants ± erect to procumbent; branches 0.5-2 dm long. Mar-Nov. Common; eroded slopes and flats. Throughout much of island. All CA Channel Islands; all Baja CA islands except San Geronimo; OR to Baja CA, e to AZ and a disjunct occurrence in NJ; S America (Chile); Australia; Asia; Atlantic Islands; native of S Africa. First collected in April 1897 and reported as "frequent on beaches" by Eastwood (1898).

TETRAGONIA L.

Annuals. Ca. 60 species; S Hemisphere. (Greek: four-angled, referring to shape of the fruit)

*T. tetragonioides (Pallas) Kuntze [*T. expansa* Murray] NEW ZEALAND SPINACH Stems 3-5 dm long. Apr-Sep. Rare; sandy slopes and flats. North side at w end of Red Eye Beach. San Miguel, Santa Rosa, Santa Cruz, Anacapa, and Santa Catalina islands; WA to Baja CA, e (disjunctly) to WI and MA; Hawaii; native of Japan, Australia, Tasmania, New Zealand, and S America (Chile). First collected at Red Eye Beach in April 1985. Fresh or cooked leaves are edible and can be used as a substitute for cultivated spinach.

ANACARDIACEAE Lindl.
SUMAC OR CASHEW FAMILY

Trees or shrubs. Ca. 75 genera and 850 species; warm temperate to tropical areas worldwide. Some genera are used ornamentally (e.g., *Rhus, Schinus*) and some species are used for wood and food (e.g., cashew, mango, pistachio). Contact with sap of several species can cause severe dermatitis in humans.

1. Leaflets 15-many, lanceolate to linear-lanceolate; stamens 10; fruit pink to red . . . .**Schinus**
1' Leaflets usually 3, oblong to ± orbicular; stamens 5; fruit whitish . . . .**Toxicodendron**

SCHINUS L. PEPPER-TREE

Trees. Ca. 30 species; tropical and warm temperate S America. (Greek: ancient name for *Pistacia lentiscus,* the mastic tree, which *Schinus* resembles in that its species can exude a mastic-like sap)

*S. molle L. PERUVIAN or CALIFORNIA PEPPER-TREE Plants 5-7 m tall. Apr-Jun. Rare; disturbed sites. Mesa at Nicktown, where it has been planted and has apparently escaped into gully w of Owens Road near transportation yard. Santa Cruz, Santa Catalina, and San Clemente islands; central CA to Baja CA, e to TX and disjunctly to FL; Hawaii; Australia; New Zealand; S Africa; native of S America (Ecuador, Peru, and Chile). First noted at Nicktown in 1988, and already established at that time. The fruits are toxic to mammals if ingested; leaves, bark, and sap have been used medicinally.

TOXICODENDRON Mill. POISON-OAK or POISON-IVY

Shrubs. Ca. 6 species; N and Central America, e Asia. (Latin: poisonous tree)

T. diversilobum (Torr. & A.Gray) Greene [*Rhus d.* Torr. & A.Gray] POISON-OAK Plants to 1 m tall. Apr-May. Rare; flats. North side on coastal flats just w of lower portion of Celery Canyon (known from a single population). All CA Channel Islands except Santa Barbara; WA to Baja CA, e to nw NV. Sap produced by this species can cause severe contact dermatitis in humans. Sap was used medicinally by Native Americans; black dye obtained from sap was also used in basketry.

APIACEAE Burnett [**Umbelliferae** Juss.]
CELERY OR CARROT FAMILY

Annual or perennial herbs. Ca. 420 genera and 3100 species; temperate and tropical mountains, ± worldwide. Many species are cultivated for food and culinary herbs (e.g., anise, caraway, carrots, celery, coriander, cumin, dill, fennel, and parsley). Some species (e.g., hemlock) are extremely toxic.

1. Ovary and fruit prickly, bristly, or hispid
    2. Stems and leaves glabrous; umbels simple . . . .**Sanicula**
    2' Stems and leaves strigose to hispidulous; umbels compound
        3. Umbels in axils of leaves, not subtended by bracts; ribs of fruit with minute prickles, tuberculate below middle . . . .**Torilis**
        3' Umbels terminal, subtended by leaf-like bracts; fruit wall minutely prickly or hispid, hispidulous between ribs, not tuberculate . . . .**Daucus**
1' Ovary and fruit glabrous (to minutely papillate in *Apiastrum*)
    4. Plants acaulescent, leaves mostly basal; mericarps dorsally compressed, marginal ribs winged . . . .**Lomatium**
    4' Plants caulescent, leaves cauline; mericarps ± cylindrical to laterally compressed
        5. Plants annual, usually less than 3 dm tall; fruit minutely papillate . . . .**Apiastrum**
        5' Plants perennial, usually more than 5 dm tall; fruit glabrous
            6. Leaves compound, leaflets lanceolate, oblong, ovate, or ± orbicular
                7. Basal leaves twice pinnately compound, leaflets lanceolate to ± orbicular, irregularly lobed or toothed . . . .**Apium**
                7' Basal leaves once compound, leaflets oblong to ovate, regularly toothed . . .**Berula**
            6' Leaves pinnately dissected, ultimate segments filiform or oblong in outline
                8. Petals white; stems purple-spotted; leaves aromatic but not like anise; fruit ± globose, 2-3 mm long and wide (highly toxic) . . . .**Conium**
                8' Petals yellow; stems not purple-spotted; leaves with a strong anise odor; fruit oblong-ovoid, mericarps laterally compressed, 3-4.5 mm long . . . . .**Foeniculum**

ANACARDIACEAE

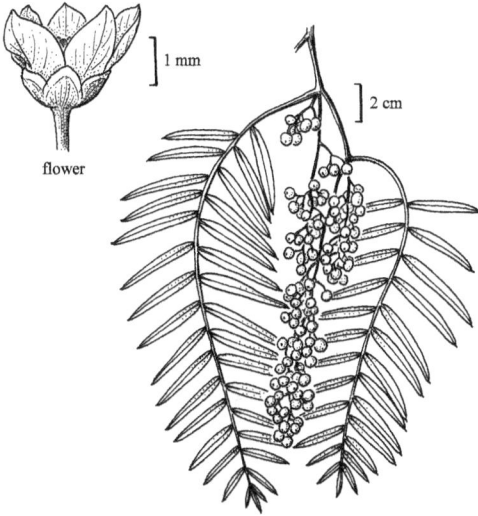

1 mm

flower

2 cm

Schinus molle

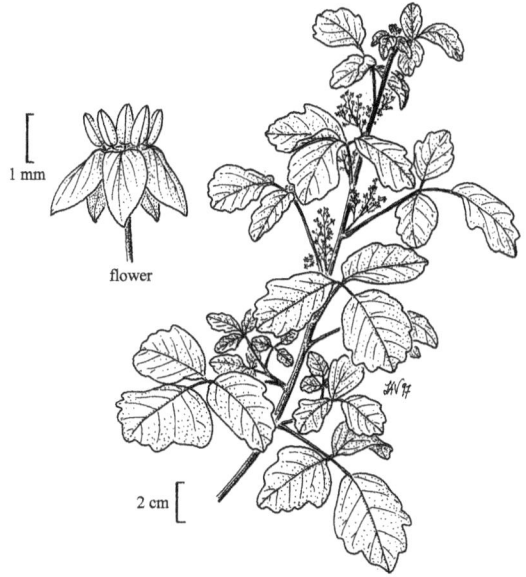

1 mm

flower

2 cm

Toxicodendron diversilobum

APIACEAE: Apiastrum-Berula

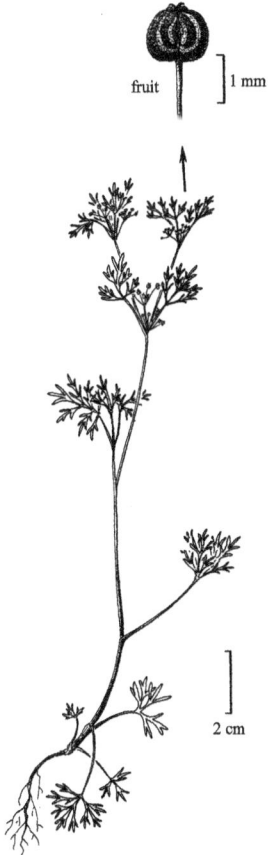

fruit

1 mm

A. graveolens

fruit

1 mm

fruit

B. erecta

2 cm

leaf

2 cm

2 cm

Apiastrum angustifolium

Apium graveolens

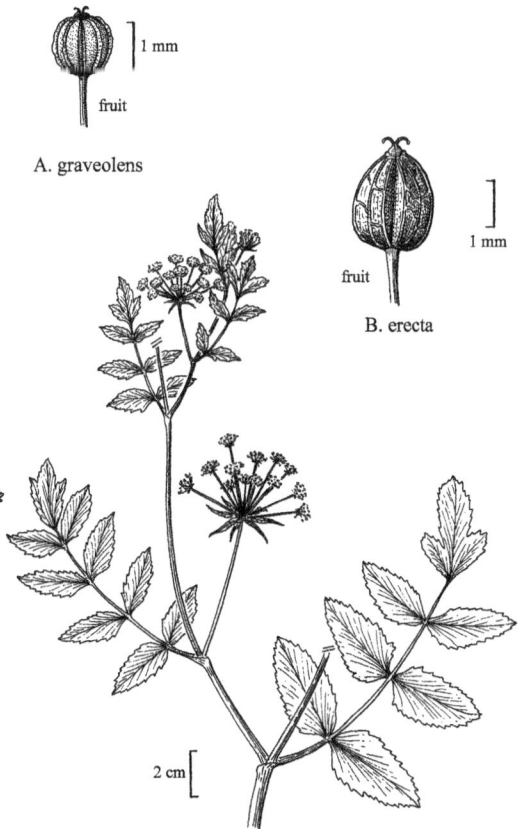

Berula erecta

APIASTRUM Nutt.

Annual herbs. One species. (Latin: wild celery)

A. angustifolium Nutt. WILD CELERY Plants 5-30 cm tall. Mar-May. Rare; n-facing slopes and flats. Northeastern coastal flats, near N and S Spur canyons. All CA Channel Islands; Los Coronados, Todos Santos, and Cedros islands; n CA (Mendocino Co.) to Baja CA, e to AZ. This inconspicuous annual has only been noted sporadically on the island; it was reportedly "rare amid cacti" in 1901, not reported again until 1932, and then not again until 1993. It is usually associated with *Opuntia* patches on the island. This species was eaten by Native Americans.

APIUM L.

Perennial herbs. Ca. 20 species; temperate regions, mostly in S Hemisphere. (classical Latin name for celery and parsley)

*A. graveolens L. CELERY Plants 5-10 dm tall. May-Jun. Scarce; n-facing slopes, coastal bluffs, canyon bottoms, and other moist sites. North side at Thousand Springs, in Tule Creek, on moist coastal bluffs at NavFac Beach, and in Celery Canyon; on mesa at Army Springs and wells area on Tufts Road. Santa Rosa, Santa Cruz, and San Clemente islands; Cedros Island; WA to Baja CA, e (disjunctly) to MA and FL; native of Europe. First collected on "moist coastal bluffs" at Corral Harbor in July 1939. Petioles are eaten as a vegetable and seeds are used for flavoring; leaves can accumulate toxic levels of nitrates.

BERULA W.D.J.Koch

Perennial herbs. One species. (Latin: water-cress)

B. erecta (Hudson) Coville CUT-LEAF WATER PARSNIP Plants 4-5 dm tall. Rare; moist flats. North side at Thousand Springs (historic collections only). San Miguel and Santa Cruz islands; WA to Baja CA, e to NY; Europe; Asia. Collected several times at Thousand Springs between June 1969 and July 1979, but apparently extirpated when its habitat was destroyed in the early 1980s. Plants may be toxic to grazing animals; an infusion of the whole plant has been used medicinally (as an external wash) by Native Americans.

CONIUM L.

Annual or biennial herbs. Ca. 6 species; Mediterranean Europe and S Africa. (Greek name for the plant and the poison derived from it)

*C. maculatum L. POISON-HEMLOCK Plants 1-2 m tall. May-Jun. Rare; moist flats and slopes. North side at Thousand Springs (historic collection only) and in Tule Creek. Santa Cruz Island; WA to Baja CA, e to ME; S America; Australia; New Zealand; native of Europe, n Africa, and Asia. First collected along Tule Creek in July 1939. This taxon was locally common near Humphrey's Sump in Tule Creek in the 1980s, but is now known only from a few scattered plants upstream from that site. An extremely toxic species, which was reportedly used to poison Socrates.

DAUCUS L.

Annual or biennial herbs. Ca. 20 species; temperate areas of N Hemisphere, Australia. (Latin name for carrot)

1. Biennial; bract segments elongate, narrowly linear to ± thread-like; central flower of umbel usually purple; fruit widest at middle . . . .**D. carota**
1' Annual; bract segments short, linear to lanceolate; central flower of umbel white; fruit widest below middle . . . .**D. pusillus**

APIACEAE: Conium-Sanicula

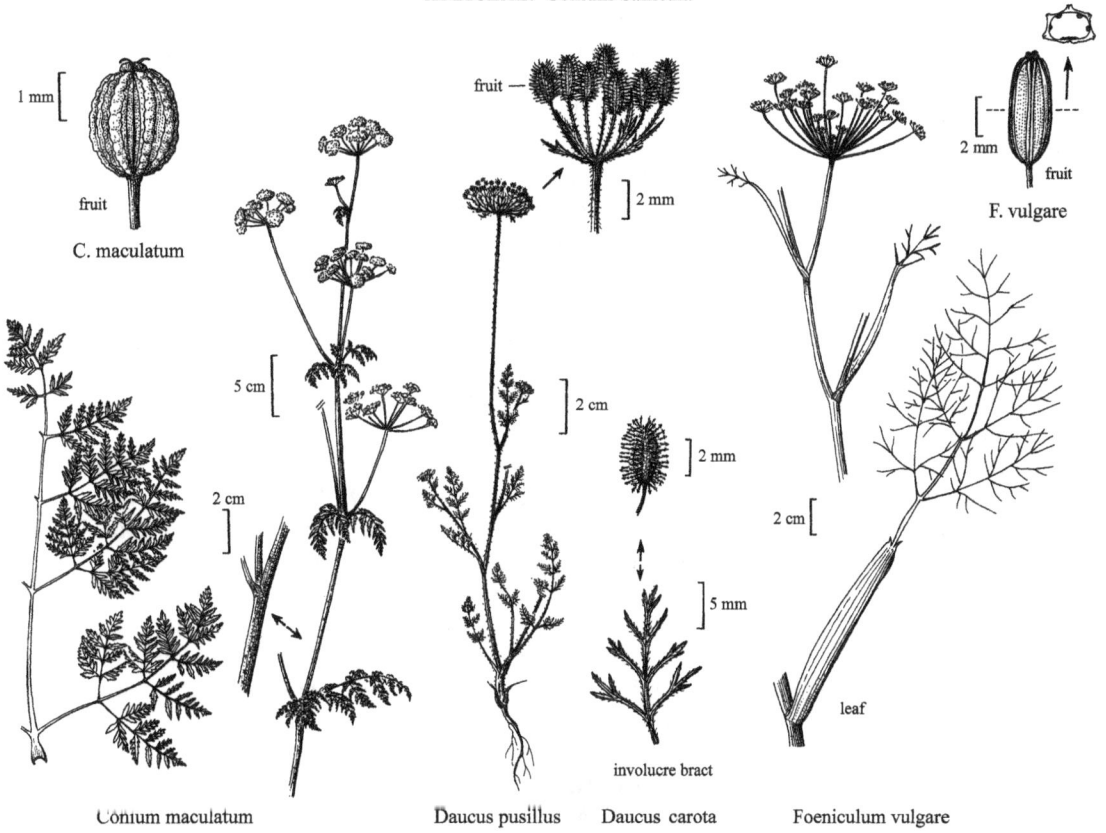

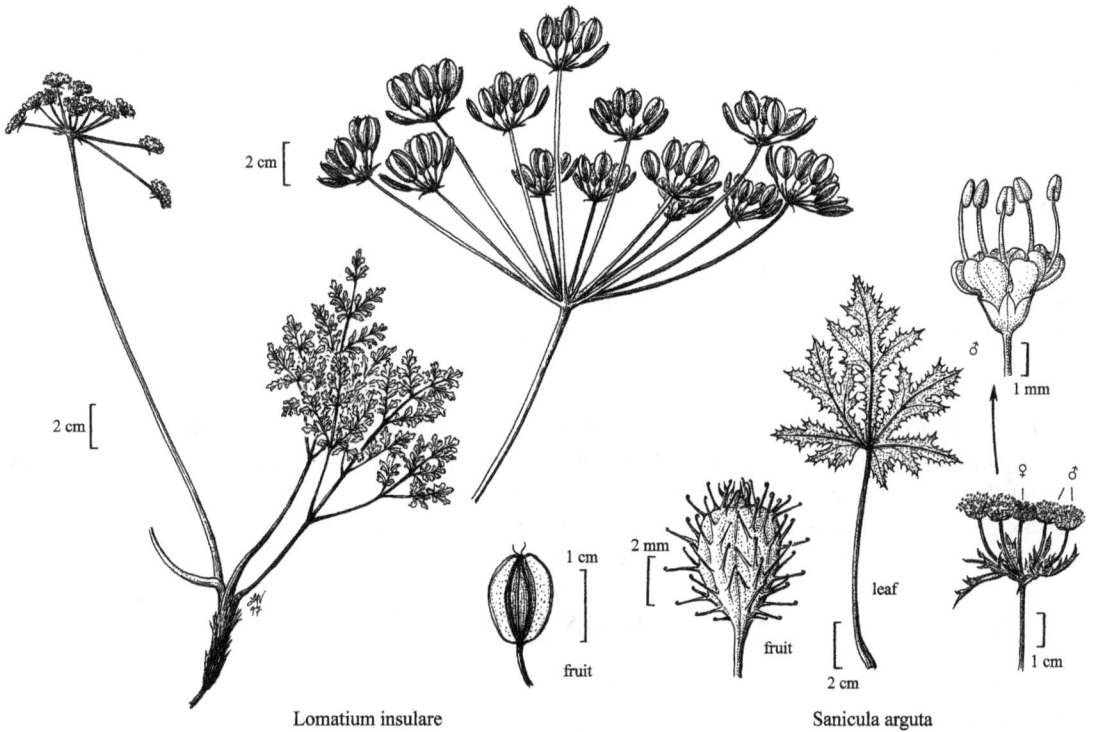

1 mm

fruit

C. maculatum

fruit —

2 mm

2 mm

2 mm

5 mm

involucre bract

5 cm

2 cm

2 cm

Conium maculatum

Daucus pusillus   Daucus carota

2 mm

fruit

F. vulgare

2 cm

leaf

Foeniculum vulgare

2 cm

2 cm

1 cm

fruit

Lomatium insulare

2 mm

fruit

leaf

2 cm

♂

1 mm

♀   ♂

1 cm

Sanicula arguta

D. carota L. WILD CARROT or QUEEN ANNE'S LACE Plants 2-10 dm tall. May-Oct. Rare; flats. Mesa along Jackson Highway just e of intersection with Shannon Road; s side near Army Springs (historic collections only). WA to Baja CA, e to ME; native of Europe. First collected near Army Springs in July 1981; last collected on mesa in October 1983. This species is cultivated for the edible root (carrot).

D. pusillus Michx. RATTLESNAKE WEED Plants 1-7 dm tall. Mar-May. Abundant; flats, stabilized sand dunes, swales, canyon bottoms, and slopes. Widespread locations throughout much of island. All CA Channel Islands; Los Coronados, Todos Santos, and Guadalupe islands; British Columbia to Baja CA, e to FL; S America. Roots are edible either raw or cooked and were used as food by some Native Americans; this species was also used medicinally.

FOENICULUM Mill.

Perennial herbs. One species. (diminutive of Latin word for hay, referring to the odor)

F. vulgare Mill. SWEET FENNEL Plants 1-2.5 m tall. Feb-Jun. Common; disturbed flats, roadsides, canyon bottoms, and dry slopes. North side between Tule Creek and Rock Jetty (including coastal flats s of Anchor Point, W Mesa Canyon, E Mesa Canyon, and coastal flats near Rock Jetty); numerous locations on mesa (including scattered sites near roads, W Mesa Canyon, E Mesa Canyon, and vicinity of airfield); s side in an unnamed canyon above Dutch Harbor. All CA Channel Islands except Anacapa and Santa Barbara; WA to Baja CA, e (disjunctly) to ME and FL; Central America; Hawaii; Australia; New Zealand; China; many other parts of the world; native of Europe. First collected at mouth of small canyon e of Nicktown in April 1961. Eradication efforts, utilizing herbicides and some manual removal, were initiated in 1982 but have yet to be successful. A noxious and persistent weed, long cultivated for its essential oils. The leaves, roots, and seeds are edible either raw or cooked; seeds are used for seasoning in Italian dishes.

LOMATIUM Raf. BISCUIT-ROOT

Perennial herbs. Ca. 75 species; temperate N America. Roots of many species were eaten by Native Americans. (Greek: bordered, referring to the winged fruits)

L. insulare (Eastw.) Munz [*Peucedanum i.* Eastw.] SAN NICOLAS ISLAND LOMATIUM Plants 3-7 dm tall. Jan-Apr (Jun). Abundant; open ridgetops, flats, canyon bottoms, and slopes. North side from Mineral Canyon to Sand Spit (especially on escarpment and in associated canyons); s side from w end of mesa to Sand Spit. Especially common and conspicuous in se portion of island. ENDEMIC to San Nicolas, San Clemente (last seen there in 1918), and Guadalupe islands. A detailed distribution map for this taxon can be found in Junak (2003c).

SANICULA L. SNAKEROOT or SANICLE

Perennial herbs. Ca. 40 species; temperate areas worldwide. Young leaves of some Asian species are boiled and eaten as a vegetable; roots of *Sanicula tuberosa* were eaten by Native Americans. (Latin: to heal)

S. arguta J.M.Coult. & Rose SHARP-TOOTHED SNAKEROOT Plants 2-7 dm tall. Feb-Apr. Occasional; north-facing slopes and swales. North side on escarpment and in canyons e of Mineral Canyon, coastal flats n of Nicktown; mesa near Nicktown; scattered locations on s side, as on n-facing slopes in upper e portion of Twin Rivers drainage and near *Opuntia* and *Lycium* thickets in se portion of island. All CA Channel Islands except Santa Barbara; central CA (Alameda Co.) to Baja CA.

TORILIS Adans. HEDGE-PARSLEY

Annual herbs. Ca. 12 species; Mediterranean Europe, e Asia, and Canary Islands. (meaning of the name is not known)

*T. nodosa (L.) Gaertn. KNOTTED HEDGE-PARSLEY Plants to 5 dm tall. Apr-May. Rare; canyon walls and slopes. Known only from mesa, in upper reaches of Mineral Canyon. San Miguel, Santa Rosa, Santa Cruz, and Santa Catalina islands; OR to Baja CA; native of Europe. First collected on the island in May 2001.

ASTERACEAE Dumort. [**Compositae** Giseke]
SUNFLOWER FAMILY

Annual herbs, perennial herbs, or shrubs. Ca. 1300 genera and 21,000 species; worldwide. The largest family of dicots. Many species are cultivated for food, ornamental use, or medicinal properties.

1. Heads ligulate; plants with milky sap . . . .**GROUP 1**
1' Heads discoid or radiate; plants with clear sap
  2. Heads discoid; corollas usually all tubular and radial
    3. Pappus absent or minute and inconspicuous . . . .**GROUP 2**
    3' Pappus present, composed of awns, bristles, or scales . . . .**GROUP 3**
  2' Heads radiate; both disk and ray corollas present
    4. Pappus absent or minute and inconspicuous . . . .**GROUP 4**
    4' Pappus present, composed of awns, bristles, or scales
      5. Pappus of awns or scales . . . .**GROUP 5**
      5' Pappus of bristles . . . .**GROUP 6**

GROUP 1. Heads composed of ligulate flowers; plants with milky sap.

1. Pappus composed of plumose bristles
  2. Flowers subtended by chaffy, scarious bracts; involucres 12-16 mm long; pappus tawny or dull white . . . .**Hypochaeris**
  2' Flowers not subtended by bracts; involucres 35-70 mm long; pappus brown . . . .**Tragopogon**
1' Pappus composed of awns, scales, or capillary bristles, or absent
  3. Pappus composed of scales, these sometimes with awned tips
    4. Pappus awns usually shorter than scales, glabrous . . . .**Uropappus**
    4' Pappus awns equal to or longer than scales, ± barbellate
      5. Pappus scales 5-8 mm long, deltate to ovate, tapered to awns; leaves all basal . . . . .
      . . . .**Microseris**
      5' Pappus scales less than 2 mm long, linear-lanceolate, awned from bifid apices; cauline leaves sometimes present . . . .**Stebbinsoseris**
  3' Pappus composed of capillary bristles or absent
    6. Achene truncate at top, pappus attached directly to fruit
      7. Achenes strongly flattened . . . .**Sonchus**
      7' Achenes cylindrical to fusiform, not flattened . . . .**Malacothrix**
    6' Achene with slender beak between fruit and pappus
      8. Achenes strongly flattened . . . .**Lactuca**
      8' Achenes cylindrical to fusiform, not flattened
        9. Cauline leaves present, hispid, some trichomes pustulate; pappus plumose . . . . .
        . . . .**Picris**
        9' Leaves basal, glabrous to pubescent, but trichomes neither hispid nor pustulate; pappus of simple, smooth bristles . . . .**Taraxacum**

GROUP 2. Heads discoid; corollas ± all tubular and radial; pappus absent or minute and inconspicuous.

1. Flowers and heads unisexual; pistillate heads with bur-like or spiny involucre; anthers free
   2. Staminate heads with phyllaries strongly fused, forming a cup-like involucre; phyllaries of pistillate heads spiny, spines not hooked . . . .**Ambrosia**
   2' Staminate heads with phyllaries ± free to base; phyllaries of pistillate heads with hooked prickles
. . . .**Xanthium**
1' Flowers and heads bisexual; involucres not bur-like or spiny; anthers fused into a tube surrounding style
   3. Woody shrubs; heads in racemes or panicles . . . .**Artemisia**
   3' Perennial herbs; heads solitary . . . .**Cotula**

GROUP 3. Heads discoid; corollas ± all tubular and radial; pappus present, composed of awns, bristles, or scales.

1. Pappus of chartaceous scales . . . .**Amblyopappus**
1' Pappus of capillary or plumose bristles
   2. Shrubs or subshrubs
      3. Corollas white or cream . . . .**Baccharis**
      3' Corollas yellow to orange . . . .**Isocoma**
   2' Annuals or herbaceous perennials
      4. Leaves with spiny teeth or lobes; receptacles bristly (also see *Centaurea*)
         5. Pappus of barbellate bristles; leaves mottled with white . . . .**Silybum**
         5' Pappus of long-plumose bristles; leaves green throughout
            6. Largest leaves toothed to deeply lobed; involucre 1-6 cm in diameter; receptacle not fleshy; phyllaries linear to ovate . . . .**Cirsium**
            6' Largest leaves often ± compound; involucre 3-15 cm in diameter; receptacle fleshy; phyllaries ovate . . . .**Cynara**
      4' Leaves entire to toothed or lobed, but not spiny; receptacles without bracts subtending disk flowers or with chaffy bracts near margin, bristly only in *Centaurea*
         7. Phyllaries prominently spine-tipped . . . .**Centaurea**
         7' Phyllaries entire to laciniate, but not spine-tipped
            8. Receptacle with chaffy bracts subtending or enfolding outer or marginal flowers
. . . .**Filago**
            8' Receptacle without bracts subtending disk flowers
               9. Phyllaries lanceolate to linear with scarious margins, in 2-3 series
                  10. Outer phyllaries conspicuously black-tipped and less than ⅓ as long as the inner ones . . . .**Senecio**
                  10' Outer phyllaries usually not black-tipped, more than ⅓ as long as the inner ones . . . .**Conyza**
              9' Phyllaries ovate to oblong, ± scarious to chartaceous throughout, in several series
                  11. Heads in subglobose, compact, axillary clusters; inflorescence with leaf-like bracts . . . .**Gnaphalium**
                  11' Heads in rounded to open, terminal panicles; inflorescence bracts much reduced . . . .**Pseudognaphalium**

GROUP 4. Heads radiate; both disk and ray corollas present; pappus absent or minute and inconspicuous

1. Disk and ray corollas white . . . .**Achillea**
1' Disk corollas yellow or brown, ray corollas yellow or orange (sometimes white in *Chrysanthemum*)

2. Receptacle without bracts subtending disk flowers
   3. Leaves 1-2 times pinnatifid; phyllaries unequal, overlapping in several series . . . .
   . . . . **Chrysanthemum**
   3' Leaves simple, entire to very shallowly lobed; phyllaries ± equal, in 2-3 series, slightly
   overlapping . . . . **Calendula**
2' Receptacle with chaffy bracts subtending flowers or in a ring between disk and ray flowers
   4. Leaves 2-3 times pinnatifid; disk corollas yellow; phyllaries in 2 unequal series, the outer
   spreading; disk achenes not enclosed by receptacular bracts . . . . **Coreopsis**

   4' Leaves simple, ± entire to undulate; disk corollas brown; phyllaries in 2-3 ± equal, overlapping series; disk achenes enclosed by receptacular bracts . . . . **Encelia**

GROUP 5. Heads radiate; both disk and ray corollas present; pappus of awns or scales.

1. Receptacle naked, disk flowers not subtended by bracts
   2. Annuals; leaves opposite . . . . **Lasthenia**
   2' Perennials; leaves alternate . . . . **Grindelia**
1' Receptacle bristly or with chaffy bracts subtending disk flowers
   3. Annuals; phyllaries often in 2-3 series; inner phyllaries not enfolding ray achenes; disk flower
   pappus of 2 deciduous scales . . . . **Helianthus**
   3' Perennials; phyllaries in 1 series, partly enfolding ray achenes; disk flower pappus of 10 ± persistent scales . . . . **Deinandra**

GROUP 6. Heads radiate; both disk and ray corollas present; pappus of bristles.

1. Ray corollas white . . . . **Conyza**
1' Ray corollas yellow
   2. Pappus of 1-6 awn-like bristles, ± V-shaped in cross-section; outer phyllaries ± recurved . . .
   . . . . **Grindelia**
   2' Pappus of many capillary bristles; outer phyllaries not recurved . . . . . Heterotheca

ACHILLEA L. YARROW or SNEEZEWORT
Perennial herbs. Ca. 85 species; temperate areas of N Hemisphere. (Greek: honoring Achilles of ancient mythology, who reportedly valued the medicinal properties of yarrow)

A. millefolium L. YARROW or MILFOIL Plants 5-10 dm tall. Feb-Jun. Common; flats, swales, canyon bottoms, and slopes. Widespread locations throughout much of island. All CA Channel Islands; Cedros Island; N America, Europe, and Asia. This species, which contains essential oils and a bitter alkaloid (achilleine), is widely used as a medicinal herb.

AMBLYOPAPPUS Hook. & Arn.
Annual herbs. One species. (Greek & Latin: blunt pappus)

A. pusillus Hook. & Arn. PINEAPPLE WEED Plants 1-4 dm tall. Mar-Jul. Common; open flats, slopes, and sandy areas. Widespread locations in sandy areas near the coast on both sides of island, on n escarpment, and in open areas on mesa. All CA Channel Islands; all Baja CA islands except San Geronimo; central CA (San Luis Obispo Co.) to Baja CA; S America (Peru and Chile).

AMBROSIA L. RAGWEED or BUR-SAGE
Perennial herbs or subshrubs. Ca. 24 species; mostly N America. Pollen of some species can cause

severe allergies. (Greek: early name for aromatic plants and for the mythic "food of the gods")

1. Fruiting heads unarmed or with a few blunt or vestigial spines, mostly on upper half of bur; stems erect; herbage greenish with short, stiff appressed or spreading hairs . . . .**A. psilostachya**
1' Fruiting heads with at least 10-20 sharp spines; stems sprawling; herbage silvery canescent . . . . . .
. . . .**A. chamissonis**

A. chamissonis (Less.) Greene SILVER BEACH-WEED or BEACH-BUR [*Franseria c.* Less.] Plants 1.5-4 dm tall and 1-3 m across. Mar-Oct. Common; sand dunes and sandy flats. Widespread locations throughout much of island, mostly in sandy soils near the coast but also common on mesa. All CA Channel Islands except Santa Barbara; San Martin and Cedros islands; British Columbia to Baja CA; S America.
A. psilostachya DC. WESTERN RAGWEED Plants 3-10 dm tall. Jul-Oct. Rare; flats. North side near Jetty Canyon (at a disturbed site); on mesa on sw side of airfield runways. San Miguel, Santa Rosa, Santa Cruz, and Santa Catalina islands; WA to Baja CA, e to ME and FL. Possibly introduced on San Nicolas Island, where it was first collected on the mesa in March 1983.

ARTEMISIA L. SAGEBRUSH, WORMWOOD or MUGWORT
Shrubs. Ca. 350-500 species; mostly in temperate areas of N Hemisphere. Many species have been used medicinally. (Greek: honoring Artemis of ancient mythology, goddess of the hunt and protector of all wild places)

1. Leaf segments generally less than 1 mm wide; corollas 0.8-1.2 mm long; pistillate flowers 6-10; bisexual flowers 18-25 . . . .**A. californica**
1' Leaf segments 1-3 (5) mm wide; corollas 1.2-1.5 mm long; pistillate flowers 0; bisexual flowers 20-50
. . . .**A. nesiotica**

A. californica Less. COASTAL SAGEBRUSH Plants 5-15 dm tall. Aug-Feb (Apr). Occasional; dry slopes and flats. North side in first canyon w of Celery Canyon; canyons of s escarpment, especially in se portions of island. All CA Channel Islands; Los Coronados, Todos Santos, Guadalupe, and Cedros islands; n CA (Mendocino Co.) to Baja CA. This species is an important medicinal plant for Native Americans.
A. nesiotica Raven [*A. californica* Less. var. *insularis* (Rydb.) Munz] ISLAND SAGEBRUSH Plants 5-15 dm tall. Sep-Feb. Occasional; gullies, swales, canyon slopes, and flats. North side near Mineral and Celery canyons, and on n coastal flats and escarpment from Airfield Grade e to Sand Spit; s escarpment near Twin Towers (Building 186) and in upper portions of canyons of se escarpment. ENDEMIC to Santa Barbara, San Nicolas, and San Clemente islands.

BACCHARIS L.
Shrubs. Ca. 350-450 species; mostly in N and S America. Leaves of several species were used medicinally by Native Americans. (Latin: honoring Bacchus, god of wine in ancient mythology)

1. Leaves oblanceolate to obovate, obtuse, 1-4 cm long; achenes with 8-10 ribs . . . .**B. pilularis**
1' Leaves lanceolate, acute, 2-14 cm long; achenes with 4-5 ribs . . . .**B. salicifolia**

B. pilularis DC. [*B. p.* subsp. *consanguinea* (DC.) C.B.Wolf] COYOTE BRUSH Plants 1-2.5 m tall. Jul-Nov. Common; flats and canyon bottoms. Widespread locations in n portion of island and on mesa; rare on s escarpment, where it is typically found only in canyons. All CA Channel Islands; OR to Baja CA, e to NM. Stems were used by Native Americans to construct dwellings.
B. salicifolia (Ruiz & Pav.) Pers. [*B. glutinosa* Pers., *B. viminea* DC.] MULE FAT Plants 2-3 m tall. Feb-Nov. Scarce; moist flats, swales, and depressions. Mesa at wells area along Tufts Road, at old bor-

ASTERACEAE: Achillea-Centaurea

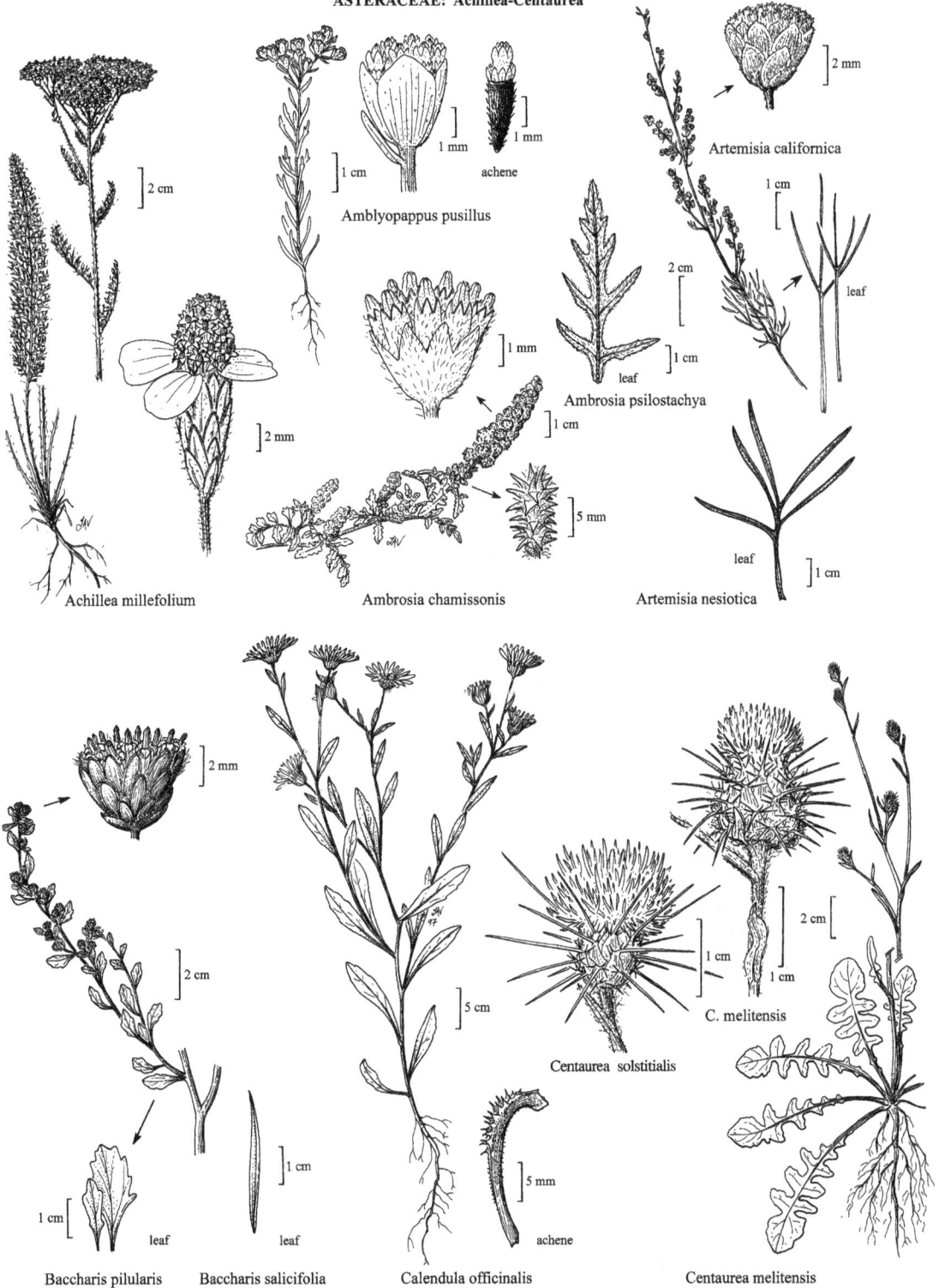

2 cm

Achillea millefolium

1 cm

Amblyopappus pusillus

1 mm

achene

1 mm

1 mm

Ambrosia psilostachya

leaf

2 cm

1 cm

1 cm

5 mm

Ambrosia chamissonis

2 mm

Artemisia californica

1 cm

leaf

Artemisia nesiotica

leaf

1 cm

2 mm

Baccharis pilularis

leaf

1 cm

Baccharis salicifolia

leaf

1 cm

2 cm

5 cm

Calendula officinalis

achene

5 mm

Centaurea solstitialis

1 cm

C. melitensis

1 cm

2 cm

Centaurea melitensis

row pit along Monroe Drive, in gully e of old borrow pit, in canyon ne of Nicktown, and at ne side of reservoir near Nicktown; s side at barge landing near Daytona Beach. All CA Channel Islands except Santa Barbara; n CA (Tehama Co.) to Baja CA, e to TX; Mexico; S America.

CALENDULA L.

Annual or perennial herbs. Ca. 15 species; Europe, n Africa, and w Asia. Some species are cultivated for ornamental use. (Latin: calendar, referring to extended flowering season of some species)

*C. officinalis L. GARDEN or POT MARIGOLD Plants 5-7 dm tall. Mar-Aug. Rare; disturbed flats and roadsides. Northeastern coastal flats near Coast Guard Beach (historic collections). WA; n CA (Napa Co.) to central CA (Santa Barbara Co.), e (disjunctly) to ME; Eurasia; Africa; Atlantic Islands; native of Mediterranean Europe. First collected on the island in July 1977; persisted without cultivation until at least the mid-1980s but not seen recently. Flowers and leaves are edible; dried flower petals can be used as seasoning and to produce yellow dye; this species also has many medicinal uses.

CENTAUREA L. KNAPWEED or STAR-THISTLE

Annual herbs. Ca. 500 species; mostly Eurasia and n Africa; widely introduced ± worldwide. Some species are cultivated for ornamental use; many are noxious or invasive weeds. (Greek: ancient plant name associated with Chiron, a centaur famous for knowledge of medicinal plants)

1. Apical spines of largest phyllaries 5-10 mm long . . . .**C. melitensis**
1' Apical spines of largest phyllaries 10-25 mm long . . . .**C. solstitialis**

*C. melitensis L. TOCALOTE Plants to 6 dm tall. Apr-Jun. Abundant; sandy swales, dry slopes and flats. Scattered on n side of island, as in dune swale w of Red Eye Beach and on ne coastal flats from NavFac area to airfield; widespread in se portion of mesa, especially s of airfield runways and w of airfield terminal; occasional in other parts of mesa, as in Radar Row area (Building 127); scattered on s side, especially n of Cormorant Rock and on coastal flats just w of Cattail Canyon. All CA Channel Islands; Todos Santos, Guadalupe, and Cedros islands; WA to Baja CA, e to TX and disjunctly to MA; Hawaii; native of Europe. First reported for island by Eastwood (1898). This invasive species has been spreading rapidly on the island in recent years. It has been used medicinally.

*C. solstitialis L. YELLOW STAR THISTLE Plants to 6 dm tall. Jun-Sep. Rare; disturbed flats. Mesa at Nicktown. Santa Rosa, Santa Cruz, and Santa Catalina islands; WA to Baja CA, e to NH and disjunctly to FL; Australia; New Zealand; native of Europe. First collected on mesa in September 1999. Eradication efforts (manual removal) were initiated in 1999. This species, toxic to horses, has also been used medicinally.

CHRYSANTHEMUM L.

Annual herbs. Ca. 2 species; Mediterranean Europe and n Africa. Several species are cultivated for ornamental use. (Greek: gold flower)

*C. coronarium L. [*Glebionis c.* (L.) Spach.] GARLAND CHRYSANTHEMUM or CROWN DAISY Plants 8-12 dm tall. Mar-Jun. Rare; sandy flats and slopes. North side on coastal flats near Jetty Canyon; mesa near wells area along Tufts Road. Santa Catalina and San Clemente islands; Todos Santos Island; central CA to nw Baja CA; native of Europe. First collected near wells area along Tufts Road in June 1977. Eradication efforts were initiated on the island in November 1983 and intensified in 1991 and 1992, but plants have persisted. Leaves and tender stems are cooked and eaten as a vegetable in China and Japan; flower petals can be eaten raw; flowers and leaves have been used medicinally.

ASTERACEAE: Chrysanthemum-Cynara

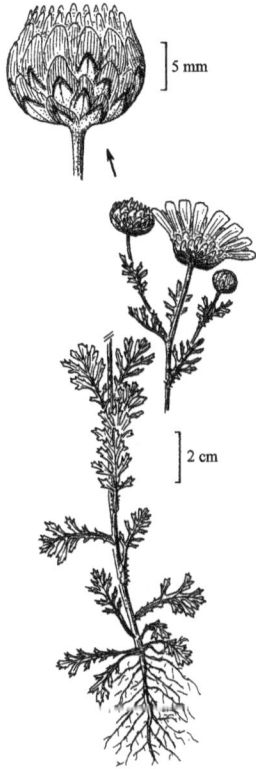

5 mm

2 cm

Chrysanthemum coronarium

2 cm

Cirsium occidentale var. occidentale

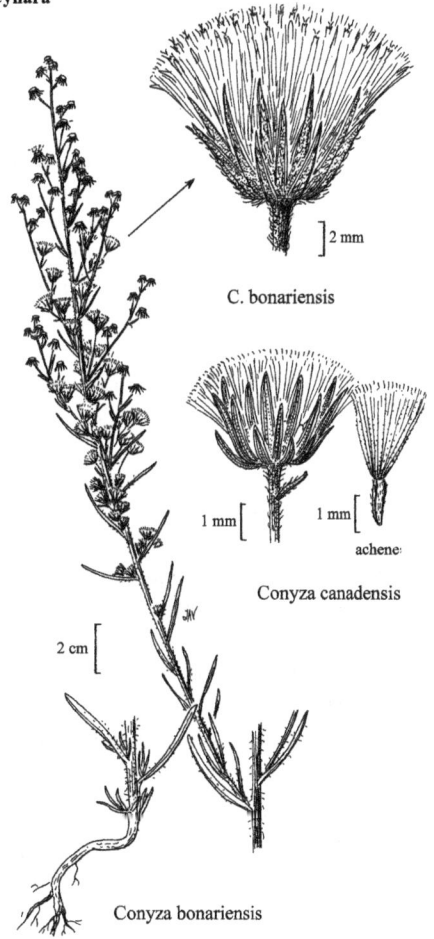

2 mm

C. bonariensis

1 mm

1 mm

achene

Conyza canadensis

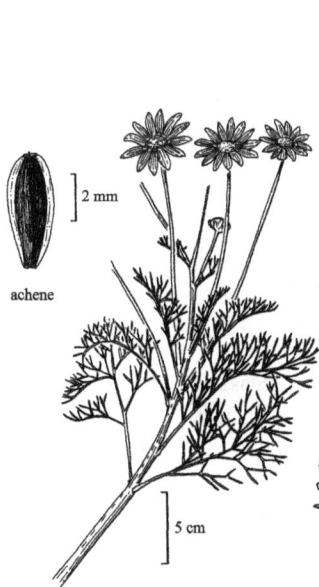

2 cm

Conyza bonariensis

achene

2 mm

Coreopsis gigantea

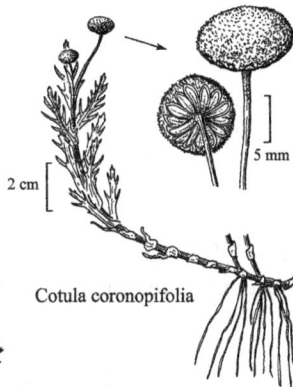

5 cm

2 cm

5 mm

Cotula coronopifolia

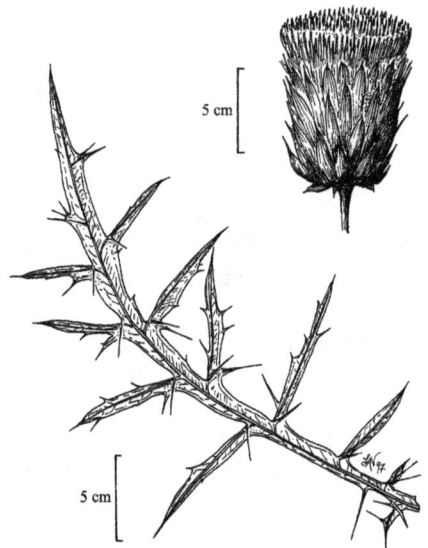

1 cm

5 cm

5 cm

Cynara cardunculus subsp. flavescens

CIRSIUM Mill. THISTLE

Annual herbs. Ca. 200 species; temperate N America and Eurasia. A number of species are invasive weeds. (Greek name for some thistle)

C. occidentale (Nutt.) Jeps. var. **occidentale** COBWEBBY or WESTERN THISTLE Plants 3-12 dm tall. Mar-Jun. Common; flats, slopes, and gully bottoms. North side on coastal flats and lower escarpment slopes between Corral Harbor and Coast Guard Beach; n edge of mesa w of airfield runways. All CA Channel Islands except Anacapa and Santa Barbara; n CA (Mendocino Co.) to Baja CA. Six other varieties of this species occur in OR, CA, and NV. Some depauperate plants on the island approach var. *compactum* Hoover. Peeled roots and young stems were eaten by Native Americans.

CONYZA Less.

Annual herbs. Ca. 25-40 species; temperate and subtropical regions worldwide. (ancient name for fleabane, perhaps alluding to powdered plant being used to repel insects)

1. Phyllaries usually hispidulous or strigose; pistillate flowers 60-150 . . . .**C. bonariensis**
1' Phyllaries ± glabrous; pistillate flowers 20-45 . . . .**C. canadensis**

*C. bonariensis (L.) Cronquist FLAX-LEAVED FLEABANE Plants 3-8 dm tall. Apr-Nov. Common; disturbed flats and eroded slopes. Widespread locations throughout much of island. All CA Channel Islands; OR to Baja CA, e to FL; native of S America. First collected at nw end of mesa in April 1966.
*C. canadensis (L.) Cronquist HORSEWEED Plants 5-12 dm tall. Jul-Nov. Occasional; flats. North side, especially on coastal flats near NavFac Grade; mesa near wells area along Tufts Road, in vicinity of Jackson Peak, and along roads between Nicktown and airfield. All CA Channel Islands; British Columbia to Baja CA, e to ME; introduced ± worldwide. Probably not native on San Nicolas Island. First collected in small canyon e of Nicktown in July 1965.

COREOPSIS L. TICKSEED

Shrubs. Ca. 35 species; mostly temperate N and S America. (Greek: bedbug-like, referring to shape of achenes)

C. gigantea (Kellogg) H.M.Hall [*Leptosyne g.* Kellogg] GIANT COREOPSIS Plants to 2.5 m tall. Feb-Apr. Abundant; n-facing slopes, canyon bottoms, swales, and flats. Widespread throughout many parts of island, especially on ne coastal flats and on mesa; scattered and localized on lower slopes of s escarpment and coastal flats w of Dutch Harbor. All CA Channel Islands; Guadalupe Island; central and s CA (coastal San Luis Obispo Co. to Los Angeles Co. and conspicuously planted in San Diego Co.).

COTULA L.

Perennial herbs. Ca. 55 species; mostly in S Hemisphere of Old World. Some species are widely naturalized. (Greek: small cup, perhaps referring to sheathing leaf bases of some species)

*C. coronopifolia L. BRASS-BUTTONS Plants 1-2 dm tall. Apr-Sep. Occasional; moist canyon bottoms and flats. North side in canyon bottoms between Celery Canyon and vicinity of Nicktown; mesa near wells area along Tufts Road and in upper w fork of W Mesa Canyon. All CA Channel Islands except Santa Barbara and San Clemente; Alaska to Baja CA, e to AZ and disjunctly to MA; S America; Australia; Europe; native of S Africa. First collected at an unspecified location on the island in September 1945. The whole plant yields a gold dye.

CYNARA L.

Perennial herbs. Ca. 8 species; Mediterranean Europe, Macaronesia, and w Asia. *Cynara cardunculus* subsp. *cardunculus* is the cultivated artichoke; young flower heads are cooked and eaten as a vegetable.

(early Latin name for artichoke)

*C. cardunculus L. subsp. **flavescens** Wiklund CARDOON or ARTICHOKE THISTLE Plants to 2 m tall. May-Jun. Rare; flats. Mesa at wells area along Tufts Road. Santa Cruz and Santa Catalina islands; WA to Baja CA; S America (especially Argentina); Australia; New Zealand; native of Mediterranean Europe. First collected in wells area in May 1985. Artichoke thistle is an extremely invasive plant (Hillyard 1985, Thomsen *et al.* 1986) and needs to be eradicated from the island. Globe artichoke (*C. cardunculus* L. subsp. *cardunculus*) has persisted without cultivation on ne coastal flats near Rock Jetty (at Building 122) but has not become naturalized. Leaves, petioles, roots, and flower receptacles of artichoke thistle can be boiled and eaten as vegetables; dried flowers are a rennet substitute, used for curdling milk; leaves have been used medicinally; plants yield a yellow dye.

DEINANDRA Greene TARPLANT
Shrubs. Ca. 21 species; sw N America; all were previously included in the genus *Hemizonia*. Five taxa are endemic to the islands off California and Baja California. (meaning of the name is not known)

D. clementina (Brandegee) B.G. Baldwin [*Hemizonia c.* Brandegee] CATALINA TARPLANT or ISLAND TARPLANT Plants 3-7 dm tall. Mar-Jul (Sep). Abundant; slopes, flats, barren ridgetops, and canyon walls. North side on escarpment and on n coastal flats between Corral Harbor and Sand Spit; widespread in central and e portions of mesa; scattered on s side on slopes of s escarpment from w end of mesa, but especially e of Twin Rivers drainage, where it is mostly restricted to siltstone substrates. ENDEMIC to Santa Cruz, Anacapa, Santa Barbara, San Nicolas, Santa Catalina, and San Clemente islands. A detailed distribution map for this taxon can be found in Junak *et al.* (1995b).

ENCELIA Adans.
Shrubs. Ca. 13 species; sw N America, Mexico, and S America. (for Christoph Entzelt, 1517-1583, German naturalist)

*E. californica Nutt. BUSH SUNFLOWER Plants to 5 dm tall. Rare; roadsides and sandy flats. Mesa along Shannon Road n of intersection with Tufts Road and in wells area along Tufts Road. All CA Channel Islands except San Miguel and Santa Rosa; Los Coronados, Todos Santos, and San Martin islands; s CA (Santa Barbara Co.) to n Baja CA. Although native to CA; this taxon was presumably introduced to San Nicolas Island as it appeared to be cultivated when first collected in June 1977. The foliage of this species causes contact dermatitis in some people.

GNAPHALIUM L. CUDWEED or EVERLASTING
Annual herbs. Ca. 38 species; ± worldwide. (Greek: name for a downy plant with soft white leaves used for stuffing cushions)

G. palustre Nutt. LOWLAND CUDWEED or WESTERN MARSH CUDWEED Plants to 1 dm tall. Apr-May. Rare; flats. Known only from a collection of extremely depauperate plants on an "upland ridge" in April 1901 and from a collection on mesa near airfield in October 2001. Santa Cruz, Santa Catalina, and San Clemente islands; WA to Baja CA, e to ND and NM. This species was used medicinally by Native Americans.

GRINDELIA Willd. GUMPLANT
Suffrutescent perennials. Ca. 30 species; central and w N America, Mexico, and S America. (for David Hieronymous Grindel, 1776-1836, Latvian botanist)

*G. hirsutula Hook. & Arn. [*G. camporum* var. *bracteosum* (J.T.Howell) M.A.Lane; *G. latifolia* Kellogg] Plants 8-12 dm tall. May-Jul. Rare; roadsides and disturbed flats. Mesa along Shannon Road just n of

ASTERACEAE: Deinandra-Heterotheca

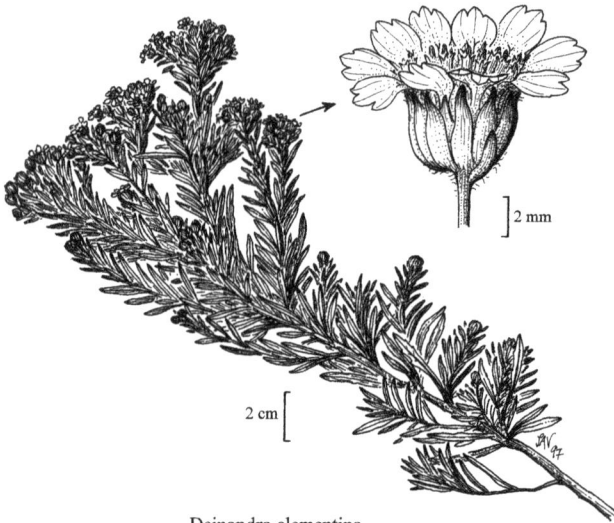

2 mm

2 cm

Deinandra clementina

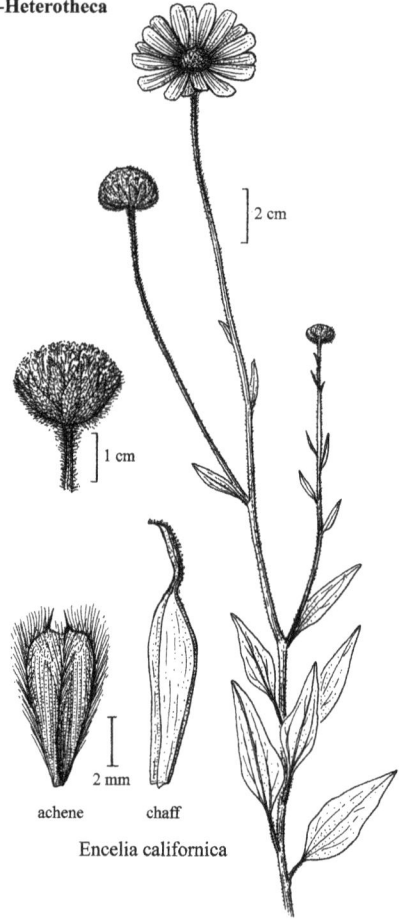

2 cm

1 cm

2 mm

achene   chaff

Encelia californica

2 cm

2 cm

Grindelia hirsutula

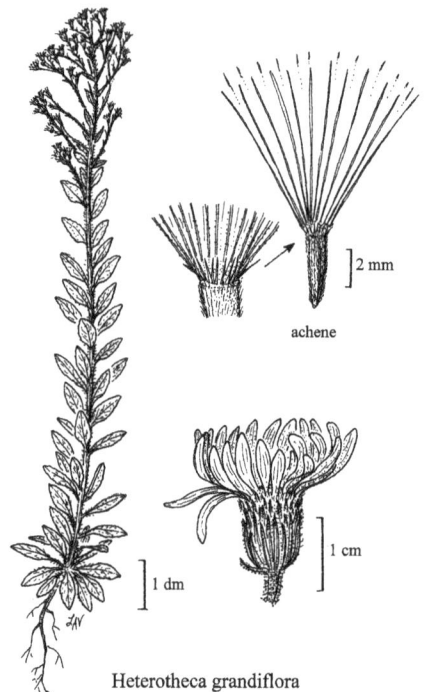

2 mm

achene

1 dm

1 cm

Heterotheca grandiflora

intersection with Tufts Road. All CA Channel Islands except Santa Barbara; British Columbia to Baja CA, e to ND and disjunctly to NY; introduced in Mexico (Yucatan). Presumably introduced to San Nicolas Island as it occurs only in disturbed areas and was not collected there until June 1987. Dried leaves and flower heads were used medicinally by Native Americans; leaves were also eaten.

HELIANTHUS L. SUNFLOWER
Annual herbs. Ca. 52 species; N America and Mexico. *Helianthus tuberosus* L., the Jerusalem artichoke, is cultivated for edible tubers. Several other species are cultivated for ornamental use or have edible seeds or tubers. (Greek: sunflower)

*H. annuus** L. COMMON SUNFLOWER Plants to 6 dm tall. Rare; roadsides and disturbed flats. North side on coastal flats near Jetty Canyon. Santa Cruz, Anacapa, Santa Catalina, and San Clemente islands; AK to Baja CA, e to ME and FL; Hawaii; introduced nearly worldwide. First collected on island in June 1996. This species is the source of sunflower seeds and oil; yellow, purple, and black dyes have been made from the flowers.

HETEROTHECA Cass.
Biennial herbs. Ca. 28 species; N America and Mexico. (Greek: different container or ovary, referring to dimorphic achenes)

*H. grandiflora** Nutt. TELEGRAPH WEED Plants to 12 dm tall. Mar-Nov. Common; roadsides, disturbed flats. North side in E Mesa Canyon; widespread locations on mesa. Santa Rosa, Santa Cruz, and San Clemente islands; n CA (Mendocino Co.) to Baja CA, e to NV, UT, and AZ (probably introduced to these 3 states); also introduced to Hawaii. Probably introduced on San Nicolas Island (and also on the other Channel Islands where it now occurs). First collected along Thousand Springs Road in July 1965. The Luiseño Tribe of southern CA used the tall stems to make arrow shafts; this species was also used medicinally by Native Americans.

HYPOCHAERIS L. CAT'S-EAR
Perennial herbs. Ca. 50 species; Europe, Asia, and S America. (Greek name used by Theophrastus for this or related genus)

*H. radicata** L. HAIRY CAT'S EAR Plants 3-5 dm tall. Jun-Aug. Rare; disturbed flats. Known from a single collection at Nicktown in July 1979; not seen recently. Santa Catalina and San Clemente islands; WA to Baja CA, e to ME and FL; Hawaii; native of Europe. The entire plant can be cooked and eaten.

ISOCOMA Nutt. COASTAL GOLDENBUSH
Shrubs. Ca. 16 species; southwestern N America (including Mexico). This group of plants was previously treated as a section of the genus *Haplopappus*. (Greek: equal hair of the head, apparently referring to its equal flowers)

1. Leaf margins entire or with 1-3 pairs of teeth or lobes near apex ......**I. menziesii** var. **menziesii**
1' At least some lower leaves with teeth along most margins ....**I. menziesii** var. **vernonioides**

I. menziesii (Hook. & Arn.) G.L.Nesom var. **menziesii** [*Haplopappus venetus* (Kunth) S.F.Blake subsp. *oxyphyllus* (Greene) H.M.Hall] Plants to 1 m tall. Aug-Jan. Common; flats and coastal bluffs. Widespread locations over most of island. Santa Catalina and San Clemente islands; n CA (San Mateo Co.) to Baja CA.
I. menziesii var. **vernonioides** (Nutt.) G.L.Nesom [*Haplopappus venetus* subsp. *vernonioides* (Nutt.) H.M.Hall] Plants to 1 m tall. Jul-Dec. Scarce; flats and coastal bluffs. Scattered with above taxon and

apparently intergrading with it. San Miguel, Santa Rosa, Santa Cruz, and Anacapa islands; coastal northern CA (San Mateo Co.) to Baja CA. Plants most closely approaching this variety are found on slopes of the n escarpment.

LACTUCA L.

Annual herbs. Ca. 100 species; temperate areas worldwide. *Lactuca sativa* L. is cultivated lettuce. (Latin: milk, referring to milky sap)

*L. serriola L. PRICKLY LETTUCE Plants to 2 m tall. (May) Jul-Nov. Widespread along roadsides on n side and in disturbed areas on mesa. All CA Channel Islands except Santa Barbara; WA to Baja CA, e to ME and FL; introduced ± worldwide; native of Europe. First collected along old runway south of air terminal in July 1965. Plants with deeply lobed leaves have been called var. *serriola,* while those bearing leaves with nearly entire margins have been called var. *integrata* Gren. & Godron.; both forms are present on the island. Leaves of this species are edible and can be cooked as greens, achenes are the source of Egyptian lettuce seed oil, and the milky sap found in most parts of the plant has been used medicinally.

LASTHENIA Cass. GOLDFIELDS

Annual herbs. Ca. 18 species; western N America, Mexico (Baja CA and Sonora), and S America (Chile). Seeds of *Lasthenia glabrata* Lindl. were ground into flour and eaten dry by Native Americans; the leaves were also eaten. (named for Greek girl who dressed as a boy in order to attend Plato's classes)

L. gracilis (DC.) Greene [*L. chrysostoma* Fisch. & C.A.Mey. subsp. *g.* (DC.) Ferris] Plants to 3 dm tall. Mar-Apr (Jul). Common; open grassy sites on flats and coastal slopes, in openings between shrubs on flats. Widespread on n coastal flats, on slopes of n escarpment, and on mesa; rare on slopes of s escarpment. All CA Channel Islands; Los Coronados, Todos Santos, Guadalupe, and Cedros islands; central CA to northwestern Mexico, e to AZ.

MALACOTHRIX DC.

Annual herbs, perennial herbs, or suffrutescent perennials. Ca. 20 species; w N America. (Greek: soft hair)

1. Annuals, stems herbaceous throughout . . . .**M. foliosa** subsp. **polycephala**
1' Perennial herbs, either suffrutescent or with woody, underground base
    2. Corollas yellow; inner phyllaries obtuse . . . .**M. incana**
    2' Corollas white; inner phyllaries attenuate
        3. Stems densely leafy; upper leaves ovate to obovate, bipinnately divided into linear segments
        . . . .**M. saxatilis** var. **implicata**
        3' Stems ± sparsely leafy; upper leaves linear to lanceolate, sharply and narrowly lobed . . . . .
    . . . .**M. saxatilis** var. **tenuifolia**

M. foliosa A.Gray subsp. **polycephala** W.S.Davis SAN NICOLAS ISLAND MALACOTHRIX Plants to 6 dm tall. Apr-Jun. Abundant; stabilized dunes and sandy sites on flats. Widespread locations throughout much of island, especially on coastal flats between Corral Harbor and Sand Spit. Less common on ne portion of mesa and on coastal flats on s side. ENDEMIC to San Nicolas Island.
M. incana (Nutt.) Torr. & A.Gray DUNE MALACOTHRIX or DUNEDELION Plants to 4 dm tall. Mar-Jul. Common; dunes, sandy terraces, and sandy flats. Widely scattered on n and sw sides of island; sandy sites on mesa. San Miguel, Santa Rosa, and Santa Cruz islands (historical occurrence only on Santa Cruz); central CA (San Luis Obispo Co.) to s CA (Ventura Co.; previously to San Diego Co.). Apparently not seen on island until 1977. Both pubescent and glabrous-leaved forms are common on the island. This taxon hybridizes freely with *M. foliosa* subsp. *polycephala* (Davis & Junak 1993).

ASTERACEAE: Hypochaeris-Malacothrix

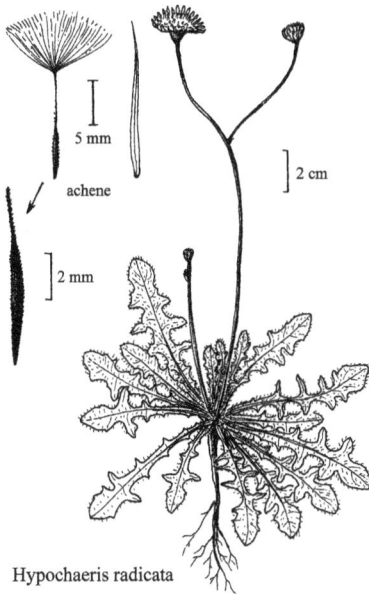

5 mm

achene

2 cm

2 mm

Hypochaeris radicata

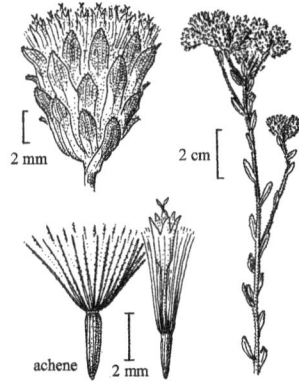

2 mm

2 cm

achene   2 mm

Isocoma menziesii var. menziesii

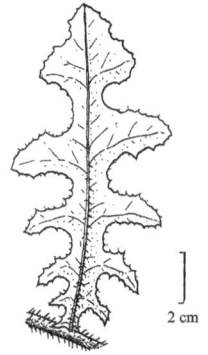

2 cm

Lactuca serriola

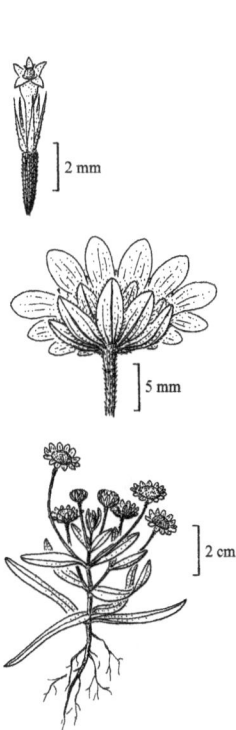

2 mm

2 mm

5 mm

2 cm

Lasthenia gracilis

2 cm

Malacothrix foliosa subsp. polycephala

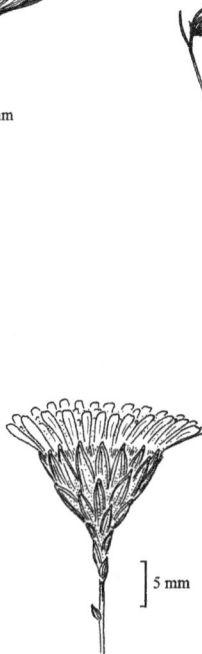

5 mm

Malacothrix incana

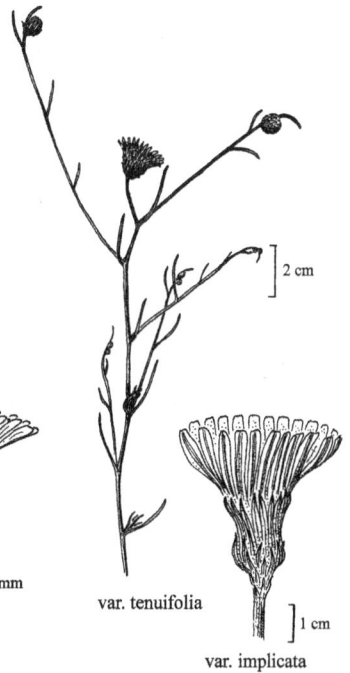

2 cm

var. tenuifolia

var. implicata

1 cm

Malacothrix saxatilis

M. saxatilis (Nutt.) Torr. & A.Gray var. **implicata** (Eastw.) H.M.Hall ISLAND CLIFF-ASTER or CLIFF MALACOTHRIX Plants to 7 dm tall. Mar-Sep. Common; canyons and slopes. Widely scattered on n side; se portion of island from ridge e of Cattail Canyon to Sand Spit. ENDEMIC to San Miguel, Santa Rosa, Santa Cruz, Anacapa, and San Nicolas islands. Other varieties occur on the CA mainland from Monterey Co. to San Diego Co.

*****M. saxatilis** var. **tenuifolia** (Nutt.) A.Gray CLIFF-ASTER Plants to 1 m tall. May-Jun. Rare; disturbed sites (especially gravel piles). North side along lower portion of Shannon Road and near Rock Jetty; mesa at Nicktown. Santa Catalina and San Clemente islands (introduced on San Clemente); central and s CA (San Luis Obispo to Orange Co. and recently introduced inland in San Bernardino Co.). First collected on island at Nicktown in May 1983. Presumably introduced to San Nicolas Island, this taxon should be removed as soon as possible, before it hybridizes with *M. s.* var. *implicata.*

MICROSERIS D.Don

Annual herbs. Ca. 14 species; western N America, S America (Chile), Australia, and New Zealand. (Greek: small chicory)

1. Achenes ± cylindrical, 1.5-3.5 mm long; pappus awns 1-4 mm long . . . .**M. elegans**
1' Achenes ± fusiform, 3-6.5 mm long; pappus awns 3-7 mm long
    2. Pappus paleae 0.5-1 mm long (or nearly obsolete) . . . .**M. douglasii** subsp. **tenella**
    2' Pappus paleae 1-6 mm long . . . .**M douglasii** subsp. **douglasii**

M. douglasii (DC.) Sch.Bip. subsp. **douglasii** Plants to 3 dm tall. Mar-May. Rare; eroded flats with clay soils. Mesa near Jackson Hill and s of airfield. OR to Baja CA.

M. douglasii subsp. **tenella** (A.Gray) K.Chambers Plants to 2 dm tall. Mar-May. Rare; eroded flats with clay soils. Known only from mesa s of airfield. San Miguel, Santa Rosa, and Santa Cruz islands; n CA (Tehama Co.) to s CA (Santa Barbara Co.).

M. elegans A.Gray Plants to 2.5 dm tall. Mar-May. Rare; eroded flats with clay soils. Known only from mesa s of airfield. San Miguel, Santa Cruz, and San Clemente islands; n CA (Butte Co.) to Baja CA.

PICRIS L. OX-TONGUE

Annual or biennial herbs. Ca. 45 species; Europe, Asia, and Africa. (Greek: bitter)

*****P. echioides** L. [*Helminthotheca e.* (L.) Holub] BRISTLY OX-TONGUE Plants to 7 dm tall. Mar-Jul. Scattered in disturbed sites; locally abundant in drainage ditches and along roadsides in south-central mesa, especially near Jackson Peak; Nicktown. Santa Cruz and Santa Catalina islands; WA to Baja CA, e (disjunctly) to ME; native of Europe. First collected on mesa near Jackson Hill in March 1983. This is an extremely invasive weed that is rapidly spreading on San Nicolas Island; it should be eliminated if possible. Young leaves can be cooked and eaten.

PSEUDOGNAPHALIUM L. CUDWEED or EVERLASTING

Annual or short-lived perennial herbs. Ca. 100 species; worldwide. Leaves and flower heads of the native species found on San Nicolas Island were used medicinally by Native Americans. (Greek: deceptively similar to *Gnaphalium*)

1. Leaves green and glandular above or grayish green and thinly tomentose, becoming glabrous in age
    2. Stems and lower surfaces of leaves thinly villous or glabrate; leaves strongly scented . . . .
. . . .**P. californicum**
    2' Stems and lower surfaces of leaves densely woolly; leaves not strongly scented . . . .**P. biolettii**
1' Leaves white to grayish, persistently woolly on both surfaces
    3. Heads 3-4 mm long; pappus bristles falling in clusters; pistillate corollas 1.5-2 mm long, lobes red
. . . .**P. luteo-album**

ASTERACEAE: Micropus-Pseudognaphalium

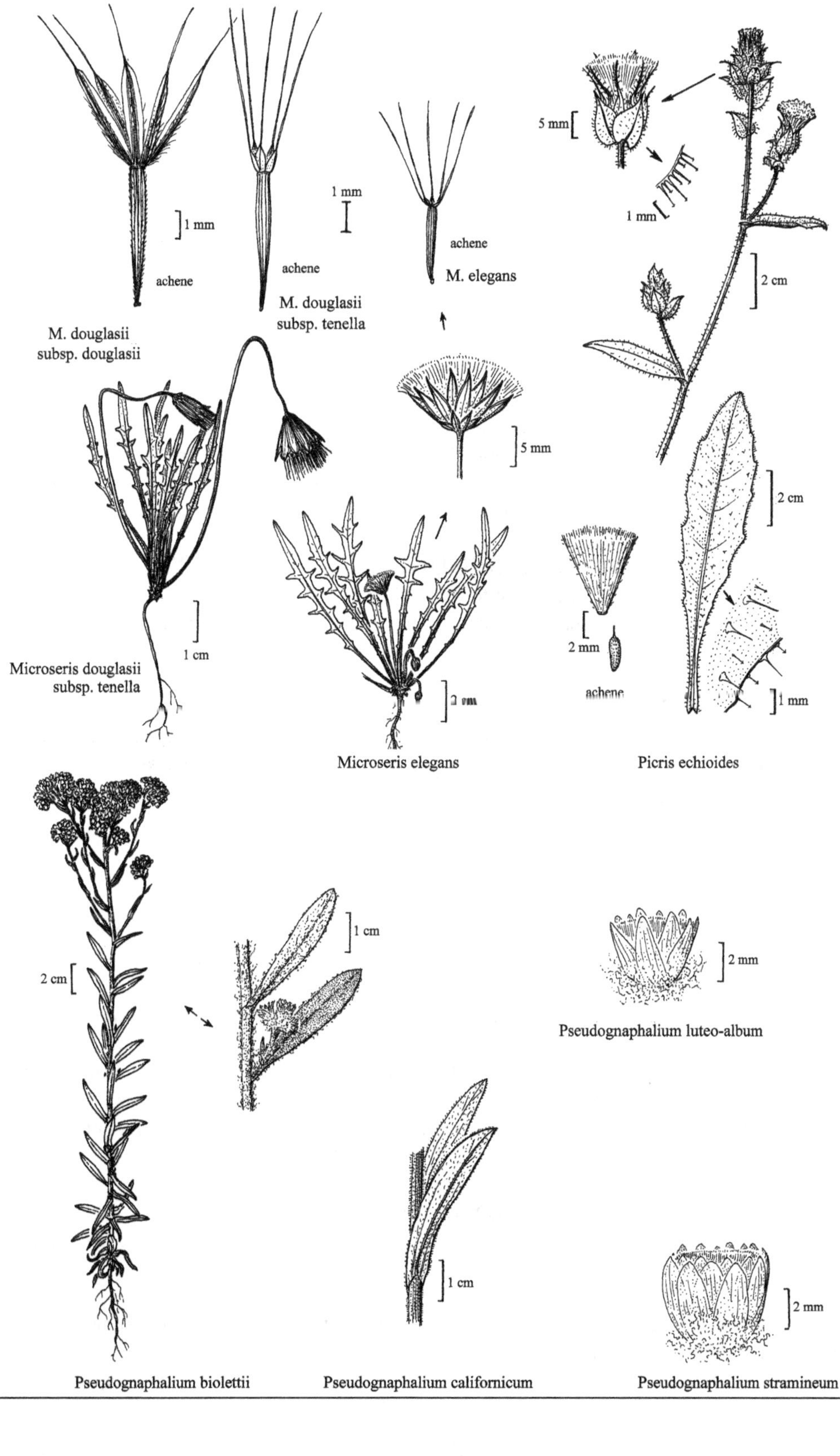

achene

M. douglasii
subsp. douglasii

achene

M. douglasii
subsp. tenella

achene
M. elegans

5 mm

1 mm

2 cm

2 cm

2 mm

achene

1 mm

Microseris douglasii
subsp. tenella

Microseris elegans

Picris echioides

5 mm

1 cm

2 cm

2 mm

1 cm

1 cm

Pseudognaphalium luteo-album

Pseudognaphalium biolettii

Pseudognaphalium californicum

Pseudognaphalium stramineum

3' Heads 4-6 mm long; pappus bristles falling separately; pistillate corollas 2-2.5 mm long, lobes yellow . . . .**P. stramineum**

P. biolettii Anderberg [*Gnaphalium bicolor* Bioletti] BICOLORED EVERLASTING Perennials to 12 dm tall. Mar-May (Jul). Scarce; dry gullies and flats. Northern coastal flats east of Rock Jetty; scattered in sw portion of mesa and more common in ne portion of mesa. All CA Channel Islands; Todos Santos, San Martin, Guadalupe, and Cedros islands; n CA (Napa Co.) to Baja CA.

P. californicum (DC.) Anderberg GREEN EVERLASTING Annuals or biennials to 13 dm tall. May-Jun. Rare; dry canyon. Known only from the ne escarpment, near Nicktown and in upper e fork of Keyhole Canyon. All CA Channel Islands; WA to Baja CA.

*P. luteo-album (L.) Hilliard & B.L.Burtt WEEDY CUDWEED Annuals 3-8 dm tall. Mar-Jul. Occasional; moist sites in gullies and canyons and in disturbed areas on flats. Widely scattered locations on n coastal flats and on mesa. All CA Channel Islands; WA to Baja CA, e to FL; Australia; New Zealand; Africa; native of Europe and Asia. First collected in Tule Canyon in April 1961. This species is used medicinally by the Australian Aborigines.

P. stramineum (Kunth) Anderberg [*Gnaphalium chilense* Spreng.] COTTON-BATTING Biennials to 12 dm tall. Feb-Jul. Scarce; dunes, canyons, and flats. North side near Thousand Springs and Celery Canyon; mesa in dunes near wells area along Tufts Road, on flats s and e of Nicktown, and w of airfield runways; s side at sw end of island. All CA Channel Islands except Santa Barbara; Guadalupe and Cedros islands; British Columbia to Baja CA, e to TX and disjunctly to NY; Mexico; S America.

SENECIO L. GROUNDSEL or RAGWORT

Annual herbs. Ca. 1000 species; temperate to tropical areas worldwide. Several species are cultivated for ornamental use. (Latin: old man or woman, referring to white pappus bristles that resemble the white hair of an elderly person)

*S. vulgaris L. COMMON GROUNDSEL Plants to 4 dm tall. Feb-Apr. Occasional; disturbed areas on flats and along roads. Coastal flats in ne portion of island; scattered locations on mesa; coastal flats in se portion of island. All CA Channel Islands except Santa Barbara; AK to Baja CA, e to ME; Hawaii; native of Europe. First collected at ne end of island in April 1982. This species has been used in folk medicine, but all parts of the plant can be toxic to many mammals (including humans).

SILYBUM Adans.

Annual herbs. Two species; Mediterranean Europe. (Greek name for a thistle-like plant)

*S. marianum (L.) Gaertn. MILK-THISTLE Plants to 15 dm tall. May-Jun. Rare; disturbed flats. North side near Anchor Point; mesa on s side of airfield. All CA Channel Islands except San Miguel (but it has been eradicated on Anacapa, Santa Barbara, and San Clemente); WA to Baja CA, e to NH and GA; S America (Chile and Argentina); Hawaii; Australia; New Zealand; S Africa; native of Europe. First collected on mesa in June 1993. Very young leaves and peeled upper stems are edible uncooked, roasted achenes have been used as a coffee substitute, and an infusion of the plant has been used in folk medicine. Achenes, which contain silymarin (a combination of three flavonoids), have been used medicinally for liver ailments and as antioxidants. Wilted plants can contain large amounts of nitrates and can be toxic to livestock.

SONCHUS L. SOW-THISTLE

Annual or biennial herbs. Ca. 55 species; Europe, Asia, and Africa. (ancient Greek name for a type of thistle)

1. Stem slender; leaves deeply lobed, the lobes ± linear and often with smaller secondary lobes; achenes only slightly flattened; ligule longer than corolla tube . . . .**S. tenerrimus**
1' Stem stout; leaves toothed or with broad lobes; achenes flattened; ligule shorter than or equal to corolla tube

ASTERACEAE: Senecio-Stebbinsoseris

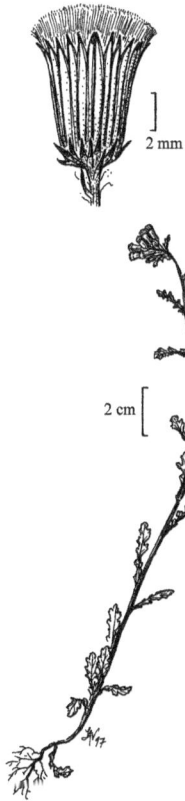

2 mm

2 cm

2 cm

5 mm

achene

Silybum marianum

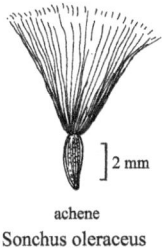

Senecio vulgaris

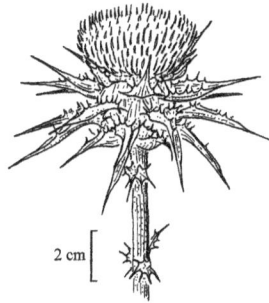

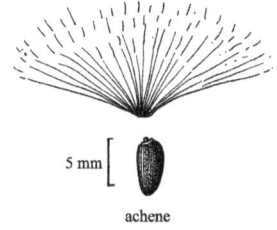

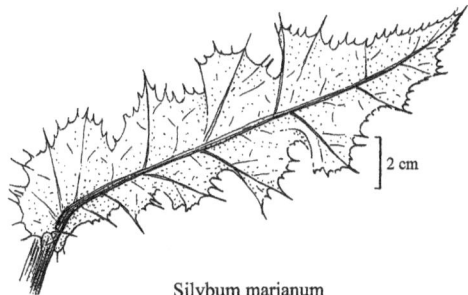

2 mm

achene
Sonchus oleraceus

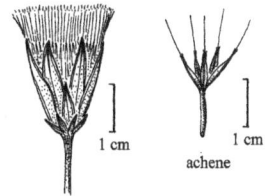

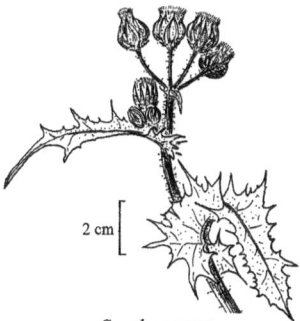

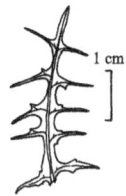

1 cm

1 cm

achene

2 cm

2 cm

2 cm

2 cm

1 cm

Sonchus asper

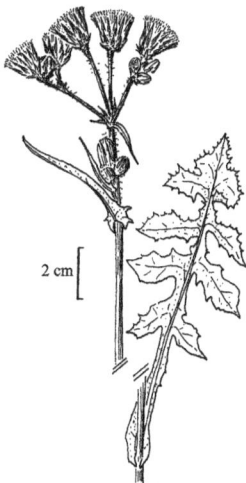

Sonchus oleraceus

Sonchus tenerrimus

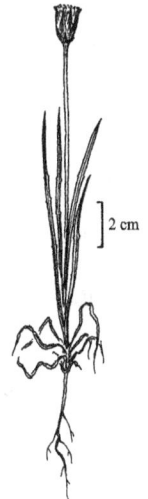

Stebbinsoseris heterocarpa

2. Lobes at leaf base rounded; leaf margins dentate and spinulose; achenes with 3 ribs on each side, otherwise smooth; ligule shorter than corolla tube . . . .**S. asper**

2' lobes at leaf base acute; leaf margins irregularly toothed; achenes with 2-4 ribs on each side and transversely wrinkled; ligule ± equal to corolla tube . . . .**S. oleraceus**

*S. asper (L.) Hill PRICKLY SOW-THISTLE Plants to 6 dm tall. Mar-May. Occasional; along roadsides and in disturbed areas in scattered localities; Thousand Springs area; gully sw of wells area on Tufts Road. All CA Channel Islands except Santa Barbara; Cedros Island; throughout N America; Hawaii; many parts of the world; native of Europe. First collected "on moist slopes" in April 1897. This species is edible and can be cooked for greens; stems can be bruised to remove bitter milky juice, peeled, and eaten.

*S. oleraceus L. COMMON SOW-THISTLE Plants to 7 dm tall. Apr-Jul. Abundant: disturbed flats and slopes. Widespread over most of island (one of the most widespread plants on the island). All CA Channel Islands; Los Coronados, Todos Santos, San Martin, Guadalupe, Cedros, and Natividad islands; throughout N America; Hawaii; nearly worldwide; native of Europe, Eurasia, and northern Africa. First collected on a "moist slope" on island in April 1897. This species is edible and can be cooked for greens.

*S. tenerrimus L. SLENDER SOW-THISTLE. Plants to 5 dm tall. Apr-May. Rare; coastal slopes and ridges. North escarpment on ridge ne of barge landing at Daytona Beach; s escarpment on ridge along e side of Grand Canyon. Santa Barbara, Santa Catalina, and San Clemente islands; all Baja CA islands except San Geronimo; n CA (Butte Co.) to Baja CA; disjunct occurrences in AL and NY; native of s Europe. First collected on "moist slopes" in April 1897.

STEBBINSOSERIS K.L.Chambers

Annual herbs. Two species; w N America. (Greek: Stebbin's chicory, honoring G. Ledyard Stebbins, Jr., 1906-2000, American botanist and geneticist)

S. heterocarpa (Nutt.) K.L.Chambers [*Microseris h.* (Nutt.) K.L.Chambers] Annual herb to 3 dm tall. Mar-Apr. Rare; flats. Known from a single collection on mesa just w of airfield runways. All CA Channel Islands except San Miguel and Santa Barbara; Guadalupe Island; n CA (Butte Co.) to Baja CA; e to AZ.

TARAXACUM F.H.Wiggers DANDELION

Perennial herbs. Ca. 60 species; N and S America, Europe, and Asia. (Arabic: bitter herb)

*T. officinale F.H.Wiggers COMMON DANDELION Plants to 1.5 dm tall. Jun-Jul. Rare; disturbed flats. Known from a single collection at Nicktown in July 1979. Santa Rosa, Santa Cruz, and Santa Catalina islands; throughout N America; Hawaii; nearly worldwide; native of Europe. This species is often cultivated for its edible leaves and as a source of wine; dried roots are used medicinally; roasted roots are sometimes used as a coffee substitute.

TRAGOPOGON L. GOAT'S BEARD

Biennial herbs. Ca. 45 species; Europe and Asia. (Greek: goat's beard, probably referring to silky pappus)

*T. porrifolius L. SALSIFY or OYSTER PLANT. Plants to 1 m tall. Apr-Jul. Common; disturbed flats. Northeastern coastal terrace near Rock Jetty; widely scattered locations in central and e portions of mesa. Santa Cruz, Santa Catalina, and San Clemente islands; throughout most of N America; Hawaii; Australia; native of Europe. First collected on mesa "near Telemetry Buildings" in June 1983. This species is widely cultivated for its edible roots; latex from the roots has been used by Native Americans in British Columbia for chewing gum.

ASTERACEAE: Taraxacum- Xanthium

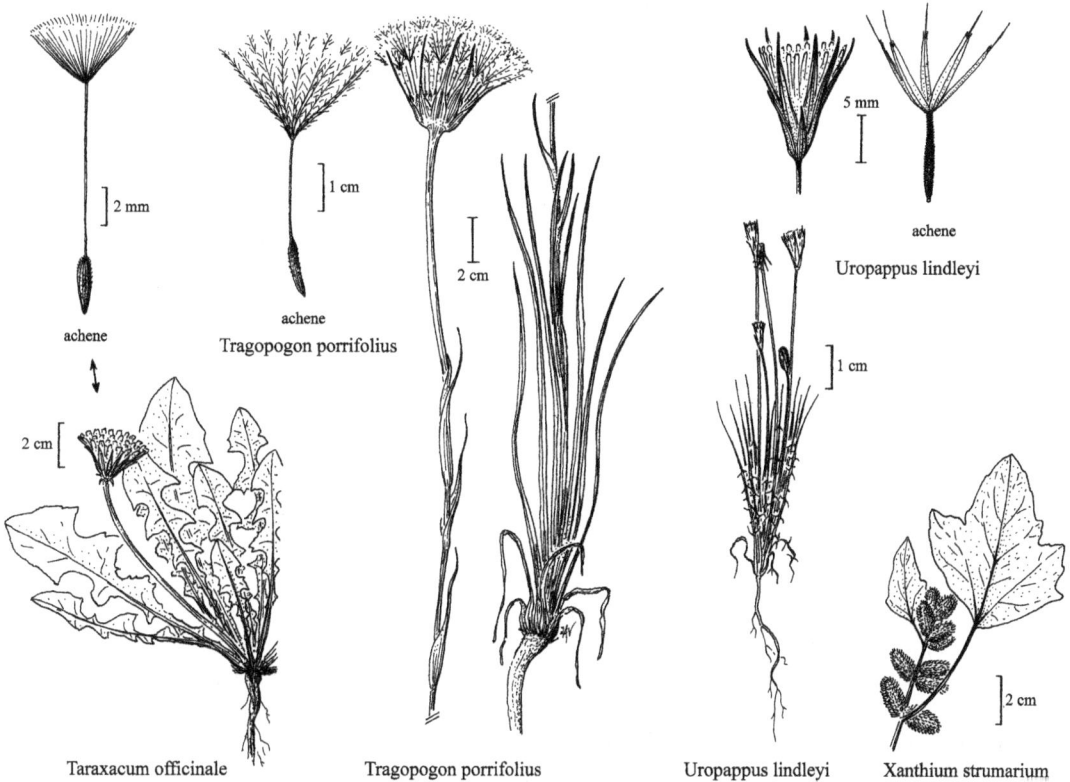

2 mm

achene

2 cm

Taraxacum officinale

1 cm

achene
Tragopogon porrifolius

2 cm

Tragopogon porrifolius

5 mm

achene

Uropappus lindleyi

1 cm

Uropappus lindleyi

2 cm

Xanthium strumarium

UROPAPPUS Nutt.

Annual herbs. One species; western N America. (Greek: tailed pappus, referring to slender terminal bristle on each pappus scale)

U. lindleyi (DC.) Nutt. [*Microseris lindleyi* (DC.) A.Gray; *M. linearifolia* (DC.) Sch.Bip.] SILVER PUFFS Plants to 3 dm tall. Apr-May. Rare; flats. Known only from a collection made at "one locality" in April 1897; not seen recently. All CA Channel Islands; Los Coronados, Todos Santos, San Martin, Guadalupe, and Cedros islands; WA to Baja CA, e to ID, UT, and TX.

XANTHIUM L. COCKLEBUR

Annual herbs. Two species; New World; introduced worldwide. (Greek: yellow, apparently alluding to an ancient name of a plant that yielded a yellow dye)

*X. strumarium L. Plants to 5 dm tall. Jul-Aug. Rare; disturbed flats. Known only from barge landing area at Daytona Beach (on and near gravel piles). It was first collected there in September 1988, when all plants seen were removed; it was not seen in 1989 but was collected again in 1990. Santa Rosa, Santa Cruz, Santa Catalina, and San Clemente islands; N America; Hawaii; nearly worldwide. Presumably introduced to San Nicolas Island. A weedy species first described from Europe, but region of origin is unknown. This species can cause contact dermatitis in some humans and can be toxic to humans and livestock; leaves, roots, and seeds have been used medicinally; yellow dye can be obtained from leaves.

BORAGINACEAE Juss.
BORAGE FAMILY

Annual or perennial herbs. Ca. 100 genera and 2000 species; temperate to tropical areas worldwide, especially w N America and Mediterranean Europe. Several genera are cultivated for ornamental or medicinal uses (e.g., *Borago, Echium, Myosotis, Symphytum*). Most genera may be toxic because of alkaloids or accumulated nitrates. Some authors would include Hydrophyllaceae (waterleaf family) in Boraginaceae (Olmstead *et al.* 2000).

1. Perennials; ovary not strongly lobed; style absent, stigma at ovary apex, disk-like . . .**Heliotropium**
1' Annuals; ovary strongly lobed; style inserted between ovary lobes
    2. Nutlets widely spreading at maturity, margins with hooked bristles; flowers axillary . . . .
    . . . .**Pectocarya**
    2' Nutlets erect at maturity, margins entire; flowers in terminal spikes
        3. Corollas yellow to orange . . . .**Amsinckia**
        3' Corollas white . . . .**Cryptantha**

AMSINCKIA Lehm. FIDDLENECK

Annual herbs. Ca. 10 species; temperate w N and S America. (honoring Wilhelm Amsinck, 1752-1831, German botanist and patron of the Hamburg Botanical Garden)

1. Calyx lobes ± equal, fused near base; stems usually erect; nutlets 2-3.5 mm long . . . .
    . . . .**A. menziesii** var. **intermedia**
1' Calyx lobes unequal, 2-3 strongly fused, others ± free; stems spreading to decumbent; nutlets 1.5-2.5
    mm long . . . .**A. spectabilis** var. **spectabilis**

*A. menziesii** (Lehm.) A.Nelson & J.F.Macbr. var. **intermedia** (Fisch. & C.A.Mey.) Ganders [*A. i.* Fisch. & C.A.Mey.] COMMON FIDDLENECK Plants to 5 dm tall; stems erect. Feb-Apr. Scarce; flats, disturbed slopes, and canyon bottoms. North side on coastal flats near Celery Canyon; n edge of mesa w of Nicktown (above NavFac buildings), along Owens Road n of fire station, and s of airfield. All CA Channel Islands; Todos Santos and Cedros islands; British Columbia to Baja CA, e to ME; Hawaii. Probably introduced to San Nicolas Island. It was first collected on mesa near fire station on Owens Road in March 1980. Seeds were pounded into flour, made into cakes, and eaten raw by Native Americans. Variety *menziesii* occurs from British Columbia to Baja CA, e to CO and disjunctly to CT.
A. spectabilis Fisch. & C.A.Mey. var. **spectabilis** [*A. sancti-nicolai* Eastw., *A. spectabilis* var. *nicolai* (Jeps.) Munz] SEASIDE FIDDLENECK Plants to 1.5 dm tall; stems sprawling to prostrate. Mar-May (Jul). Scarce; stabilized dunes and sandy flats. North side at Vizcaino Point, e end of Red Eye Beach, Tender Beach, near Thousand Springs, and se of Corral Harbor. All CA Channel Islands except Santa Catalina; British Columbia to Baja CA. Variety *microcarpa* (Greene) Jeps. & Hoover occurs in coastal Santa Barbara and San Luis Obispo cos.

CRYPTANTHA Lehm. ex G.Don

Annual herbs. Ca. 160 species; temperate w N and S America. (Greek: hidden flowers, referring to minute corollas of the type species)

1. Nutlets usually 1-2 per calyx, heteromorphic; back of 1 nutlet usually tubercled, the other shiny and
    smooth . . . .**C. maritima** var. **maritima**
1' Nutlets usually 4 per calyx, homomorphic; nutlet backs fine-granular to minutely tubercled above the
    middle, fine-granular to smooth below the middle . . . .**C. traskiae**

C. maritima (Greene) Greene var. **maritima** [*C. m.* var. *pilosa* I.M.Johnst.] GUADALUPE ISLAND CRYPTANTHA Plants to 2 dm tall; rounded. Mar-Jun. Scarce; sandy flats, ridges, arroyos, and slopes. North side on ridgetop ne of barge landing at Daytona Beach; se edge of mesa se of Peak 606; s coastal

BORAGINACEAE

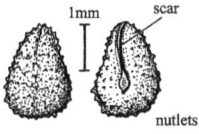

1mm   scar

nutlets

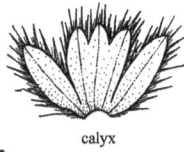

calyx

2 mm

flower

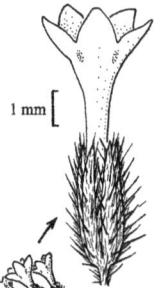

1 mm

flower

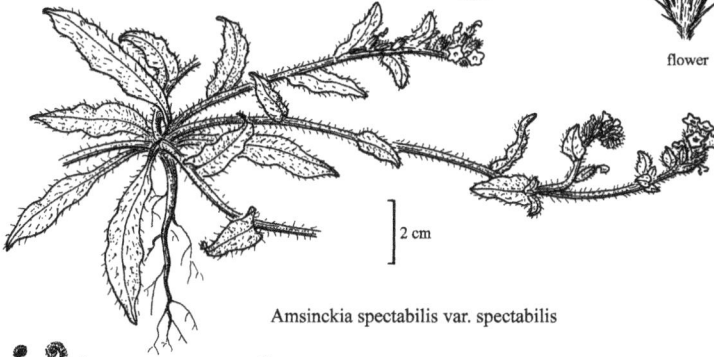

2 cm

Amsinckia spectabilis var. spectabilis

2 cm

Amisinckia menziesii var. intermedia

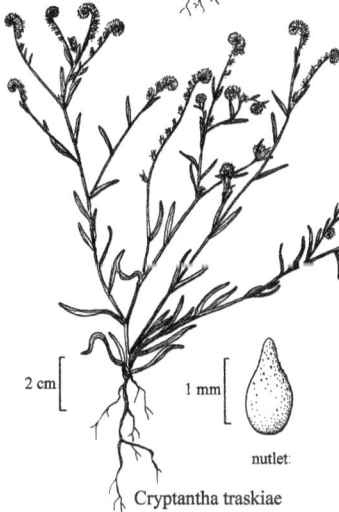

2 cm   1 mm

nutlet

Cryptantha traskiae

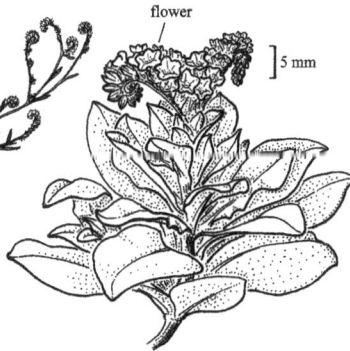

flower

5 mm

H. curassavicum

1 mm

nutlets

fruit & calyx

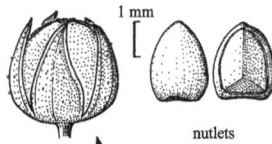

5 mm

2 cm

1 cm

1 mm

nutlets

Cryptantha maritima
var. maritima

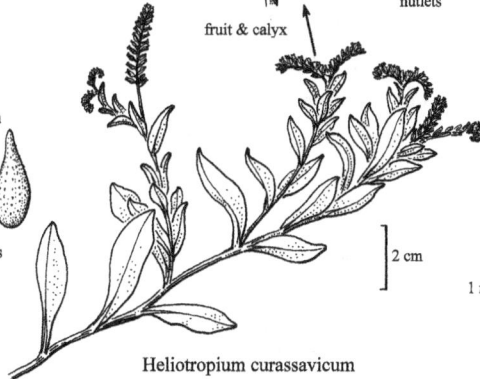

2 cm

Heliotropium curassavicum

1 mm   nutlets

Pectocarya linearis subsp. ferocula

flats and escarpment slopes between Cattail Canyon and Daytona Beach. Santa Barbara, Santa Catalina, and San Clemente islands; Los Coronados, Guadalupe, San Benito, and Natividad islands; e CA (Inyo Co.) to Baja CA, e to NV and AZ. Abundance of this taxon is extremely variable from year to year. Variety *cedrosensis* (Greene) I.M.Johnst. is endemic to Cedros Island.

C. traskiae I.M.Johnst. TRASK'S CRYPTANTHA Plants to 2 dm tall; stems sprawling. Mar-May. Occasional; stabilized dunes and sandy flats. Northern coastal flats and escarpment near Red Eye Beach, between Thousand Springs and Corral Harbor, and e of Celery Canyon; mesa near wells area along Tufts Road and near w end of airfield; s escarpment in upper e fork of Twin Rivers drainage. ENDEMIC to San Nicolas and San Clemente islands. Habitat for this species may be threatened by the extremely invasive sahara mustard *(Brassica tournefortii)*.

HELIOTROPIUM L. HELIOTROPE

Perennial herbs. Ca. 250 species; temperate and tropical areas worldwide. (Greek: sun-turning, either because some species flower at summer solstice or referring to an early belief that flowers of some species "followed" the sun)

H. curassavicum L. [*H. c.* var. *oculatum* (A.Heller) I.M.Johnst.] Plants to 3 dm tall; stems sprawling. Apr-Aug. Scarce; saline flats and seasonally damp areas. North side on peninsula at w end of Red Eye Beach, Thousand Springs, and along roadside near NavFac buildings; s side at S Range Marker poles w of Dutch Harbor. All CA Channel Islands except Santa Barbara; San Martin and Cedros islands; OR to Central America and w coast of S America, e to ME; Hawaii; Puerto Rico; Virgin Islands; naturalized in s Europe. Seeds were eaten by Native Americans, who also used the leaves and roots medicinally.

PECTOCARYA DC. ex Meisn. COMB-SEED

Annual herbs. Ca. 15 species; temperate w N and S America. (Greek: comb nut, referring to dentate nutlet margins in some species)

P. linearis (Ruiz & Pav.) DC. subsp. **ferocula** (I.M.Johnst.) Thorne SLENDER COMB-SEED Plants to 0.5 dm tall. Mar-Apr. Rare; canyon rims and sandy flats. Mesa near upper reaches of w and e forks of E Mesa Canyon and on n side of Beach Road n of airfield runways; se escarpment above Daytona Beach. All CA Channel Islands except San Miguel and Santa Barbara; Guadalupe and Cedros islands; central CA (San Joaquin Co.) to Baja CA. Subspecies *linearis* occurs in S America.

BRASSICACEAE Burnett [**Cruciferae** Juss.]
MUSTARD FAMILY

Annual or perennial herbs. Ca. 300 genera and 3000 species; temperate to tropical areas worldwide. Some genera are cultivated for ornamental use or for food (e.g., broccoli, brussel sprouts, cabbage, cauliflower, mustard, radish, turnip, and water cress). Some authors would include Capparaceae (the caper family) in Brassicaceae (Rodman *et al.* 1993, Rollins 1993).

1. Fruit a silicle, 1-6 mm long, length usually less than 3 times width
  2. Fruit deeply notched at tip and base between ± rounded halves; valves strongly flattened and disklike . . . .**Dithyrea**
  2' Fruit sometimes notched at tip but not deeply lobed; valves not disklike
    3. Silicles subterete . . . .**Cakile**
    3' Silicles flat
      4. Silicle apex notched; seeds 2 per silicle . . . .**Lepidium**
      4' Silicle apex obtuse to truncate, entire; seeds 3 or more per silicle

     5. Pubescence of appressed 2-branched hairs attached at the middle; silicles orbicular with flat margins . . . . **Lobularia**

     5' Pubescence stellate or forked, but not as above; silicles not margined

          6. Stems usually solitary, erect; leaves mostly basal; silicles usually stellate-pubescent, smooth or with inconspicuous veins . . . . **Draba**

          6' Stems many, spreading to weakly erect; leaves basal and cauline; silicles usually glabrous, usually with conspicuously raised veins . . . . **Hornungia**

1' Fruit a silique, 8-90 mm long, 3 to many times longer than width

    7. Siliques transversely jointed, indehiscent lengthwise (breaking transversely into 2 segments in *Cakile*)

        8. Leaves succulent; siliques with 2 segments, each with 1 seed; apical beak, if present, flat; plants of sandy beaches and dunes . . . . **Cakile**

        8' Leaves not succulent; siliques usually with 3 or more segments; beak terete, 1-2 cm long; plants usually in disturbed sites (roadsides, flats, etc.) . . . . **Raphanus**

    7' Siliques dehiscent lengthwise, not transversely jointed

        9. Siliques with apical beaks (elongated, seedless segment between ovary and style) 3-20 mm long

            10. Silique valves with 1 prominent midvein; plants annual . . . . **Brassica**

            10' Silique valves with 3-7 prominent veins; plants perennial . . . . **Hirschfeldia**

        9' Silique apices without beaks; styles sometimes persistent, beak-like, but less than 3 mm long

            11. Basal and lower cauline leaves simple, margins entire to dentate . . . . **Draba**

            11' Basal and lower cauline leaves deeply and pinnately lobed

                12. Petals yellow; stems hispidulous to hirsute; stems usually erect; plants of dry habitats . . . . **Sisymbrium**

                12' Petals white; stems glabrous to sparsely pubescent; stems prostrate to decumbent; plants of aquatic habitats . . . . **Nasturtium**

BRASSICA L. MUSTARD

Annual herbs. Ca. 35 species; temperate areas of Europe and Asia, with some species naturalized ± worldwide. Many species are cultivated for food or for oil obtained from seeds. (Latin name for cabbage)

1. Pedicels mostly shorter than sepals; siliques appressed against stem . . . . **B. nigra**
1' Pedicels longer than sepals; siliques divergent . . . . **B. tournefortii**

*B. nigra** (L.) Koch BLACK MUSTARD Plants to 1 m tall. Apr-May. Rare; flats. Known from a single collection near Army Springs in April 1966. All CA Channel Islands; Guadalupe Island; WA to Baja CA, e to ME; Hawaii; many temperate areas; native of Europe and Asia. Seeds of this species are used medicinally and to produce commercial mustard; leaves are edible raw or cooked. If eaten in large quantities, seeds can be toxic to livestock.

*B. tournefortii** Gouan SAHARA MUSTARD Plants to 8 dm tall. (Nov) Feb-May. Scarce; sandy flats and unstabilized dunes. North side near Thousand Springs and Corral Harbor; sw end of mesa and adjacent s escarpment. San Clemente Island; central CA (Santa Barbara and Kern cos.) to CA deserts and Baja CA; e to TX; Australia; New Zealand; native to Mediterranean Europe, N Africa, and Middle East. First collected near Building 176 in March 1999. This species is invading the habitat of the endemic Trask's cryptantha *(Cryptantha traskiae)* and should be eliminated from the island if possible.

CAKILE J.Mill. SEA-ROCKET

Annual herbs. Ca. 7 species; N America, Europe, and Africa. (Arabic name for the genus)

1. Lower fruit segment without lateral lobes or horns; petals absent or less than 3 mm wide, usually white . . . . **C. edentula** subsp. **edentula**

BRASSICACEAE: Brassica-Draba

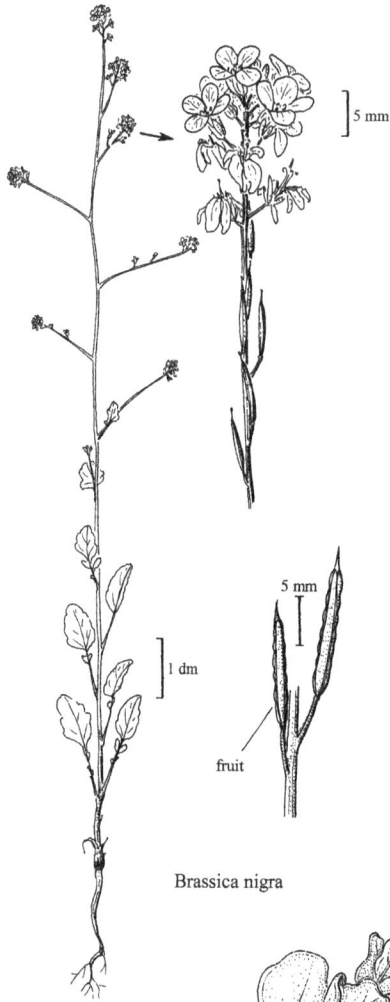

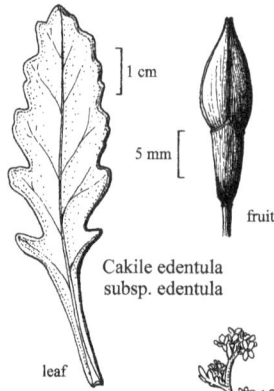

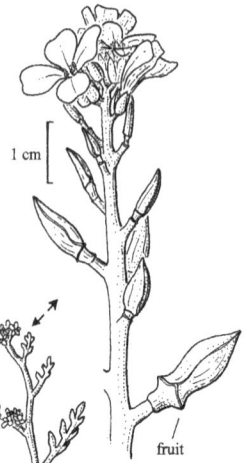

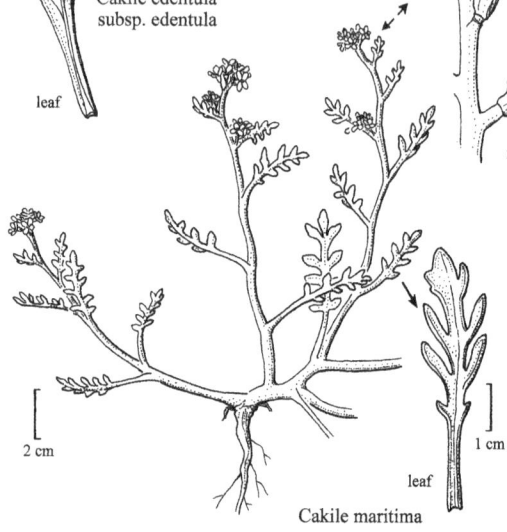

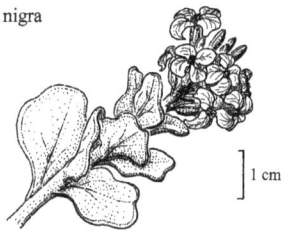

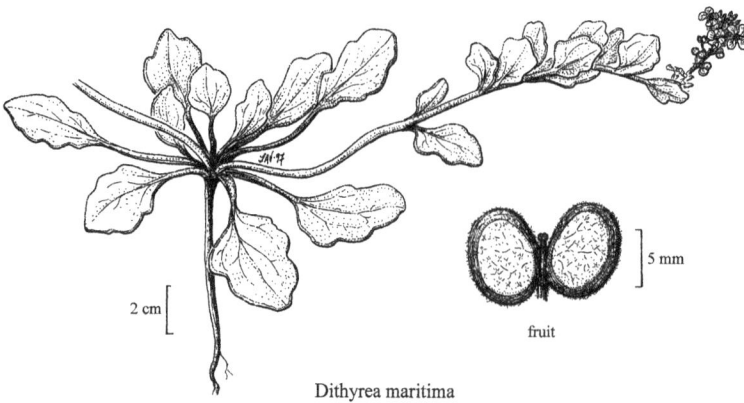

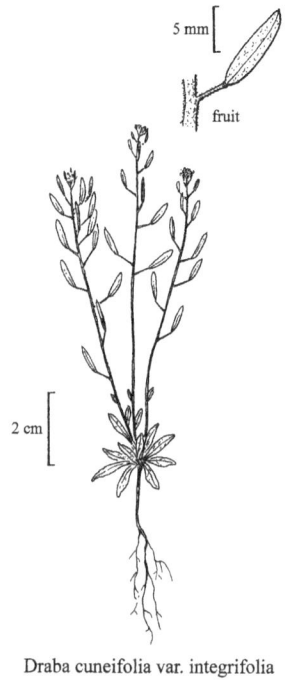

5 mm

1 cm

5 mm

fruit

Cakile edentula
subsp. edentula

leaf

1 cm

fruit

5 mm

fruit

2 cm

1 cm

leaf

Cakile maritima

1 dm

5 mm

fruit

Brassica nigra

1 cm

5 mm

fruit

Dithyrea maritima

2 cm

5 mm

fruit

2 cm

Draba cuneifolia var. integrifolia

1' Lower fruit segment with 2 lateral lobes or horns; petals 3-6 mm wide, usually lavender . . . .
. . . . .C. maritima

*C. edentula** (Bigelow) Hook. subsp. **edentula** [*C. e.* subsp. *californica* (A.Heller) Hultén] Plants to 7 dm tall. Apr. Rare; sandy beaches and dunes. Northwestern end of island. San Miguel and Santa Rosa islands; AK to Baja CA; Great Lakes and e coast of N America; Australia; native to Atlantic Coast. First collected in sand dunes at unspecified location in March 1945. Last known collection was made in dune area near nw end of island in April 1961. Putative hybrids with *C. maritima* have been collected in nw and s portions of island.

*C. maritima** Scop. Plants to 7 dm tall. Mar-Jul. Common and widespread; coastal strand and stabilized dune areas, particularly on sw and se sides of island. All CA Channel Islands except Santa Barbara; Todos Santos, San Martin, San Benito, Cedros, and Natividad islands; British Columbia to Baja CA; S America; Australia; native of Europe. First collected on ne side of island "near tank farm and Navy pier" in April 1961. Entire plant is edible after cooking; young plants can be eaten raw.

DITHYREA Harv. SPECTACLE-POD
Perennial herbs. Two species; western N America. (Greek: with 2 shields, referring to twin locules of fruit)

D. maritima Davidson BEACH SPECTACLE-POD Plants to 2 dm tall. Mar-Jul. Scarce; unstabilized sand dunes. North side on ridge east of Vizcaino Point and near Red Eye Beach and vicinity; s side e of Rock Crusher, near Cormorant Rock, and sw escarpment near Dizon's Ravine. San Miguel Island; central CA (San Luis Obispo Co.) to Baja CA.

DRABA L.
Annual herbs. Ca. 350 species; temperate areas worldwide. Some species are edible. (Greek: acrid)

D. cuneifolia Torr. & A.Gray var. **integrifolia** S.Watson Plants to 1.5 dm tall. Mar-Apr. Rare; flats. North side on coastal flats near W and E Mesa canyons. Santa Cruz, Santa Catalina, and San Clemente islands; Cedros Island; western N America.

HIRSCHFELDIA Moench
Perennial herbs. Two species; temperate areas of Europe. (honoring Christian Caius Lorenz Hirschfeldt, 1742-1792, German horticulturist)

*H. incana** (L.) Lagr.-Foss. [*Brassica geniculata* (Desf.) Ball] SHORT-PODDED MUSTARD Plants to 12 dm tall. May-Sep. Common; disturbed flats and sandy areas. North side on coastal flats e of Thousand Springs; mesa near Army Springs, wells area along Tufts Road, intersection of Shannon Road and Jackson Highway, Nicktown, and s of airfield runways; canyons of s escarpment (e.g., Grand Canyon). All CA Channel Islands except Anacapa and Santa Barbara; Cedros Island; OR to Baja CA, e to NV; Hawaii; native of Europe. First collected near Nicktown in April 1961. Seeds and young leaves are edible.

HORNUNGIA Reichenb.
Annual herbs. Three species; temperate areas of N America, Europe, and Asia. (honoring Ernst Gottfried Hornung, 1795-1862, German pharmacist, botanist, and entomologist)

H. procumbens (L.) Hayek [*Capsella divaricata* Walp., *Hutchinsia p.* (L.) Desv., *Hymenolobus p.* (L.) Nutt.] Plants to 3 cm tall. Feb-May (Jul). Rare; sandy flats, slopes, gullies, and sand dunes. North side near Vizcaino Point, Red Eye Beach, Tender Beach, Thousand Springs, and Corral Harbor; mesa near upper portion of E Mesa Canyon; sw side near Seal Beach. All CA Channel Islands except Santa Catalina; Guadalupe Island; WA to Baja CA, e to CO; N America; Australia (introduced); Europe.

BRASSICACEAE: Hirschfeldia-Lobularia

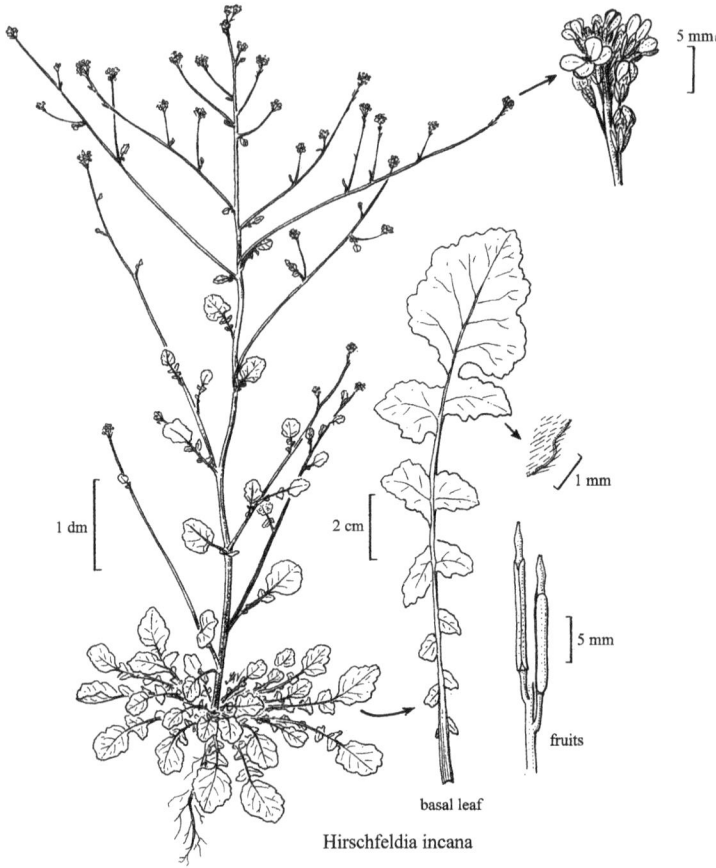

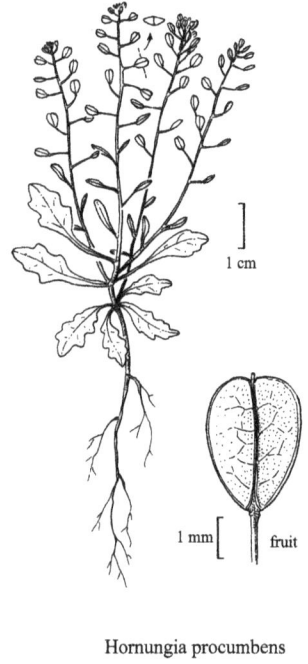

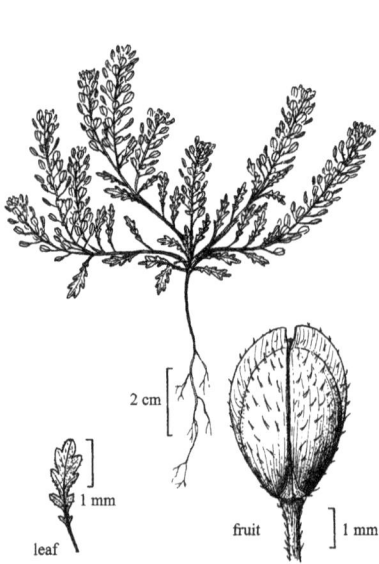

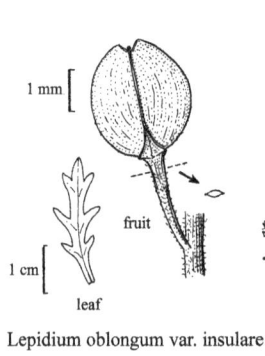

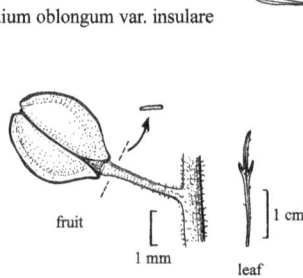

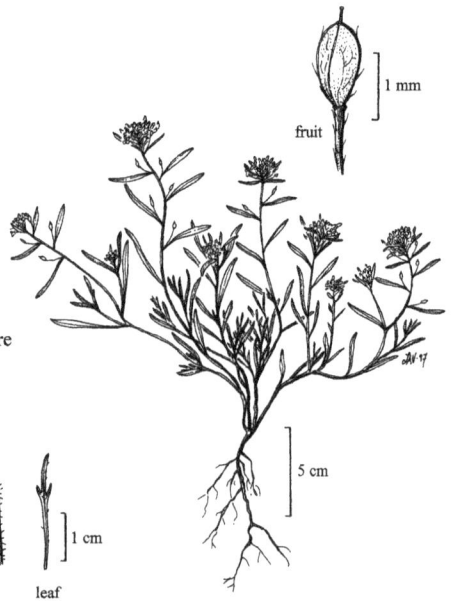

5 mm

1 mm

1 dm

2 cm

5 mm

fruits

basal leaf

Hirschfeldia incana

1 cm

1 mm fruit

Hornungia procumbens

1 mm

fruit

1 cm

leaf

Lepidium oblongum var. insulare

1 mm

fruit

2 cm

leaf

1 mm

fruit 1 mm

Lepidium lasiocarpum var. lasiocarpum

fruit

1 mm

leaf

1 cm

Lepidium nitidum var. nitidum

1 mm

fruit

5 cm

Lobularia maritima

LEPIDIUM L. PEPPERGRASS or PEPPERWORT

Annual herbs. Ca. 175 species; temperate areas worldwide. Seeds and leaves were eaten by Native Americans. The Maidu used leaves for medicinal purposes and washing hair. (Greek: little scale, referring to shape of the silicle)

1. Pedicels in fruit spreading, not crowded; silicles not strongly overlapping; cauline leaves linear and entire or with a few linear lobes; stems solitary to branched above base, usually erect . . . .
. . . .**L. nitidum** var. **nitidum**
1' Pedicels in fruit ascending, crowded; silicles strongly overlapping; cauline leaves dentate to pinnatifid; stems often branched at base, usually decumbent to prostrate
    2. Silicles 3-4.5 mm long, hirsute; plants usually growing in sand dunes . . . .
. . . .**L. lasiocarpum** var. **lasiocarpum**
    2' Silicles 2.5-3 mm long, ± glabrous, often with sparsely puberulent margins; plants usually growing on clay flats . . . .**L. oblongum** var. **insulare**

L. lasiocarpum Torr. & A.Gray var. **lasiocarpum** SAND PEPPERGRASS Plants to 1 dm tall. Feb-May (Jul). Common; usually on sandy flats and dunes, but occasionally on clay soils on n escarpment. Widely scattered locations on n and s sides of island; se edge of mesa w of Peak 606. Santa Catalina and San Clemente islands; central CA (San Luis Obispo Co.) to nw Mexico, e to CO and AZ. Three other varieties are found in Mexico and sw U.S., including var. *latifolium* C.L.Hitchc., which occurs on Guadalupe, San Benito, Cedros, and Natividad islands.
L. nitidum Torr. & A.Gray var. **nitidum** SHINING PEPPERGRASS Plants to 2 dm tall. Feb-Apr. Rare; flats and swales. North side on coastal flats near E Mesa Canyon and between N and S Spur canyons; mesa near Nicktown, n side of airfield runways, and se end of mesa. All CA Channel Islands; Los Coronados, Todos Santos, San Martin, and Guadalupe islands; WA to Baja CA. Two other varieties occur in w N America.
L. oblongum Small var. **insulare** C.L.Hitchc. LENTEJILLA Plants to 1 dm tall. Mar-Apr. Rare; sandy alluvial flats, clay flats, gullies, eroded slopes, and open ridgetops. North side on coastal flats near mouth of W Mesa Canyon; mesa near Shannon Road, s of Nicktown, and se of Peak 606; s side between Cattail Canyon area and Daytona Beach. All CA Channel Islands except Santa Barbara; all Baja CA islands except San Geronimo; central CA (Santa Barbara Co.) to Baja CA. Variety *oblongum,* with more elliptic silicles, occurs on the CA mainland.

LOBULARIA Desv.

Perennial herbs. Six species; Mediterranean area of Europe. (Latin: small lobe, referring to the small silicles)

*L. maritima (L.) Desv. SWEET ALYSSUM Plants to 8 dm tall. Jan-Jul. Occasional; disturbed sandy flats and in canyons. Mesa at wells area along Tufts Road, antenna field n of Jackson Hill, at southern edge of mesa near Jackson Hill, and near Nicktown; canyons of s escarpment near Jackson Peak. San Miguel, Santa Catalina, and San Clemente islands; WA to Baja CA, e to ME; Hawaii; native of Europe. First collected on mesa near wells area in June 1977. This species has been used medicinally; young leaves, stems, and flowers can be used as a spicy condiment. Eradication efforts were initiated in 1983 but this taxon is still spreading on the island.

NASTURTIUM Scop. CRESS

Perennial herbs. Ca. 50 species; temperate areas worldwide. (Latin: twisted or wry nose, referring to pungent qualities)

*N. officinale R.Br. [*Rorippa nasturtium-aquaticum* (L.) Hayek] WATER-CRESS Plants to 8 dm tall. May-Jul. Rare; moist flats and stream margins. North side in Tule Creek. Santa Cruz and Santa Catalina islands; WA to Baja, e to ME; N America; Europe. First collected in March 1984. Widely cultivated for

BRASSICACEAE: Nasturtium-Sisymbrium

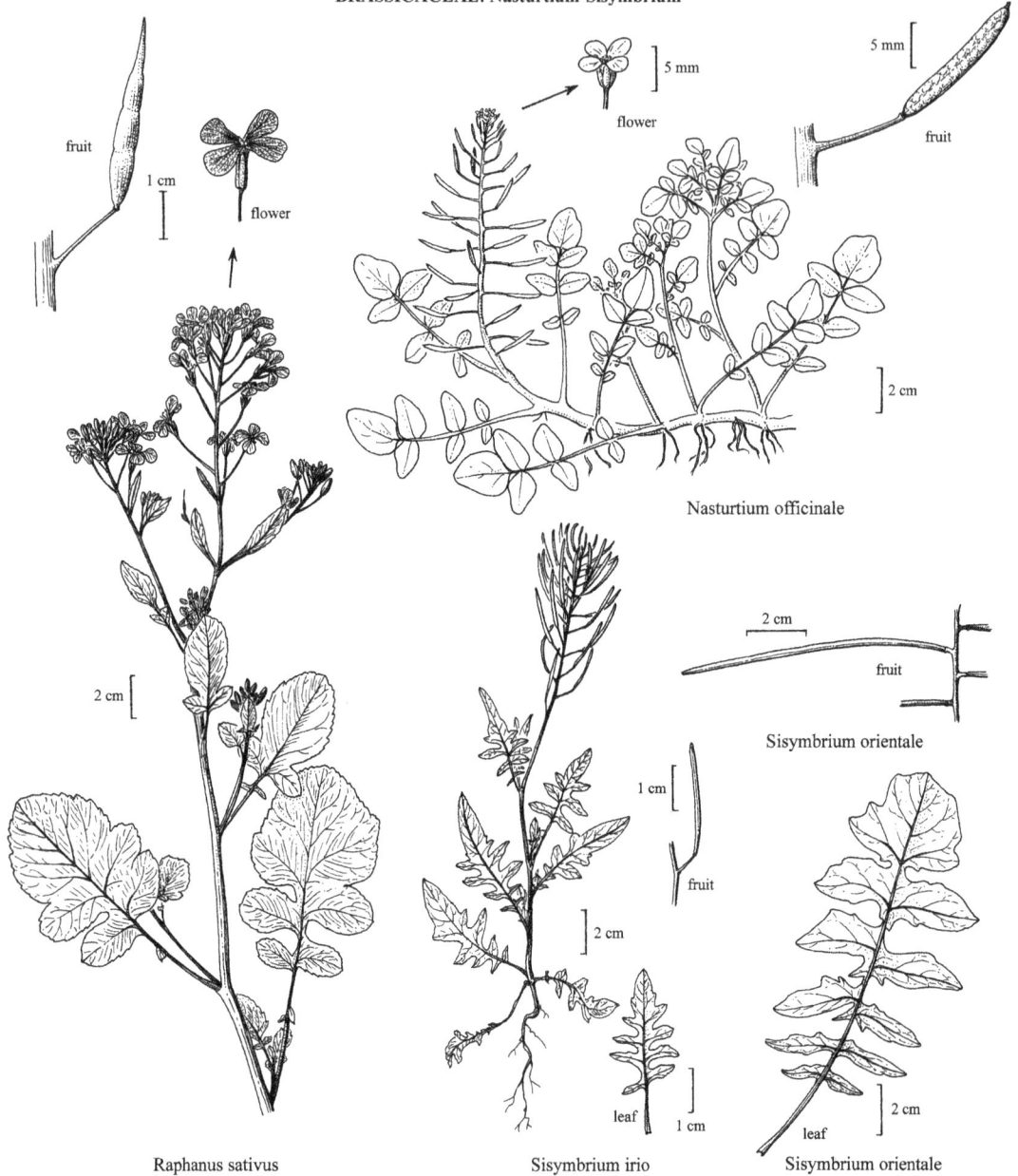

fruit

1 cm

flower

5 mm

flower

5 mm

fruit

2 cm

Nasturtium officinale

2 cm

fruit

Sisymbrium orientale

1 cm

fruit

2 cm

2 cm

leaf

1 cm

leaf

2 cm

Raphanus sativus

Sisymbrium irio

Sisymbrium orientale

edible stems and leaves; seeds can be sprouted and eaten or ground to prepare a spicy mustard.

RAPHANUS L. RADISH

Annual herbs. Three species; Mediterranean area of Europe. (Greek: appearing rapidly, referring to seed germination rate)

*R. sativus L. WILD RADISH Plants to 7 dm tall. Apr-May. Rare; disturbed areas. North side near Red Eye Beach and near Rock Jetty; s side at Daytona Beach. Not seen recently on island. All CA Channel Islands except Anacapa and Santa Barbara; Guadalupe and Cedros islands; WA to Baja CA, e to ME;

Hawaii; probably native of Mediterranean Europe and Africa. First collected near Red Eye Beach in April 1978. Entire plant is edible but roots can be fibrous and tough; young siliques are very tasty. Seeds and roots have been used medicinally.

SISYMBRIUM L. HEDGE-MUSTARD

Annual herbs. Ca. 90 species; temperate areas worldwide. Leaves and seeds of some species are edible. (Latin form of Greek name for some plant in the mustard family)

1. Width of fruiting pedicels less than width of siliques; siliques 3-4.5 cm long; petals 3-4 mm long . . . .**S. irio**
1' Fruiting pedicels ca. as wide as siliques; siliques 4-9 cm long; petals 6-9 mm long . . . .**S. orientale**

*S. irio L. LONDON ROCKET Plants to 5 dm tall. Mar-Apr. Rare; sandy coastal flats and disturbed sites. Known only from sw portion of island. Anacapa, Santa Catalina, and San Clemente islands; Todos Santos, Guadalupe, and Cedros islands; northern CA (Butte Co.) to Baja CA, e to TX and disjunctly to MA and FL; Hawaii; native of Europe. First collected in May 1992. Young leaves, flowers, and seeds can be eaten raw or cooked.

*S. orientale L. INDIAN HEDGE-MUSTARD Plants to 6 dm tall. Feb-May (Jul). Occasional; disturbed sites. North side near Vizcaino Point, Corral Harbor, Nicktown, and Rock Jetty; mesa at wells area on Tufts Road, near Nicktown, and s of airfield runways; s side at Army Springs and near Daytona Beach. Anacapa, Santa Catalina, and San Clemente islands; Guadalupe and Cedros islands; OR to Baja CA, e to TX and disjunctly to MA; Australia; native of Europe. First collected near Nicktown in June 1969.

CACTACEAE Juss.
CACTUS FAMILY

Shrubs. Ca. 90 genera and 2000 species; N and S America. Many species are cultivated for ornamental use.

OPUNTIA J.Mill.

Ca. 200 species; N and S America. Stem segments were cooked and eaten by Native Americans; fruits were eaten raw; seeds also were eaten by the Luiseño Tribe. (Greek name for a different plant found near the ancient town of Opus)

1. Stem segments cylindrical, tuberculate; perianth parts purple to reddish purple . . . .**O. prolifera**
1' Stem segments flat, not tuberculate; perianth parts yellow to orange
    2. Stems erect, segments mostly 25-55 cm long; areoles usually 3-4.5 cm apart . . . .**O. ficus-indica**
    2' Stems erect to sprawling, segments mostly 15-25 cm long; areoles usually 1.5-2.5 cm apart
        3. Stem segments elliptic to obovate, their length 1.7-2.3 times the width; unweathered spines bone-white, sometimes darkened at base; innermost perianth parts 1.2-1.7 times longer than width; style pink . . . .**O. littoralis**
        3' Stem segments orbicular to broadly elliptic, their length 1-1.5 times the width; mature but unweathered spines yellow and mostly translucent; innermost perianth parts 1.8-2.6 times longer than width; style red . . . .**O. oricola**

*O. ficus-indica (L.) J.Mill. MISSION CACTUS Plants to 2.5 m tall. May-Jun. Rare; disturbed flats. Mesa near Nicktown. Santa Cruz, Anacapa, Santa Catalina, and San Clemente islands; San Martin and Todos Santos islands; cultivated and naturalized ± worldwide; probably native of Mexico or Central America. First collected in October 1988.

O. littoralis (Engelm.) Cockerell COASTAL PRICKLY PEAR Plants to 1 m tall. Apr-Jul. Abundant;

CACTACEAE

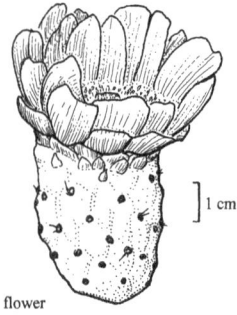

flower

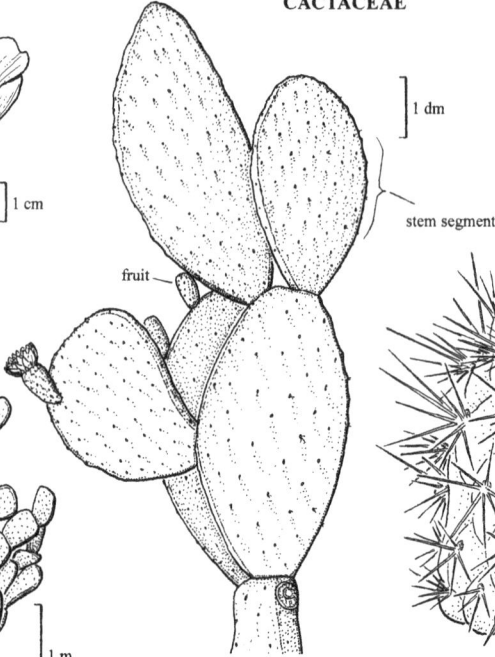

1 dm

stem segment

fruit

flower

O. prolifera

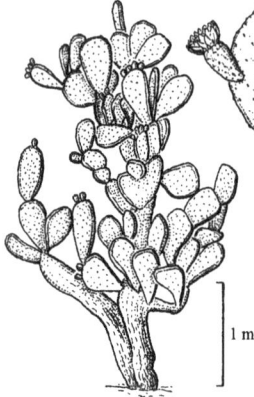

1 m

Opuntia ficus-indica

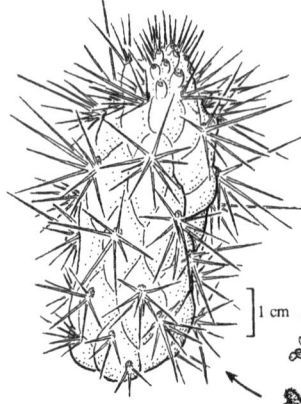

stem segment

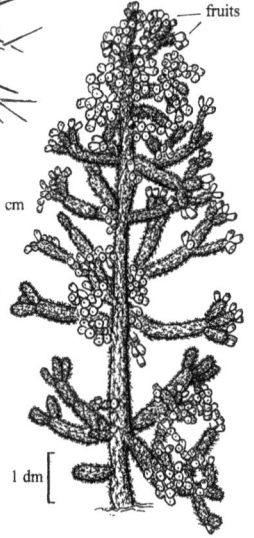

fruits

1 dm

Opuntia prolifera

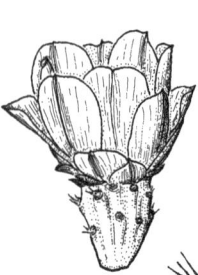

flower

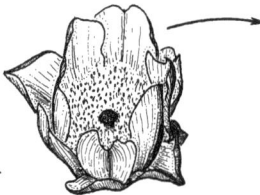

2 cm

flower

petal

O. littoralis

petal

O. oricola

1 cm

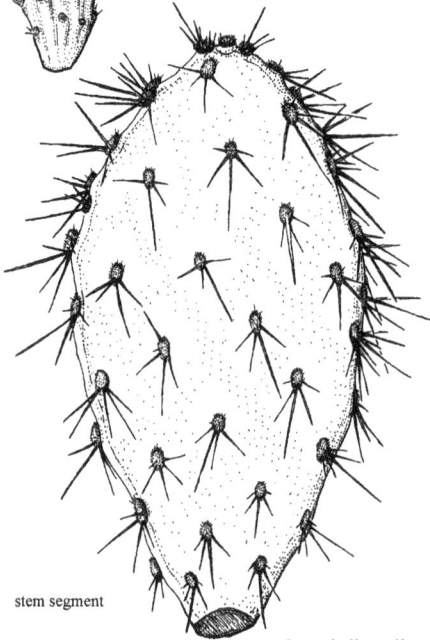

2 cm

stem segment

Opuntia littoralis

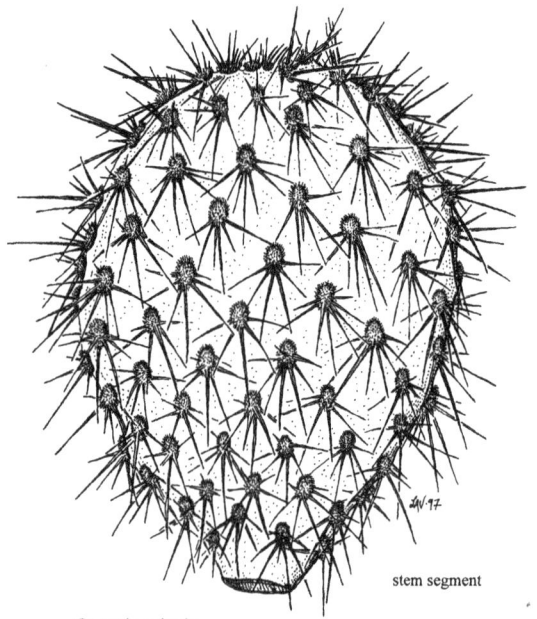

stem segment

Opuntia oricola

slopes, gullies, swales, and flats. North side between Live-forever Canyon and Sand Spit; mesa near W Mesa Canyon; s side between Grand Canyon and Daytona Beach, especially near head of e fork of Twin Rivers. All CA Channel Islands; Los Coronados and Todos Santos islands; central CA (Santa Barbara Co.) to Baja CA. Three other varieties occur on the CA mainland from Ventura Co. to Baja CA, e to UT and AZ. Hybridizes freely with *O. oricola*; hybrids have white styles (as opposed to pink or red) and stem segments intermediate in size and shape between parents (Philbrick 1963). Native *Opuntia* species, along with boxthorn *(Lycium californicum),* provide habitat for the endemic island night lizard *(Xantusia riversiana).* Detailed distribution maps for *O. littoralis, O. oricola,* and *O. prolifera* on the island can be found in Junak (2003a).

O. oricola Philbrick TALL PRICKLY PEAR Erect shrub to 1 m tall. Apr-Jul. Abundant; slopes, gullies, swales, and flats. North side between Live-forever Canyon and Sand Spit; mesa near Nicktown and airfield; s side between Cattail Canyon and Sand Spit. All CA Channel Islands; Los Coronados, Todos Santos, San Martin, and Cedros islands; central CA (Santa Barbara Co.) to Baja CA.

O. prolifera Engelm. [*Cylindropuntia p.* (Engelm.) Kunth] COASTAL CHOLLA Erect shrub to 2 m tall. Apr-Jun. Abundant; slopes, gullies, swales, and flats. North side between mouth of E Mesa Canyon and Sand Spit; mesa near airfield; s side between Cattail Canyon area and Sand Spit. All CA Channel Islands except San Miguel; Los Coronados, Todos Santos, San Martin, San Geronimo, and Cedros islands; central CA (Ventura Co.) to Baja CA.

CARYOPHYLLACEAE Juss.
PINK FAMILY

Annual or perennial herbs. Ca. 85 genera and 3000 species; arctic and temperate areas ± worldwide, mostly in N Hemisphere. Several genera are cultivated for ornamental use (e.g., *Dianthus, Gypsophila,* and *Silene*).

1. Fruit a 1-seeded indehiscent utricle . . . . **Herniaria**
1' Fruit a capsule with several to many seeds
    2. Sepals fused into a cylindrical or cup-like calyx; petals clawed . . . .**Silene**
    2' Sepals free; petals, if present, usually not clearly clawed
        3. Stipules absent . . . .**Sagina**
        3' Stipules present
            4. Flowers axillary, 1-2; sepals awn-tipped; petals absent; stipules bristle-like .**Loeflingia**
            4' Flowers in cyme, few-many; sepals awnless; petals present; stipules ovate to lanceolate
            . . . .**Spergularia**

HERNIARIA L. HERNIARY or RUPTUREWORT

Annual herbs. Ca. 45 species; S America, Europe, central and western Asia, and Africa. (Latin: rupture, alluding to use of one species in treatment of hernias)

*H. hirsuta L. var. **cinerea** (DC.) Loret & Barrandon [*H. c.* DC.] HAIRY RUPTUREWORT Plants to 7 cm across. Apr. Rare; sandy flats at top of coastal bluffs. Known from a single collection on coastal flats near mouth of W Mesa Canyon in April 1989; not seen recently. San Miguel, Santa Rosa, Santa Barbara, Santa Catalina, and San Clemente islands; Guadalupe Island; WA (not seen there recently) to Baja CA, e to AZ and disjunctly to MD; native of s Europe, n Africa, and sw Asia. Another variety occurs on the CA mainland. This taxon has been used medicinally.

LOEFLINGIA L.

Annual herbs. Seven species; N America, Mediterranean Europe, sw Asia, and Africa. (honoring Peter Loefling, 1729-1756, Swedish botanist and explorer)

CARYOPHYLLACEAE

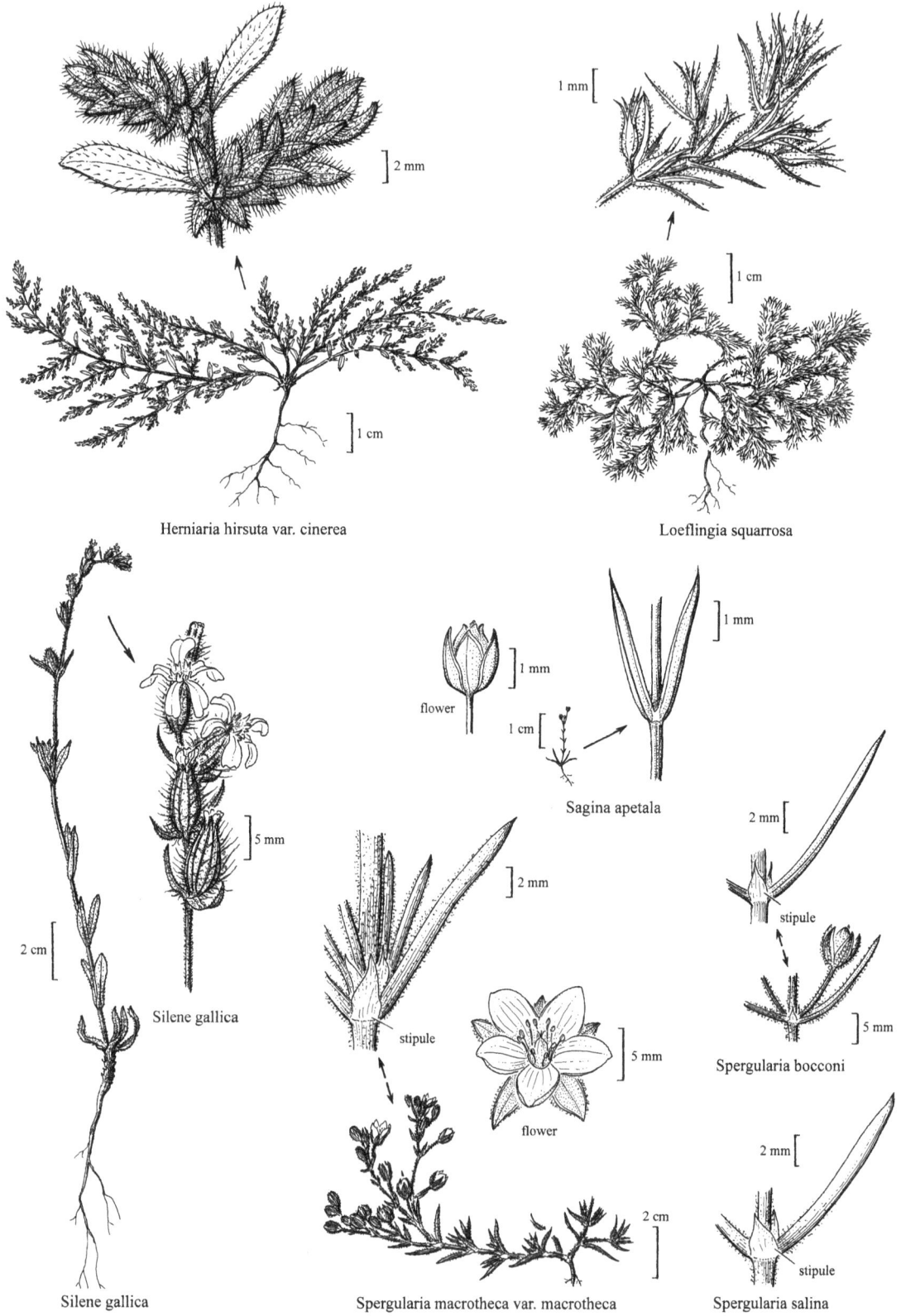

Herniaria hirsuta var. cinerea

Loeflingia squarrosa

Silene gallica

Silene gallica

flower

Sagina apetala

stipule

flower

Spergularia bocconi

Spergularia macrotheca var. macrotheca

stipule

Spergularia salina

L. squarrosa Nutt. SPREADING PYGMY-LEAF Plants to 1 dm across. Apr. Rare; disturbed sandy flats. Known only from two collections near n rim of mesa w of airfield runways in April 1989; not seen recently. WA to Baja CA, e to AR. Possibly introduced on San Nicolas Island. First record for the California Channel Islands.

SAGINA L. PEARLWORT

Annual herbs. Ca. 20 species; temperate areas in N Hemisphere and on some tropical mountains. [Latin: nourishing or fattening, referring to the qualities of *Spergula sativa (Sagina spergula),* on which sheep thrive]

*S. apetala Ard. DWARF PEARLWORT Plants to 3 cm tall. Mar-May. Rare; in pockets of sandy soil on caliche flats, and on open flats. Mesa near head of Mineral Canyon and n of airfield runways. San Miguel, Santa Rosa, Santa Cruz, and Santa Catalina islands; British Columbia to Baja CA, e (disjunctly) to KS; Australia; Europe. First collected near head of Mineral Canyon in March 1993.

SILENE L. CATCHFLY or CAMPION

Annual herbs. Ca. 700 species; mainly N Hemisphere. Some species are grown horticulturally. (Greek: probably from mythological Silenus, intoxicated companion of Bacchus who was described as often covered with foam, referring to the sticky stems of many species)

*S. gallica L. WINDMILL PINK Plants to 3 dm tall. Mar-May. Occasional; slopes and flats. North side near W and E Mesa canyons and in vicinity of "L" Canyon; mesa on flats near head of Mineral Canyon and n of airfield runways; s side on coastal flats at se end of island, just inland from e end of Daytona Beach. All CA Channel Islands; Los Coronados, Todos Santos, Guadalupe, and Cedros islands; British Columbia to Baja CA, e to ME; Hawaii; native of Europe. It was first collected on the island in April 1897, when it was seen "in one locality, a cliff overhanging a brackish stream". It was "seldom seen" in April 1901 (herbarium specimen at New York Botanical Garden) and was not noted again on San Nicolas Island until April 1992. It may have been introduced to the island on two separate occasions. This species has been used medicinally.

SPERGULARIA (Pers.) J.Presl & C.Presl SAND-SPURREY

Annual or perennial herbs. Ca. 60 species; nearly worldwide. (Latin: derivative of *Spergula*)

1. Perennial herbs, with a thick, fleshy root; cauline leaves usually with axillary leaf clusters; stipules 4-8 mm long . . . .**S. macrotheca** var. **macrotheca**
1' Annuals, with slender roots; cauline leaves usually without axillary leaf clusters; stipules 1.5-4 mm long
    2. Stems with glabrous lower internodes; stamens 6-10; stipules short-acuminate . . . . . .**S. bocconi**
    2' Stems with glandular-puberulent lower internodes; stamens 2-5; stipules acute . . . .**S. salina**

*S. bocconi (Scheele) Graebner Plants to 2 dm across. Apr-May. Rare; disturbed flats. Mesa near airfield and at se end of mesa. Santa Cruz, Anacapa, Santa Barbara, Santa Catalina, and San Clemente islands; OR to Baja CA; native of Mediterranean Europe. First collected near se end of airfield runways in May 1986.
S. macrotheca (Cham. & Schlecht.) Heyn. var. **macrotheca** LARGE-FLOWERED SAND-SPURREY Plants to 5 dm across. Mar-Jun (Nov). Common; saline flats, sandy flats, grassy flats, moist banks, and coastal bluffs. Scattered locations throughout much of the island. All CA Channel Islands; Los Coronados and Guadalupe islands; British Columbia to Baja CA. Two other varieties occur on the CA mainland.
*S. salina J.Presl & C.Presl [*S. marina* (L.) Griseb.] SALTMARSH SAND-SPURREY Plants to 7 dm across. Mar-Jul (Nov). Scarce; disturbed flats, moist sites, canyon bottoms. North side near NavFac buildings, mouth of W Mesa Canyon, and Rock Jetty; mesa near Nicktown and airfield. All CA Channel Islands except San Miguel; Guadalupe Island; N and S America, Europe, and Asia. Although this taxon

is native to CA, it was probably introduced to San Nicolas Island. First collected on mesa s of airfield terminal in November 1971.

CHENOPODIACEAE Vent.
GOOSEFOOT FAMILY

Annual herbs, perennial herbs, or shrubs. Ca. 100 genera and 1500 species; temperate to subtropical areas worldwide, often in dry climates and often in alkaline or saline habitats. Some species are cultivated for food (e.g., beets, chard, quinoa, and spinach). Some authors would expand Amaranthaceae (amaranth family) to include Chenopodiaceae.

1. Leaves scale-like, opposite; stems and inflorescence usually fleshy ....**Salicornia**
1' Leaves not scale-like, usually alternate (except in *Atriplex watsonii*); stems and inflorescence not fleshy
  2. Leaves usually fleshy, either terete and ± linear or thin, rigid, and spine-tipped; ovary pear-shaped
    3. Leaves tipped with a spine or bristle; calyx keeled and winged in fruit, but not fleshy .....
    ....**Salsola**
    3' Leaves not spine-tipped; calyx fleshy, plane, and wingless in fruit ....**Suaeda**
  2' Leaves with wide, flat blades, neither terete nor rigid, usually not fleshy; ovary usually ovoid to globose, sometimes ± compressed, but not pear-shaped
    4. Flowers unisexual, pistillate ones enclosed by 2 ± accrescent bracts ....**Atriplex**
    4' Flowers bisexual, not enclosed by bracts
      5. Calyx lobes usually 3; stamen 1 ....**Aphanisma**
      5' Calyx lobes usually 5; stamens 4-5
        6. Calyx lobes armed with stout hooked spines; plants pilose ....**Bassia**
        6' Calyx lobes unarmed; plants glabrous, mealy, or glandular-pubescent ....
    ....**Chenopodium**

APHANISMA Nutt.

Annual herbs. One species. (Greek: inconspicuous)

A. blitoides Nutt. Plants with weak stems to 1 m long. Mar-Jun. Rare; dry slopes; usually in cactus patches. North side in vicinity of "L" Canyon and in N Spur Canyon; s side on s escarpment below Jackson Hill and below Twin Towers (Building 186). All CA Channel Islands except San Miguel; all Baja CA islands except San Geronimo; central CA (Santa Barbara Co.) to Baja CA. Collected at "one locality" in April 1897 and "among cacti" in April 1901; not seen again until May 1992.

ATRIPLEX L. SALTBUSH or SALTSCALE

Annual herbs, perennial herbs, or shrubs. Ca. 250 species; subarctic, temperate, and subtropical areas worldwide. Many species have edible leaves or seeds, most notably orache (*A. hortensis* L.), which is widely cultivated for its leaves. (Greek name for *Atriplex halimus* L.)

1. Shrubs, with woody stems and branches, usually 1-3 m tall; plants dioecious
  2. Fruiting bracts with conspicuous extra wings or crests arising at middle of face ...**A. canescens**
  2' Fruiting bracts without extra lateral wings ....**A. lentiformis**
1' Annual or perennial herbs, sometimes woody at base, usually less than 1 m tall; plants monoecious
  (except *A. watsonii*)
    3. Leaves sparsely scurfy when young, becoming green and glabrous with age ......**A. prostrata**
    3' Leaves scurfy, often densely so, grayish to white
      4. Fruiting bracts broadest below middle
        5. Fruiting bracts thick, fleshy, red with age, with 3-5 teeth above middle ..**A. semibaccata**
        5' Fruiting bracts ± compressed, neither fleshy nor red, margins mostly entire

CHENOPODIACEAE: Aphanisma-Bassia

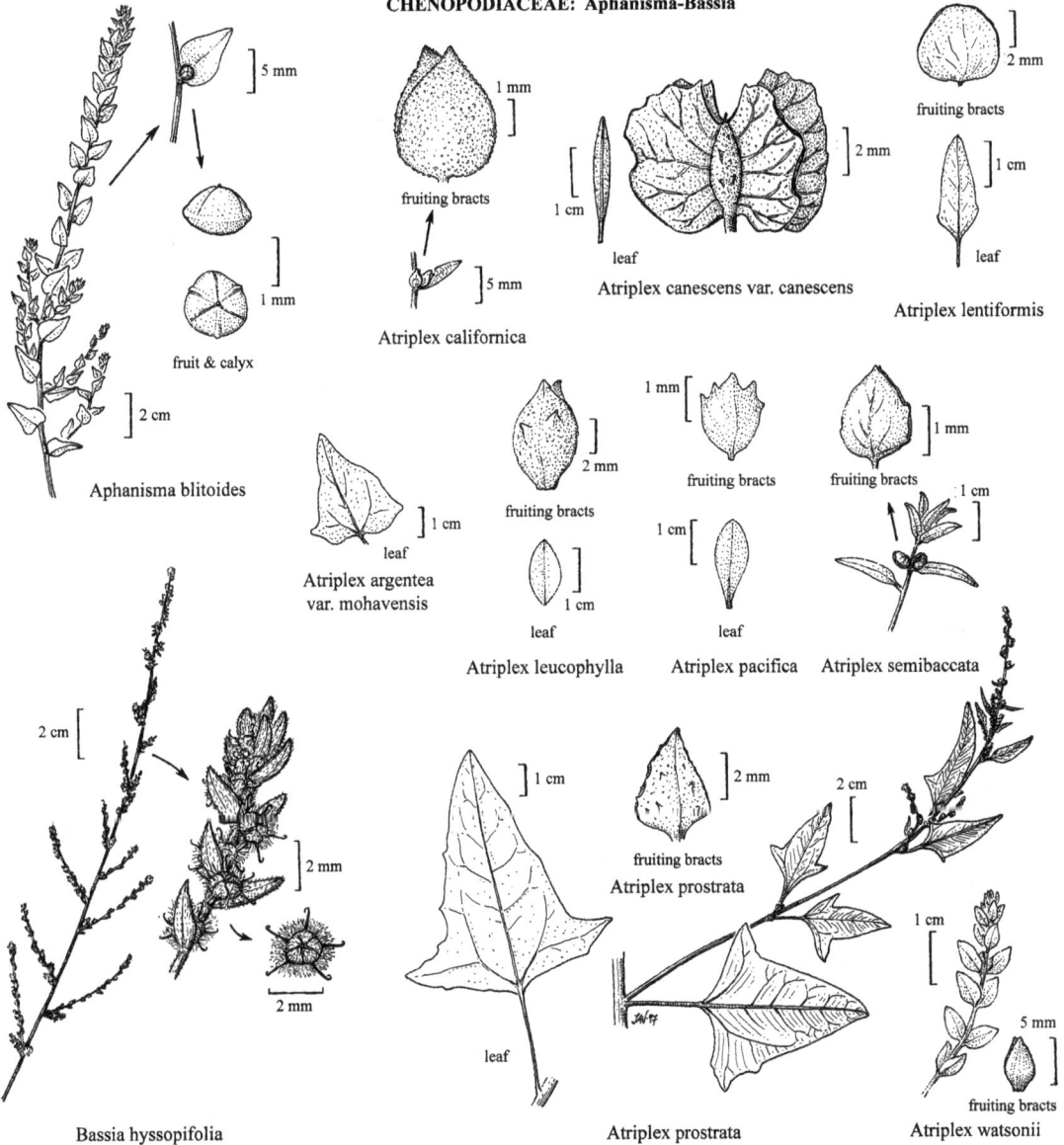

5 mm

fruiting bracts

1 mm

fruit & calyx

2 cm

Aphanisma blitoides

1 mm

fruiting bracts

5 mm

Atriplex californica

1 mm

2 mm

1 cm

leaf

Atriplex canescens var. canescens

2 mm

fruiting bracts

1 cm

leaf

Atriplex lentiformis

1 cm

leaf

Atriplex argentea
var. mohavensis

2 mm

fruiting bracts

1 cm

leaf

Atriplex leucophylla

1 mm

fruiting bracts

1 cm

leaf

Atriplex pacifica

1 mm

fruiting bracts

1 cm

Atriplex semibaccata

2 cm

2 mm

2 mm

Bassia hyssopifolia

1 cm

leaf

Atriplex prostrata

2 mm

fruiting bracts
Atriplex prostrata

2 cm

1 cm

5 mm

fruiting bracts
Atriplex watsonii

6. Leaves mostly alternate, lanceolate to oblanceolate; fruiting bracts mostly free, ovate, 2-4 mm long . . . .**A. californica**

6' Leaves mostly opposite, broadly elliptic to ovate; fruiting bracts fused to above middle, 3-7 mm long . . . .**A. watsonii**

4' Fruiting bracts broadest at or above middle

   7. Plants erect; leaves narrowly ovate to hastate; staminate and pistillate flowers usually mixed in same clusters . . . .**A. argentea**

  7' Plants prostrate; leaves lanceolate to ovate, elliptic, or oblong; staminate and pistillate flowers usually in different clusters

     8. Fruiting bracts globose to 4-angled, 4-8 mm long; seeds 2.5-3 mm long . . . .
. . . .**A. leucophylla**

    8' Fruiting bracts ± compressed, 1-3 mm long; seeds 1-2 mm long

9. Perennial herbs; leaves elliptic, lanceolate, or ovate; fruiting bracts 2-3 mm long, free margins deeply dentate . . . .**A. coulteri**

9' Annuals; leaves lanceolate to elliptic or spatulate; fruiting bracts 1-1.5 mm long, margins with 3-5 teeth at apices . . . .**A. pacifica**

*A. argentea Nutt. var. **mohavensis** (M.E.Jones) S.L.Welsh SILVERSCALE Plants to 5 dm tall. Jul-Aug. Rare; disturbed areas and flats. Mesa just e of ne end of airfield runways. Santa Rosa, Santa Cruz, Santa Catalina, and San Clemente islands; n CA (Butte Co.) to Baja CA, e to TX. First collected on island in September 2000; probably introduced to San Nicolas Island although native to CA.

A. californica Moq. CALIFORNIA SALTBUSH Plants to 12 dm tall. Mar-Aug. Common; flats, slopes, canyon bottoms, and coastal bluffs. Widespread locations on slopes and canyons of n escarpment and adjacent coastal flats, especially in w portion of island (as at Red Eye and Tender beaches); mesa near wells area on Tufts Road; on lower slopes of s escarpment and near canyon mouths (as at Cattail and Grand canyons, slopes at east end of Dutch Harbor). All CA Channel Islands; Los Coronados, Todos Santos, Guadalupe, and Cedros islands; n CA (Sonoma Co.) to Baja CA. Leaves and seeds can be cooked and eaten.

A. canescens (Pursh) Nutt. var. **canescens** FOUR-WING SALTBUSH Rare; flats. Plants to 1 m tall. Rare; flats. May-Jul. Mesa s of Nicktown and at edge of old borrow pit along Monroe Drive. WA to Baja CA; e to TX. Three other varieties occur in CA and UT. Native to CA, but probably introduced to the island. First collected at Borrow Pit in January 1991.

A. coulteri (Moq.) D.Dietr. COULTER'S SALTSCALE Plants to 4 dm across. Rare; flats. Known from a single collection made on "moist upland" in April 1901; not seen recently. All CA Channel Islands except Santa Barbara; San Benito, Cedros, and Natividad islands; central CA (Santa Barbara Co.) to Baja CA.

*A. lentiformis (Torr.) S.Watson [*A. lentiformis* subsp. *breweri* (S.Watson) H.M.Hall & Clem.] QUAIL-BUSH or BREWER'S SALTBUSH Plants to 15 dm tall. May-Jun. Rare; disturbed sites. Mesa on gravel pile in Nicktown and along Owens Road near fire station. Santa Cruz, Anacapa, Santa Catalina, and San Clemente islands; n CA (Butte Co.) to Baja CA, e to AZ; Hawaii. Reported for the island by Eastwood (1941) and native to CA, but presumably introduced as it presently occurs only in disturbed sites. First known collection was made at Nicktown in May 1983. Flowers, stems, and leaves were used medicinally by Native Americans.

A. leucophylla (Moq.) D.Dietr. SEA SCALE Plants to 3 dm tall. Apr-Jul. Scarce; sandy beaches. Widely scattered locations around perimeter, especially in ne and se portions of island. All CA Channel Islands except Santa Barbara; San Martin, Cedros, and Natividad islands; n CA (Humboldt Co.) to Baja CA.

A. pacifica A.Nelson [*A. microcarpa* D.Dietr.] SOUTH COAST SALTSCALE Plants to 6 dm across. Mar-May. Rare; sandy flats and eroded slopes. North side in lower portion of "L" Canyon and on coastal flats near N and S Spur canyons. All CA Channel Islands except San Miguel; Los Coronados, San Martin, San Benito, Cedros, and Natividad islands; s CA (Ventura Co.) to n Baja CA.

*A. prostrata DC. [*A. triangularis* Willd.] SPEAR-LEAF SALTBUSH Plants to 1 m tall and 2 m across. (Apr) Sep-Oct. Scarce; gully bottoms and sandy flats. Widely scattered locations in n side canyons between Nicktown and Rock Jetty; mesa at Nicktown and at old borrow pit along Monroe Drive; s side at Dutch Harbor and in salt marsh area near Sand Spit. San Miguel, Santa Cruz, and Santa Catalina islands; WA to Baja CA, e to ME; native to Eurasia. First collected near Nicktown in April 1961. Leaves are edible and can be used as a substitute for spinach.

*A. semibaccata R.Br. AUSTRALIAN SALTBUSH Plants to 1 m across. Apr-Jul. Common; disturbed sites, flats, slopes, ravines, and coastal bluffs. Widely scattered locations throughout much of island. All CA Channel Islands; Los Coronados, Todos Santos, San Martin, and Cedros islands; throughout much of CA to n Mexico, e to UT and TX; native of Australia. First collected on coastal bluffs at Corral Harbor in July 1939. Fruits are edible raw.

A. watsonii A.Nelson MATSCALE Plants to 1 m across. Apr-Oct. Occasional; coastal flats and bluffs. Widely scattered locations around perimeter of island. All CA Channel Islands except Santa Barbara; San Martin Island; central CA (Santa Barbara Co.) to Baja CA.

BASSIA All.

Annual herbs. Ca. 10 species; Europe, Asia, and Africa. (honoring Ferdinando Bassi, 1710-1774, Italian naturalist)

*B. hyssopifolia (Pallas) Kuntze FIVE-HOOK BASSIA Plants to 1 m tall. May-Jul. Occasional; disturbed flats and roadsides. Coastal flats in ne portion of island; mesa at Nicktown and near airfield. Santa Catalina and San Clemente islands; WA to Baja CA, e to IA and TX and disjunctly to ME; Hawaii; native of Eurasia. First collected just nw of Coast Guard Beach in June 1977. This species can be toxic to sheep.

CHENOPODIUM L. GOOSEFOOT or PIGWEED

Ca. 150 species; temperate areas worldwide. Some species are cultivated for food (edible leaves and seeds), including the protein-rich quinoa (*C. quinoa* Willd.) of the South American Andes. Other species are cultivated for ornamental or medicinal uses. Some species are segregated into the genus *Dysphania*. (Greek: goose foot, referring to the shape of the leaves of some species)

1. Leaves glandular-puberulent, strongly aromatic ....C. ambrosioides
1' Leaves glabrous to minutely powdery, generally not aromatic
   2. Perennials; calyx tube longer than lobes; most leaves broadly deltate ....C. californicum
   2' Annuals; calyx tube much shorter than lobes or absent; most leaves rhombic-ovate, upper ones often lanceolate
      3. Upper surfaces of leaves usually shiny, bright green ....C. murale
      3' Upper surfaces of leaves usually farinose; dull green ....C. berlandieri

*C. ambrosioides L. [*Dysphania a.* (L.) Mosyakin & Clemants] EPAZOTE or MEXICAN TEA Plants to 8 dm tall. Jun-Sep. Rare; disturbed sites. Mesa at Nicktown, at tank farm along Owens Road just south of Fire Station, and near se end of airfield runways (Building 121). Santa Rosa, Santa Cruz, Santa Catalina, and San Clemente islands; WA to Baja CA; e to ME; native of N and S America. Introduced to the island; first collected at tank farm along Owens Road in September 1988. This species has been used medicinally and is a source of essential oils; leaves are also used as a condiment in soups and bean dishes.
*C. berlandieri Moq. Plants to 5 dm tall. Mar-May. Rare; disturbed areas. North side on coastal flats n of airfield and on gravel piles near Rock Jetty; s side near barge landing area at Daytona Beach. Santa Rosa, Santa Cruz, Anacapa, and Santa Catalina islands; throughout much of N America (as 6 varieties). Probably introduced to the island; first collected on ne coastal flats in May 1986. Identification of the taxon on San Nicolas Island is tentative and needs further study. Leaves and seeds are edible.
C. californicum (S.Watson) S.Watson SOAP PLANT Plants with stems to 1 m long. Mar-Apr. Rare; cactus-filled swale. Known only from large swale on ne escarpment just e of Nicktown and near a small canyon n of airfield. All CA Channel Islands; Los Coronados and Todos Santos islands; n CA (Tehama Co.) to Baja CA. Collected on "moist flat" in April 1901. Roots can be dried and grated to make soap; Native Americans ate the seeds, used leaves and roots medicinally, and made gum from the milky sap.
*C. murale L. NETTLE-LEAF GOOSEFOOT Plants to 8 dm tall. Apr-Jul. Rare; disturbed flats and slopes. Widely scattered locations on n side between Thousand Springs area and Sand Spit; on mesa near air terminal; in *Opuntia littoralis* patch on s escarpment, s of Twin Towers (Building 186). All CA Channel Islands; all Baja CA islands; WA to Baja CA, e to ME; Hawaii; Puerto Rico; ± worldwide in subtropical and warm-temperate areas; native of Europe, Asia, and northern Africa. First collected on the island in April 1897 and noted as "frequent" in April 1901. Leaves and seeds are edible.

CHENOPODIACEAE: Chenopodium-Suaeda

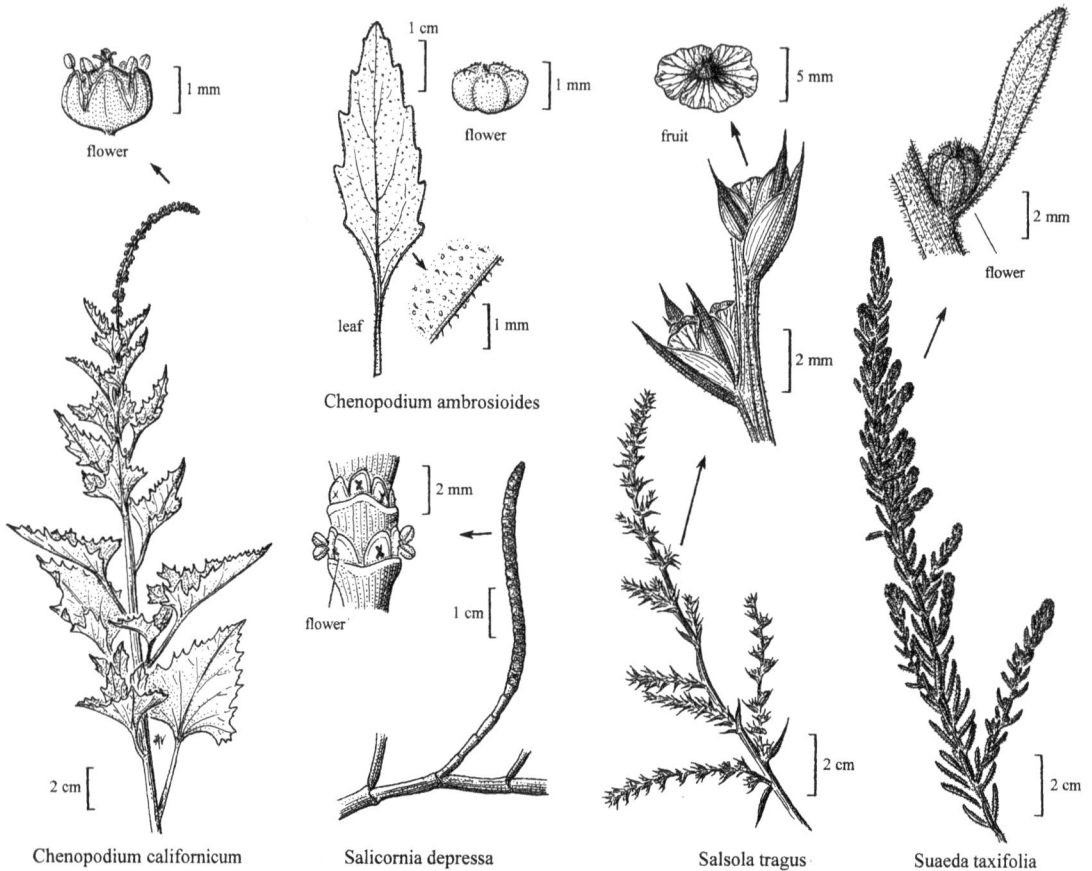

flower

1 mm

flower

1 mm

fruit

5 mm

2 mm

flower

1 cm

flower

1 mm

leaf

1 mm

Chenopodium ambrosioides

flower

2 mm

1 cm

2 mm

2 cm

2 cm

2 cm

Chenopodium californicum

Salicornia depressa

Salsola tragus

Suaeda taxifolia

SALICORNIA L. GLASSWORT or PICKLEWEED

Perennial herbs. Ca. 10 species; N Hemisphere and s Africa. (Latin: salt horn, referring to the horn-like branches and the saline habitats where the plants generally occur)

S. depressa Standl. [The name *S. virginica* L. has been misapplied to this species] PICKLEWEED Plants with stems 2-6 dm long. Jun-Jul. Rare; saline flats. Known only from salt flats on Sand Spit. All CA Channel Islands except Santa Barbara; San Martin Island; British Columbia to Baja CA; TX to ME; Hawaii. Seeds were ground into a meal and eaten by Native Americans; the fleshy stems can be eaten raw, cooked, or pickled.

SALSOLA L. SALTWORT or RUSSIAN-THISTLE

Annual herbs. Ca. 130 species; temperate and tropical areas worldwide, often in saline or alkaline habitats. (Latin: salty, referring to the habitat of many species)

1. Leaves fleshy, ± gradually narrowed into rather firm apical spine; bracts reflexed at maturity . . . . . .
S. kali subsp. **pontica**
1' Leaves usually not fleshy, usually abruptly narrowed into weak apical spine; bracts reflexed or . . . .
appressed at maturity . . . .**S. tragus**

*S. kali L. subsp. **pontica** (Pall.) Mosyakin [*S. k.* var. *p.* Pall.] Plants to 7 dm tall. Sep-Oct.

Occasional; disturbed flats and slopes. Northeastern and southeastern portions of island near Sand Spit. OR (historic collection), TX, Atlantic Coast, s to Mexico and S America; native of Europe, n Africa, and sw Asia. First collected on slopes near Sand Spit in July 1965. San Nicolas Island is the only location in CA cited for this taxon (Mosyakin 1996).

*S. tragus L. [*S. australis* R.Br., *S. iberica* (Sennen & Pau) Czerepanov, *S. kali* L. var. *tenuifolia* Moq.-Tandon, *S. pestifer* A.Nelson] RUSSIAN-THISTLE or TUMBLEWEED   Plants to 1 m tall. Sep-Oct. Occasional; roadsides, disturbed areas, and eroded slopes. North side near Celery Canyon and on coastal flats near Sand Spit; scattered locations on mesa, especially near Nicktown and airfield; s escarpment and coastal flats near Sand Spit. Santa Rosa, Santa Cruz, Anacapa, Santa Catalina, and San Clemente islands; Los Coronados, Todos Santos, and Cedros islands; British Columbia to Baja CA, e to ME; Mexico; Central and S America; Australia; S Africa; native of Eurasia. First collected near Sand Spit in April 1966. Tender branch tips can be cooked and eaten.

SUAEDA Forskk. ex J.F.Gmelin SEA-BLITE or SEEPWEED

Shrubs. Ca. 110 species; worldwide, usually in saline or alkaline habitats. (ancient Arabic name for the genus)

S. taxifolia (Standl.) Standl. [*S. californica* S.Watson var. *t.* (Standl.) Munz, *S. c.* var. *pubescens* Jeps.] WOOLLY SEA-BLITE Plants to 12 dm tall and 2 m across. Mar-Jul. Occasional at low elevations around the perimeter of the island; locally common at Red Eye Beach and in Dutch Harbor area; slopes of ne and se escarpments. All CA Channel Islands; Los Coronados, San Martin, Guadalupe, and Cedros islands; central CA (San Luis Obispo Co.) to n Baja CA. Typically found below 20 m elevation, this taxon has been seen up to elevations of 100 m on the island's escarpments. Seeds and boiled leaves were eaten by Native Americans; stems and leaves were steeped in water to produce a durable black dye.

CONVOLVULACEAE Juss.
MORNING-GLORY FAMILY

Ca. 50 genera and 1000 species; ± worldwide, especially warm temperate and tropical areas. Several genera (e.g., *Calystegia, Convolvulus, Ipomoea*) are cultivated as ornamentals. Many species of *Ipomoea*, including the sweet potato [*I. batatas* (L.) Lam.], have edible tubers and some are widely cultivated; others have edible leaves or are sources of hallucinogenic drugs.

1. Styles 2, distinct or fused near base; corolla lobes much longer than tube . . . .**Cressa**
1' Style 1, entire or with 2 minute lobes; corolla lobes usually much shorter than tube
     2. Calyx enclosed by 2 large bracts; stigmas oblong to elliptic, slightly or not at all flattened . . .
. . . .**Calystegia**
     2' Calyx not enclosed by 2 large bracts; stigmas linear, ± flattened . . . .**Convolvulus**

CALYSTEGIA R.Br. MORNING-GLORY

Perennial herbs or vines. Ca. 150 species; temperate areas worldwide. Steamed or boiled roots of several species are edible. (Greek: concealing calyx, referring to the large bracts of some species)

1. Leaves reniform; corolla white to pink, rose, or purple; plants of sandy soils near beach . . . .
C. soldanella
1' Leaves triangular-ovate to triangular-lanceolate; corolla white to cream; plants of dry places
     2. Plants strongly climbing; stems > 1 m long; leaves green . . . .**C. macrostegia**
     2' Plants not or only weakly climbing; stems usually 1-4 dm long; leaves grey . . . .
. . . .**C. malacophylla**

C. macrostegia (Greene) Brummitt subsp. **amplissima** Brummitt SOUTHERN ISLAND MORN-

ING-GLORY Stems to 2.5 m long. Mar-Sep. Abundant; sandy flats, dunes, hillsides, flats, and canyon walls. Widespread localities over much of the island. ENDEMIC to Santa Barbara, San Nicolas, and San Clemente islands. Subspecies *macrostegia* is endemic to the Northern Channel Islands, Santa Catalina, San Martin, and Guadalupe islands.

*C. malacophylla (Greene) Munz subsp. **pedicellata** (Jepson) Munz Stems to 4 dm long. May. Rare; gravel piles. Known only from 2 collections on n coastal flats near Rock Jetty. Central CA (Contra Costa Co.) to s CA (Los Angeles Co.). Although native to the CA mainland, this taxon was clearly introduced to island as it was found only on imported gravel piles. First collected on the island in March 1991; all plants seen were removed in May 1991 to prevent establishment.

C. soldanella (L.) R. Br. BEACH MORNING-GLORY Plants to 4 dm long. Mar-Jun. Rare; rocky beaches and sand dunes. North side at w and e ends of Red Eye Beach and in dunes just inland from Tender Beach. All CA Channel Islands except Anacapa and Santa Barbara; oceanic beaches ± worldwide.

CONVOLVULUS L. MORNING-GLORY

Perennial vines. Ca. 250 species; temperate areas worldwide. Some species are cultivated for ornamental uses; others are invasive weeds. (Latin: to entwine, referring to the habit of some species)

*C. arvensis L. BINDWEED Stems to 12 dm long. Mar-Sep. Rare; disturbed flats. Known only from mesa at Nicktown. Santa Rosa, Santa Cruz, and Santa Catalina islands; WA to Baja CA, e to ME; Hawaii; native of Europe. First collected on island in September 1988. Flowers, leaves, and roots have been used medicinally; the entire plant yields a green dye.

CRESSA L. ALKALI WEED

Perennial herbs. Ca. 5 species; temperate and tropical areas worldwide. (Greek: a Cretan woman)

C. truxillensis Kunth [*C. t.* var. *vallicola* (A.A. Heller) Munz] SPREADING ALKALI WEED Plants to 2 dm tall. May-Sep. Rare; sandy flats just inland from seashore. Known only from se end of island, e of Daytona Beach. All CA Channel Islands except Santa Barbara; OR to Baja CA, e to TX; Hawaii.

CRASSULACEAE DC.
STONECROP FAMILY

Annual or perennial herbs. Ca. 30 genera and 1500 species; temperate and subtropical areas ± worldwide, especially S Africa. Many genera (e.g., *Aeonium, Crassula, Dudleya, Echeveria, Kalanchoe, Sedum*) are cultivated for ornamental use.

1. Annuals, less than 7 cm tall; leaves cauline, opposite, less than 3 mm long . . . .**Crassula**
1' Perennial herbs, usually taller than 10 cm; leaves mostly basal, in rosettes, 5 or more cm long . . . **Dudleya**

CRASSULA L.

Annual herbs. Ca. 300 species; ± worldwide, especially Africa. Some species (e.g., *C. tetragona* L.) cultivated, sometimes persisting without cultivation. (Latin: diminutive of thick)

C. connata (Ruiz & Pav.) A.Berger [*C. erecta* (Hook. & Arn.) A.Berger, *Tillaea e.* Hook. & Arn] PYGMY-WEED Plants to 8 cm tall. Mar-Apr. Occasional; open flats. Widely scattered locations throughout much of island. All CA Channel Islands; all Baja CA islands except San Geronimo and Natividad; OR to S America, e to TX.

DUDLEYA Britton & Rose LIVE-FOREVER

Perennial herbs. Ca. 45 species; sw N America. Many species are cultivated for ornamental use. Leaves

CONVOLVULACEAE

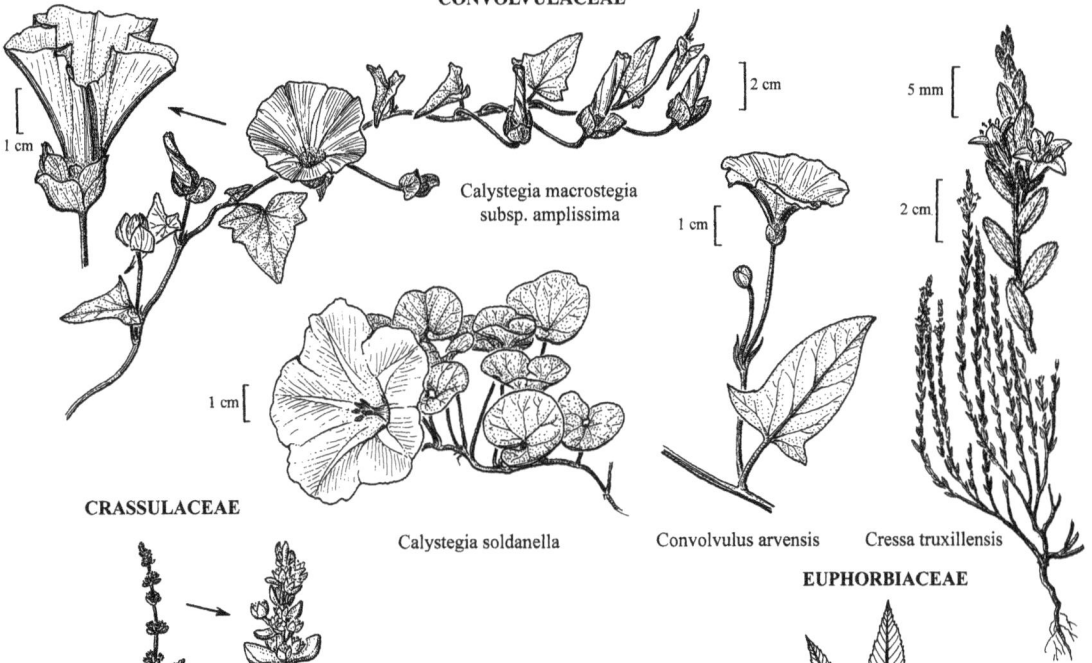

1 cm

2 cm

Calystegia macrostegia
subsp. amplissima

1 cm

5 mm

2 cm

1 cm

Calystegia soldanella

Convolvulus arvensis

Cressa truxillensis

CRASSULACEAE

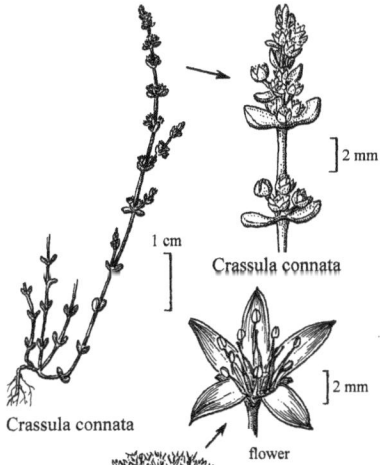

2 mm

1 cm

Crassula connata

Crassula connata

2 mm

flower

EUPHORBIACEAE

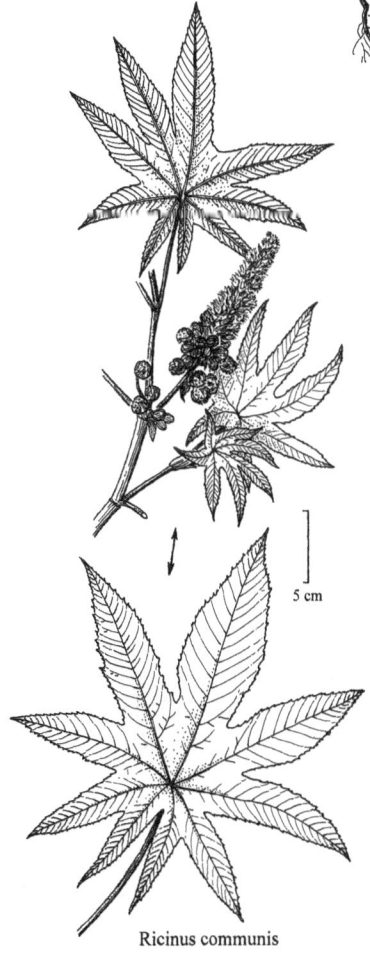

CUCURBITACEAE

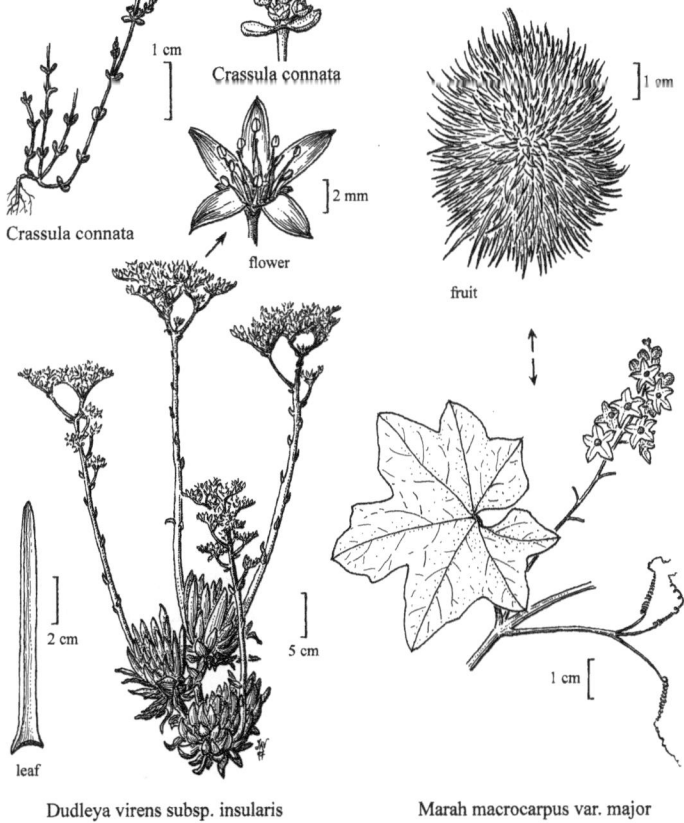

1 cm

fruit

2 cm

5 cm

5 cm

1 cm

leaf

Dudleya virens subsp. insularis

Marah macrocarpus var. major

Ricinus communis

of many species were eaten by Native Americans. (honoring William Russel Dudley, 1849-1911, first professor of systematic botany at Stanford University)

D. virens (Rose) Moran subsp. **insularis** (Rose) Moran PALOS VERDES LIVE-FOREVER Plants to 1 m tall. May-Jun. Common; flats, canyon bottoms and walls, slopes, flats, and stabilized dunes. North side in vicinity of Live-forever Canyon, coastal flats and n escarpment between E Mesa Canyon and Jetty Canyon area; mesa between Celery Canyon and airfield runways; s side near Twin Rivers drainage, coastal flats e of Dutch Harbor, Daytona Beach, and Sand Spit. Santa Catalina Island; s CA (Palos Verdes peninsula). A detailed distribution map for this taxon on San Nicolas Island can be found in Junak (2003c). Subspecies *virens* is endemic to San Clemente Island, subsp. *hassei* (Rose) Moran is restricted to Santa Catalina Island, and subsp. *extima* Moran is found only on Guadalupe Island.

CUCURBITACEAE Juss.
GOURD FAMILY

Perennial vines. Ca. 100 genera and 700 species; worldwide, mostly in tropical and warm temperate areas. Some species cultivated for fruit (e.g., chayote, gourds, melons, squashes).

MARAH Kellogg MAN-ROOT or WILD CUCUMBER

Ca. 7 species; w N America. (Latin: bitter, referring to the taste of all parts of the plant)

M. macrocarpus (Greene) Greene var. **major** (Dunn) Stocking WILD-CUCUMBER Stems to 2.5 m long. Feb-Mar. Scarce; canyon bottoms and moist slopes. Widely scattered locations on n escarpment between Nicktown and airfield runways. All CA Channel Islands; Los Coronados, Todos Santos, San Martin, and Cedros islands. This taxon may be an island endemic and needs further study. Variety *macrocarpus* occurs on CA mainland from Santa Barbara Co. to Baja CA. Seeds and roots were widely used for medicinal purposes by Native Americans; oil from seeds was mixed with pigments to make paints; roots were crushed and put in streams to stupify fish.

EUPHORBIACEAE Juss.
SPURGE FAMILY

Ca. 300 genera and 7500 species; ± worldwide, especially in tropical areas. Some species are cultivated for castor oil, tung oil, rubber, or as a source of starch (cassava, tapioca).

RICINUS L. CASTOR BEAN

Shrubs. One species. (Latin: tick, referring to appearance of the seeds)

***R. communis** L. Plants to 1.5 m tall. May-Jul. Rare; disturbed slopes and gravel piles. Near top of s escarpment overlooking Dutch Harbor (near Building 182). Santa Catalina and San Clemente islands; northern CA (Napa Co.) to Baja CA, e to AZ and disjunctly to NH; Hawaii; warm-temperate and tropical areas worldwide; native of Europe. First collected on the island in July 1979. Castor oil is obtained from the seeds, which are highly toxic if eaten raw. Ricin, a toxic protein contained in this species, is one of the most poisonous naturally-occurring substances known.

FABACEAE Lindl. [Leguminosae Juss.]
LEGUME OR PEA FAMILY

Annual herbs, perennial herbs, suffrutescent perennials, shrubs, or trees. Ca. 642 genera and 18,000 species; worldwide. Many species are important foods and sources of protein (e.g., alfalfa, beans, peanuts, peas, and soybeans). Many species are sources for oils, dyes, gums, insecticides, and other natural products. Many other species are cultivated for ornamental use.

1. Plants usually leafless or nearly so; leaves simple if present . . . . **Spartium**
1' Plants with leaves; leaves trifoliolate, pinnately compound, or palmately compound
    2. Most leaves pinnately compound with 4 or more leaflets
        3. Trees; stipules spinelike . . . . **Robinia**
        3' Annuals or herbaceous perennials, stems sometimes suffrutescent; stipules herbaceous or membranous
            4. Leaves apparently even-pinnately compound, terminated by a bristle or tendril, stems vine-like, twining, or climbing . . . . **Vicia**
            4' Leaves odd-pinnately compound, terminated by a leaflet, stems erect to prostrate, not conspicuously twining, vine-like, or climbing
                5. Inflorescence a raceme . . . . **Astragalus**
                5' Inflorescence an umbel . . . . **Lotus**
    2' Most leaves either palmately compound or with 3 leaflets (trifoliolate)
      6. Shrubs
        7. Leaves palmately compound, leaflets 5 or more; all stamens fused . . . . **Lupinus**
        7' Leaves trifoliolate (or pinnately compound); 9 stamens fused, 1 free . . . . **Lotus**
      6' Annuals or herbaceous perennials
        8. All 10 stamens fused; leaves palmately compound with 5-8 leaflets . . . . . . **Lupinus**
        8' Nine stamens fused, 1 free; leaves trifoliolate or pinnately compound with 5-6 leaflets
          9. Flowers in racemes
             10. Fruit clearly coiled 2-5 times, sometimes sickle-shaped; corolla 4-10 mm long . . . . **Medicago**
             10' Fruit ovoid, straight; corolla 2.5-5 mm long . . . . **Melilotus**
         9' Flowers in umbels or capitate clusters
             11. Fruit clearly coiled . . . . **Medicago**
             11' Fruit straight or nearly so
                12. Leaflets usually with entire margins; fruits usually dehiscent, much exceeding shriveled perianth parts . . . . **Lotus**
                12' Leaflets usually with wavy or toothed margins; fruits usually indehiscent, ± equal to and sometimes included in persistent corolla . . . . **Trifolium**

ASTRAGALUS L. LOCOWEED, MILKVETCH, or RATTLEPOD

Annual or perennial herbs. Ca. 1750 species; ± worldwide, mostly in temperate areas of N America, Europe, and Asia. Some species are grown as ornamentals and some are used medicinally. Most species are toxic to livestock, especially horses. Commercially important gum of tragacanth is obtained from several Asian species. (Greek: ankle-bone, an early name for some leguminous plant)

1. Annuals; leaflets 11-19 per leaf; calyx 2-4 mm long; corolla 3-6 mm long (including keel) . . . . . .
A. didymocarpus
1' Herbaceous perennials; leaflets 19-25 per leaf; calyx 5-8 mm long; corolla 12-16 mm long (including keel) . . . . **A. traskiae**

A. didymocarpus Hook. & Arn. var. **didymocarpus** [*A. catalinensis* Nutt.] COMMON DWARF LOCO-WEED or TWO-SEEDED MILKVETCH Plants to 2 dm tall. Apr. Rare; "on cliffs of a briny stream". Known only from collections made in April 1897; not seen recently. All CA Channel Islands except Santa Barbara; n CA (Contra Costa Co.) to Baja CA, e to NV. Three other varieties occur in sw U.S.
A. traskiae Eastw. TRASK'S LOCOWEED Plants to 8 dm tall. Mar-Aug. Abundant; sandy slopes, stabilized dunes, coastal bluffs, eroded caliche slopes, sandy canyon bottoms, and flats. Widespread

FABACEAE: Astragalus-Melilotus

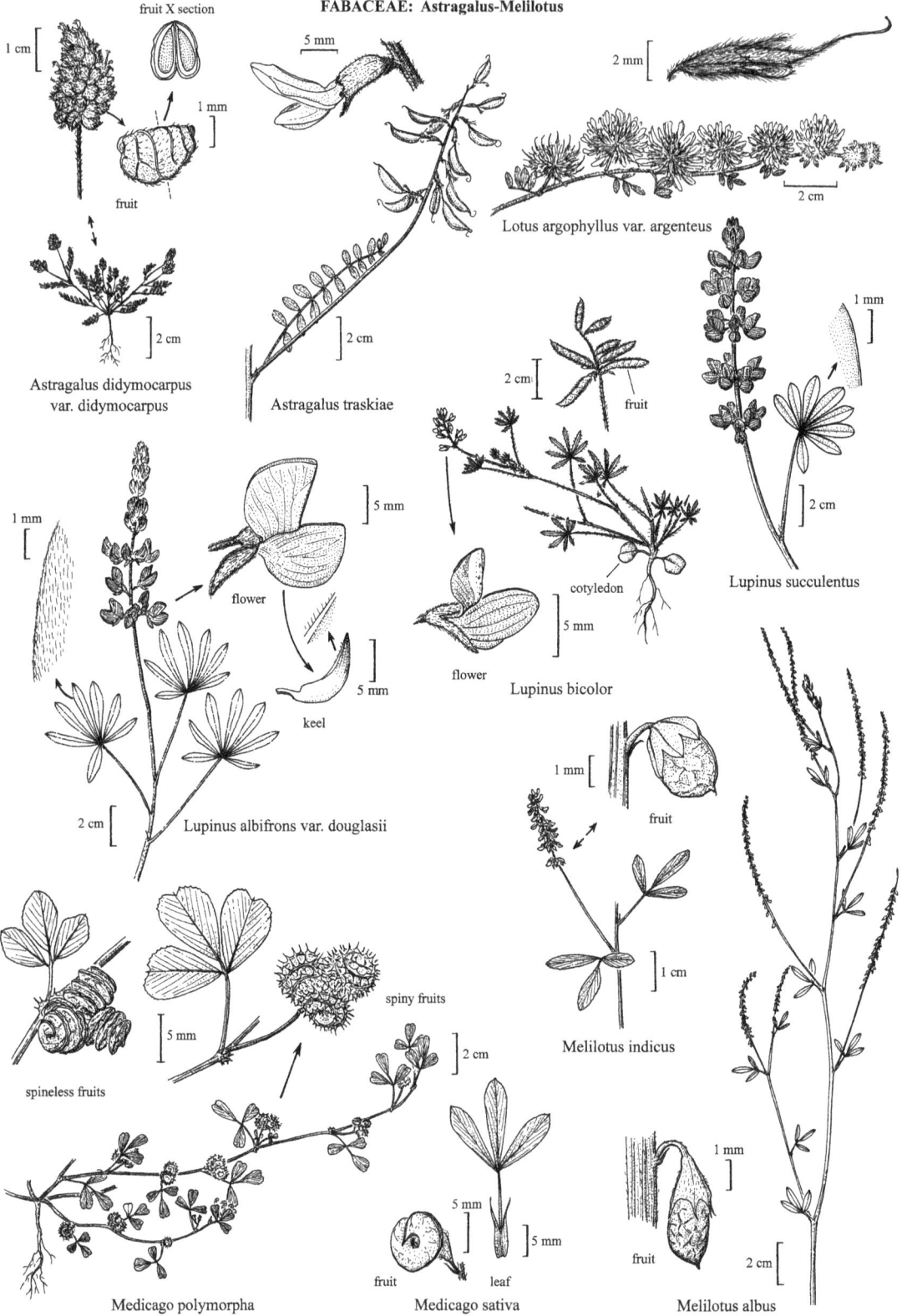

1 cm

fruit X section

1 mm

fruit

5 mm

2 mm

Lotus argophyllus var. argenteus

2 cm

Astragalus didymocarpus
var. didymocarpus

2 cm

Astragalus traskiae

2 cm

fruit

2 cm

1 mm

2 cm

Lupinus succulentus

1 mm

flower

5 mm

keel

5 mm

cotyledon

flower

5 mm

Lupinus bicolor

2 cm

Lupinus albifrons var. douglasii

1 mm

fruit

1 cm

Melilotus indicus

spiny fruits

spineless fruits

5 mm

2 cm

5 mm

fruit

leaf

5 mm

1 mm

fruit

2 cm

Medicago polymorpha

Medicago sativa

Melilotus albus

locations around perimeter of island, especially on s side; mesa s of Nicktown. ENDEMIC to San Nicolas and Santa Barbara islands. A detailed distribution map for this taxon can be found in Junak (2003c).

LOTUS L. TREFOIL

Annuals or suffrutescent perennials. Ca. 100 species; ± worldwide, mostly in temperate areas of N America, Europe, and Asia. (ancient Greek name applied to several plants)

1. Annuals; stems to 3 dm long; flowers solitary or 2-4 per umbel . . . .**L. salsuginosus**
1' Suffrutescent perennials; stems to 8 dm long; flowers 6-18 per umbel . . . .**L. argophyllus**

L. argophyllus (A.Gray) Greene var. **argenteus** Dunkle SOUTHERN ISLAND SILVER LOTUS Stems to 8 dm long. Mar-Aug. Abundant; stabilized dunes, sandy slopes, flats, coastal bluffs, and shallow swales. Widespread locations throughout much of island, especially in stabilized dunes near coast. ENDEMIC to Santa Barbara, San Nicolas, Santa Catalina, and San Clemente islands. Variety *adsurgens* Dunkle is restricted to San Clemente Island, var. *niveus* (Greene) Ottley is known only from San Cruz Island, and var. *ornithopus* (Greene) Ottley occurs only on Guadalupe Island.

*L. salsuginosus Greene var. **salsuginosus** COASTAL LOTUS Stems to 4 dm long. Mar-Jul. Rare; gravel piles. North side on coastal flats near Jetty Canyon; mesa on s side of airfield runways. All CA Channel Islands except Santa Barbara; central CA (Monterey Co.) to Baja CA. First collected on island in June 1996 (at both locations listed above). Variety *brevivexillus* Ottley occurs on San Benito and Cedros islands and in interior deserts of sw N America.

LUPINUS L. LUPINE

Annual herbs or shrubs. Ca. 200 species; temperate areas of N and S America, Europe, and Africa. Several species are cultivated for ornamental uses; many are toxic to livestock. (Latin: wolf, from the mistaken idea that plants rob nutrients from the soil)

1. Shrubs . . . .**L. albifrons**
1' Annuals
    2. Keel 4-6 mm long, upper margin ciliate near apex; leaflets villosulous . . . .**L. bicolor**
    2' Keel 10-14 mm long, upper margin glabrous near apex, ciliate near claw; leaflets glabrous to
       sparsely strigose . . . .**L. succulentus**

L. albifrons Benth. var. **douglasii** (J.Agardh) C.P.Sm. SILVER LUPINE Plants to 12 dm tall. Mar-Aug. Abundant; stabilized dunes, sandy flats, swales. Widespread locations throughout much of island, especially on slopes of the n escarpment, n portion of mesa, and in coastal dunes around perimeter of island. All CA Channel Islands except Santa Barbara and San Clemente; n CA (Marin Co.) to s CA (Santa Barbara Co.). Two to three other varieties occur from s OR to s CA. Eastwood (1898) reported that it was "infrequent on San Nicolas, growing on bare stretches of sand". A medicinal tea was made from the species by the Karok tribe.

L. bicolor Lindl. DOVE LUPINE Plants to 2 dm tall. Mar-Apr. Rare; flats with clay soil. Northeast end of mesa on n side of airfield runways and on mesa w of airfield terminal. All CA Channel Islands except Santa Barbara; Guadalupe Island; British Columbia to Baja CA, e to AZ. Eastwood (1898) reported that it grew "on the borders of a small lake, 1000 feet above the sea".

*L. succulentus Koch SUCCULENT LUPINE Plants to 4 dm tall. Mar-May. Rare; gravel piles. North side near Rock Jetty; mesa along Monroe Drive; s side near barge landing area at Daytona Beach. All CA Channel Islands except Santa Barbara; Los Coronados Islands; n CA (Tehama Co.) to Baja CA, e to AZ. Although native to much of CA, this species was apparently introduced to the island with gravel imported for construction projects. First collected on mesa in March 1983.

MEDICAGO L.

Annual or perennial herbs. Ca. 85 species; Eurasia, especially Mediterranean area. (classical Greek name for a crop plant which came to Greece from Media)

1. Corollas yellow, 3-6 mm long; annuals with prostrate to ascending stems . . . . **M. polymorpha**
1' Corollas violet to purple, 8-10 mm long; perennial herbs with erect stems . . . .**M. sativa**

*M. polymorpha** L. [*M. hispida* Gaertn., *M. p.* var. *brevispina* (Benth.) Heyn.] BUR-CLOVER Stems to 4 dm long. Mar-May (Jul). Abundant; flats, canyon bottoms, slopes, and disturbed areas. Widespread locations throughout most of island, especially on mesa and adjacent ne escarpment and coastal flats. All CA Channel Islands; Guadalupe and Cedros islands; WA to Baja CA, e to FL and disjunctly to ME; Hawaii ± worldwide; native of Mediterranean Europe. First collected on "two slopes" on San Nicolas Island in April 1897. It was reportedly "abundant on uplands" and was "the main sheep feed on the island" in July 1939 (label on herbarium specimen at Santa Barbara Botanic Garden).

*M. sativa** L. ALFALFA Plants to 1 m tall. Apr-Dec. Rare; flats and disturbed areas. Mesa at old landfill site and nw of Jackson Hill (near Building 113). All CA Channel Islands except Santa Barbara; Cedros Island; WA to Baja CA, e to FL and disjunctly to ME; Hawaii; native of Europe and Asia. First reported on the island "in one locality, a flat above a brackish stream" in April 1897 (Eastwood 1898). It was collected "in old corral" in April 1901, where it was reportedly "luxuriant" in 1897 but "scarce" in 1901 (label on herbarium specimen at Pomona College). This species was not collected on San Nicolas Island between 1902 and 1991. Alfalfa can be toxic to livestock.

MELILOTUS Miller SWEET-CLOVER

Annual or biennial herbs. Ca. 20 species; warm temperate to subtropical areas of Europe and Asia. (Greek: honey-lotus)

1. Corollas white, 3.5-6 mm long; inflorescences (including peduncles) 6-12 cm long . . . .**M. albus**
1' Corollas yellow, 2-3 mm long; inflorescences (including peduncles) 2-5 cm long . . . .**M. indicus**

*M. albus** Medik. WHITE SWEET-CLOVER Plants to 13 dm tall. Mar-Jul (Nov). Common; disturbed areas, roadsides, flats, and canyon bottoms. Widespread locations on island, especially in gullies on s escarpment. Santa Cruz, Anacapa, Santa Catalina, and San Clemente islands; British Columbia to Baja CA, e to ME; native of Eurasia. First collected in Tule Canyon in April 1961. This species is the original source of the anticoagulant drug dicoumarol; improperly dried plants in hay can be toxic to livestock.

*M. indicus** (L.) All. [*M. parviflorus* Desf.] YELLOW SWEET-CLOVER Plants to 4 dm tall. Mar-Aug. Abundant; disturbed areas, roadsides, flats, slopes, canyon bottoms, and sand dunes. Widespread locations over much of island. All CA Channel Islands; Guadalupe, San Benito, and Cedros islands; WA to Baja CA, e to FL and disjunctly to ME; Hawaii; native of Mediterranean Europe. First collected in "canyon bottom on north side" in July 1939. This species can be toxic to livestock.

ROBINIA L. LOCUST

Trees. Ca. 5 species; N America. (named for Jean Robin, 1550-1629, and Vespasian Robin, 1579-1662, French royal gardeners who introduced plants to Europe)

*R. pseudoacacia** L. BLACK LOCUST Plants to 6 m tall. Apr. Rare; canyon bottom. Known only from n side in Mineral Canyon. Santa Cruz Island; WA to Baja CA, e to ME; native of e N America. First collected on San Nicolas Island in July 1979. Seeds, leaves, and bark can be toxic to humans and livestock if ingested.

SPARTIUM L.

Shrubs. One species. (Greek name for *Stipa tenacissima* L., which, like *Spartium,* is a source of fibers used in weaving and cordage)

*S. junceum L. SPANISH BROOM Shrub. Rare; disturbed flats. Mesa near intersection of Tufts and Shannon roads (where it may be naturalized) and Nicktown (where it is clearly planted). Santa Cruz and Santa Catalina islands; WA to Baja CA; Hawaii; native to Mediterranean Europe and Canary, Madeira, and Azores islands. First collected on San Nicolas Island in April 1980. Stems are used as a source of durable fiber; flowers yield a yellow dye.

TRIFOLIUM L. CLOVER

Annual herbs. Ca. 240 species; temperate to subtropical areas, ± worldwide. (Latin: three leaves)

1. Spike not subtended by involucre of bracts
    2. Flowers sessile, not reflexed; calyx lobes plumose-villous . . . .**T. albopurpureum**
    2' Flowers pedicellate, often reflexed with age; calyx lobes glabrous
        3. Leaflets obovate to obcordate, 1.5-2.5 times longer than wide; apex emarginate . . . .
        . . . .**T. gracilentum**
        3' Leaflets narrowly elliptic or lanceolate, 3.5-5 times longer than wide; apex acute . . . .
        . . . .**T. palmeri**
1' Spike subtended by involucre of bracts (bracts sometimes inconspicuous in *T. depauperatum*)
    4. Banner conspicuously inflated in fruit; involucral bracts ± free . . . .**T. depauperatum**
    4' Banner not inflated in fruit; involucral bracts fused
        5. Involucre flat, rotate . . . .**T. willdenovii**
        5' Involucre campanulate to bowl-shaped
            6. Calyx lobes more than ½ the length of tube, bristle-tipped with entire margins . . . .
            . . . .**T. microcephalum**
            6' Calyx lobes less than ½ the length of tube, not bristle-tipped, margins ciliate . . . .
            . . . .**T. microdon**

T. albopurpureum Torr. & A.Gray [*T. insularum* P.B.Kenn.] RANCHERIA CLOVER Plants to 3 dm tall. Mar-May. Scarce; flats. Northeastern coastal flats near first canyon w of "L" Canyon; mesa e of Nicktown, sw of airfield, and w of Building 121. Santa Rosa, Santa Cruz, and Santa Catalina islands; British Columbia to Baja CA, e to AZ. Reported for island by Eastwood (1898) as *T. dichotomum* Hook. & Arn.
T. depauperatum Desv. var. **truncatum** (Greene) Isely [*T. amplectens* Torr. & A.Gray var. *truncatum* (Greene) Jepson] DWARF BLADDER CLOVER. Plants to 2 dm tall. Mar-Apr. Scarce; flats and slopes. Mesa near nw and s sides of airfield and w of Building 121; se escarpment near Sand Spit. All CA Channel Islands except Santa Barbara; n CA (Shasta Co.) to Baja CA. Three other varieties occur in w N America.
T. gracilentum Torr. & A.Gray PINPOINT CLOVER Plants to 2 dm tall. Mar-Apr. Rare; flats. Mesa n of airfield. All CA Channel Islands; Todos Santos and Guadalupe islands; WA to Baja CA, e to AZ. Population sizes apparently vary significantly from year to year; not usually seen in dry years. This species was documented on the island in April 1897, but not collected again until May 1993.
T. microcephalum Pursh SMALL-HEADED CLOVER Plants to 2 dm tall. Mar-May. Occasional; flats, slopes, and canyon bottoms. Northeastern escarpment from vicinity of W Mesa Canyon to Sand Spit area; ne portion of mesa between Nicktown and airfield; se portion of mesa w of Building 121. All CA Channel Islands except Santa Barbara; Guadalupe Island; AK to Baja CA, e to MT and AZ.
T. microdon Hook. & Arn. [*T. m.* var. *pilosum* Eastw.] VALPARAISO CLOVER Plants to 2.5 dm tall. Mar-Apr. Rare; moist slopes and flats. Coastal flats and lower slopes of n escarpment near W Mesa Canyon; mesa s of Nicktown and near w end of airfield. Santa Catalina Island; British Columbia to s CA (Orange Co.); S America (Chile). Population sizes apparently vary significantly from year to year; not usually seen in dry years. This species was first noted in "two localities on moist slopes" in April 1897,

FABACEAE: Robinia-Vicia

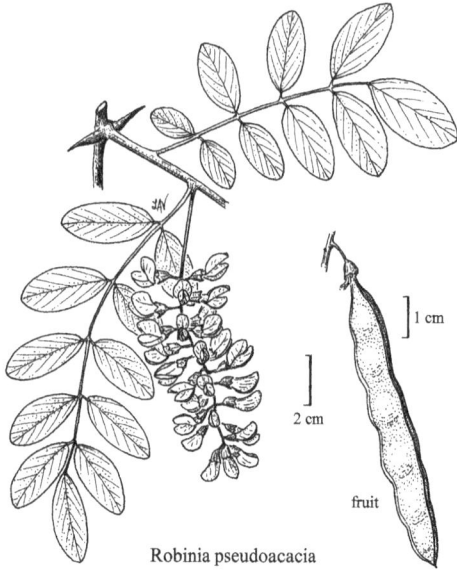

Robinia pseudoacacia

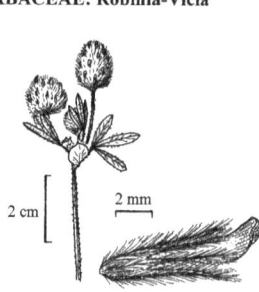

Trifolium albopurpureum

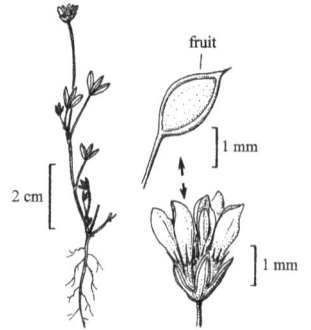

Trifolium depauperatum var. truncatum

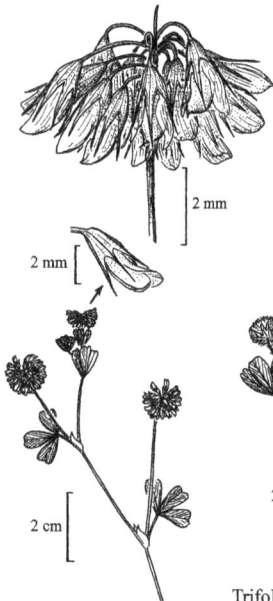

fruit

2 cm

2 mm

1 cm

fruit

1 mm

1 mm

2 mm

2 mm

2 mm

1 cm

2 mm

1 mm

1 mm

2 mm

2 mm

2 cm

Trifolium willdenovii

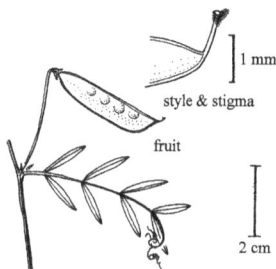

Trifolium palmeri

2 mm

2 cm

2 cm

2 cm

2 cm

Trifolium var. gracilentum

Trifolium microcephalum

Trifolium microdon

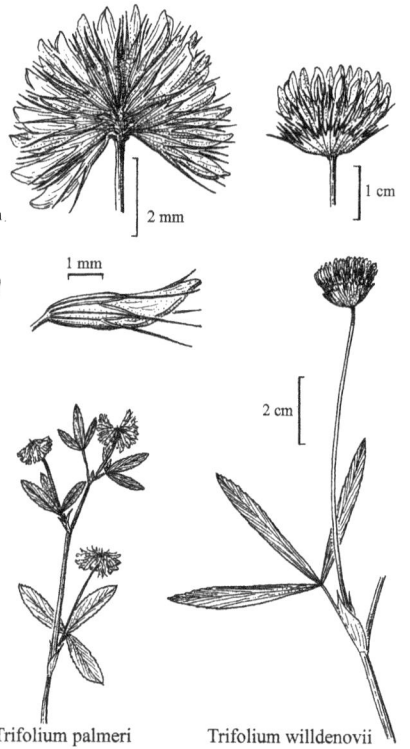

1 mm
style & stigma

fruit

2 cm

fruit

fruit

1 cm

2 cm

Vicia hassei

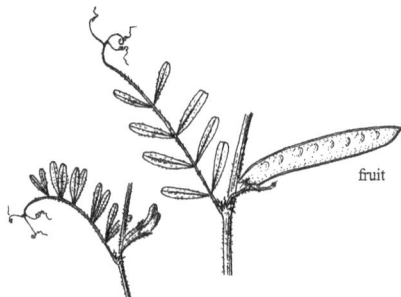

Vicia sativa subsp. sativa

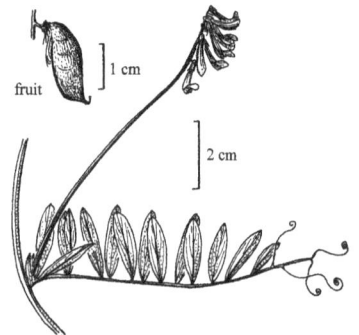

Vicia villosa var. varia

but not collected again until April 1992. Detailed distribution maps for this taxon can be found in Junak *et al.* (1996) and Junak (2003c).

T. palmeri S.Watson SOUTHERN ISLAND CLOVER Plants to 2 dm tall. Mar-May. Occasional; moist slopes and flats. Scattered locations on coastal flats and ne escarpment between vicinity of "L" Canyon and Sand Spit area; mesa near nw end of airfield and w of Building 121. ENDEMIC to Anacapa, Santa Barbara, San Nicolas, Santa Catalina, San Clemente, and Guadalupe islands. Population sizes apparently vary significantly from year to year; not usually seen in dry years. This species was documented at one locality in April 1897, but not collected again until April 1978. Detailed distribution maps for this taxon can be found in Junak *et al.* (1996) and Junak (2003c).

T. willdenovii Spreng. [*T. tridentatum* Lindl.] TOMCAT CLOVER Plants to 3 dm tall. Mar-Apr. Occasional; moist slopes and flats. Coastal flats e of Corral Harbor and scattered locations in canyons and on slopes of n escarpment from W Mesa Canyon area to airfield; mesa e of Nicktown. All CA Channel Islands; Los Coronados and Todos Santos islands; British Columbia to Baja CA, e to ID and AZ. Leaves and seeds of this taxon were eaten by Native Americans.

VICIA L. VETCH

Annual herbs. Ca. 140 species; N America, Europe, and Asia. (Latin name for the genus)

1. Inflorescence a sessile or barely peduncled cluster of 1-4 pedicelled flowers . . . .**V. sativa**
1' Inflorescence a peduncled raceme
   2. Flower 7-8 mm long . . . .**V. hassei**
   2' Flower 10-18 mm long . . . .**V. villosa**

V. hassei S.Watson [*V. exigua* Nutt. var. *h.* (S.Watson) Jeps.] SLENDER VETCH Stems to 7 dm long. Apr. Rare; flats in or near cactus patches. Mar-Apr. Northeastern coastal flats near S Spur Canyon. All CA Channel Islands except Santa Barbara; Todos Santos and Guadalupe islands; OR to Baja CA. Population sizes apparently vary significantly from year to year; not usually seen in dry years. This taxon was documented "about cacti" in April 1901, but was not collected again until March 2003.

*V. sativa L. subsp. **sativa** COMMON VETCH or SPRING VETCH Stems to 7 dm long. Apr. Rare; disturbed flats. North edge of mesa near w end of airfield runways. Santa Cruz, Anacapa, and Santa Catalina islands; WA to Baja CA, e (disjunctly) to ME and FL; native of Eurasia. First collected on mesa in April 1995. Subspecies *nigra* (L.) Erhart occurs on Santa Rosa Island and is widespread throughout much of N America.

*V. villosa Roth subsp. **varia** (Host) Corbiere [*V. dasycarpa* Ten.] WINTER VETCH Stems to 1 m long. Apr-Jun. Rare; moist flats, swales, and canyon bottoms. North side at Thousand Springs and in Tule Creek. Santa Catalina Island; WA to Baja CA, e (disjunctly) to ME and FL; native of Europe. First collected at Tule Creek in July 1965. Subspecies *villosa* occurs throughout much of N America.

FRANKENIACEAE L.
FRANKENIA FAMILY

Suffrutescent perennials. Two genera and ca. 81 species; temperate and subtropical areas worldwide, mostly in saline or gypseous habitats.

FRANKENIA L.

Ca. 80 species; temperate and subtropical areas worldwide, mostly in saline or gypseous habitats. Some species are used medicinally or as a source of salt. (honoring Johan Frankenius, 1590-1661, professor of anatomy, medicine, and botany at Uppsala, Sweden)

F. salina (Molina) I.M.Johnst. [*F. grandifolia* Cham. & Schltdl.] ALKALI HEATH or YERBA REUMA Plants to 3 dm tall. Apr-Aug. Occasional; flats with sandy or clay soils, especially in saline areas.

FRANKENIACEAE

flower

] 2 mm

2 cm

Frankenia salina

GENTIANACEAE

flower

] 5 mm

2 cm

Zeltnera venusta

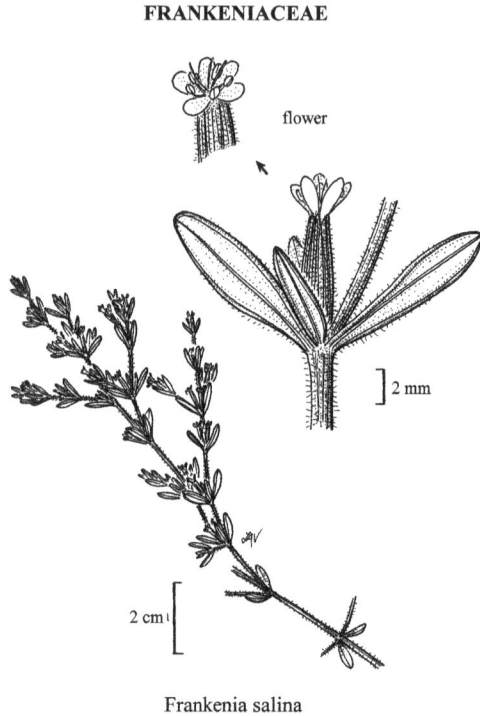

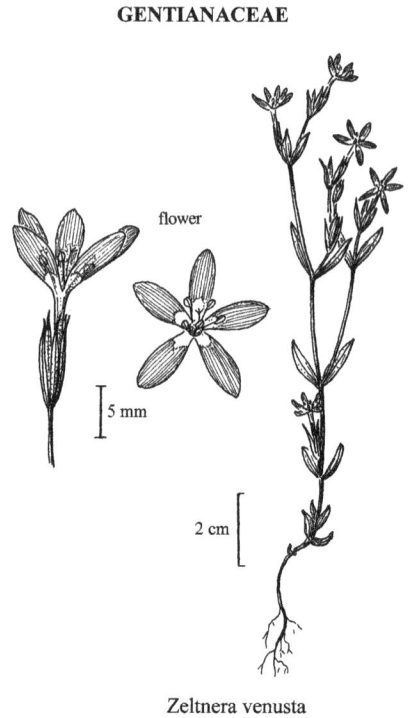

Widespread locations around perimeter of island; scattered locations on mesa, especially at se end. All CA Channel Islands except Santa Barbara; Todos Santos and Guadalupe islands; n CA (Butte Co.) to Baja CA, e to NV; S America (Chile); Hawaii (introduced on Tern Island but not seen recently). This species was used medicinally by Native Americans.

GENTIANACEAE Juss.
GENTIAN FAMILY

Annual herbs. Ca. 78 genera and 1225 species; temperate to tropical areas worldwide. Some genera (e.g., *Eustoma, Gentiana*) are cultivated as ornamentals.

ZELTNERA Mansion

Ca. 25 species; temperate areas of N America, Central America, and S America. Stems and leaves of several species were used medicinally by Native Americans. (honoring Louis Zeltner, b.1938, and his wife Nicole, b.1934, botanists from Switzerland)

Z. venusta (A.Gray) Mansion [*Centaurium v.* (A.Gray) B.L.Rob.] CANCHALAGUA Plants to 1.5 dm tall. Apr-Jun. Occasional; moist slopes and canyon bottoms. Northern escarpment between Live-forever Canyon and S Spur Canyon. Santa Catalina Island; central CA to Baja CA.

GERANIACEAE Juss.
GERANIUM FAMILY

Annual or perennial herbs. Ca. 11 genera and 700 species; temperate to subtropical areas worldwide. Some genera (e.g., *Geranium, Pelargonium,* and *Sarcocaulon*) are cultivated for essential oils or as ornamentals.

1. Flowers radial; petals all alike, alternating with nectar glands; sepals without nectar spur; annuals . . . .
. . . .**Erodium**
1' Flowers ± bilateral; petals unequal in length, without basal nectar glands; upper sepal with a nectar spur
   adnate to pedicel; perennial herbs . . . .**Pelargonium**

ERODIUM L'Her. FILAREE or STORKSBILL

Annual herbs. Ca. 60 species; temperate areas ± worldwide. Some species are cultivated for forage or for dyes. (Greek: heron, referring to bill-like fruit)

1. Lower leaves lobed to ± pinnatifid; style column 8-12.5 cm long . . . .**E. botrys**
1' Lower leaves pinnately compound; style column 2.5-4 cm long
   2. Leaflets serrate to coarsely toothed; sepal tips with 1-2 white bristles; petal claws ciliate . . . .
   . . . .**E. cicutarium**
   2' Leaflets pinnately lobed or divided; sepal tips without bristles; petal claws glabrous . . . .
   . . . .**E. moschatum**

*E. botrys (Cav.) Bertol. BROAD-LEAF FILAREE Stems to 3 dm long. Feb-May. Scarce; flats. Eastern end of mesa, especially near airfield. Santa Rosa, Santa Cruz, Santa Catalina, and San Clemente islands; OR to Baja CA, e (disjunctly) to TX and ME; native of s Europe. First collected just w of fire station at airfield in March 1984.
*E. cicutarium (L.) Aiton REDSTEM FILAREE Stems to 1.5 dm long. Feb-Jun. Occasional; flats and slopes. Widespread locations throughout much of island, especially in disturbed sites. All CA Channel Islands; Los Coronados, Todos Santos, San Martin, Guadalupe, and Cedros islands; AK to Baja CA, e to ME; Hawaii; native of Eurasia. First collected on "one moist flat" in April 1897.
*E. moschatum (L.) Aiton WHITESTEM FILAREE Stems to 4 dm long. Feb-Jun. Occasional; flats, slopes, and sand dunes. Widespread locations throughout much of island, especially in disturbed sites. All CA Channel Islands, Los Coronados, Todos Santos, San Martin, and Guadalupe islands; WA to Baja CA, e to AZ and disjunctly to ME; native of Europe. First collected at "southeast corner" of island in March 1932.

PELARGONIUM L'Her. GARDEN GERANIUM

Suffrutescent perennials. Ca. 280 species; mostly in temperate areas of s Africa and Australia. Many species are cultivated for ornamental use or for aromatic oils. Leaves of several species are used for food. (Greek: stork, referring to the beaked fruit)

1. Leaf peltate, shallowly 5-lobed . . . .**P. peltatum**
1' Leaf round to reniform, base deeply notched . . . .**P. x hortorum**

*P. x hortorum L.H.Bailey FISH GERANIUM Plants to 6 dm tall. Mar-Sep (Nov). Rare; flats. Cultivated in Nicktown and occasionally naturalizing on surrounding mesa. San Miguel, Santa Rosa, Santa Cruz, Santa Catalina, and San Clemente islands; Todos Santos Island; s CA (Santa Barbara Co. to Los Angeles Co.). Naturalized plants were first noted on mesa in the 1980s. This taxon is a fertile hybrid derived from *P. inquinans* (L.) L'Her. and *P. zonale* (L.) L'Her., both of which are natives of s Africa.
*P. peltatum (L.) L'Her. IVY-LEAVED GERANIUM Plants to 3 dm tall. Apr-Jul. Rare; flats and slopes. Scattered locations on mesa; s side at Army Springs. Southern CA (Ventura Co.); native of s Africa. First collected near Army Springs in July 1965. Leaves have been used medicinally; flowers yield a blue dye.

HYDROPHYLLACEAE R.Br.
WATERLEAF FAMILY

Annual or perennial herbs. Ca. 18 genera and 270 species; worldwide, but mostly N America. Several

GERANIACEAE

5 mm

1 mm

leaf   2 cm

fruit tip

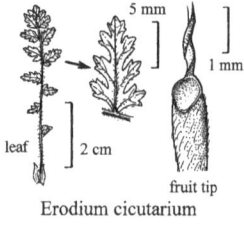

Erodium cicutarium

1 mm

.5 mm

sepal

fruit tip

2 cm

1 mm

2 cm

fruit

leaf

fruit tip

Erodium botrys   Erodium moschatum

2 cm

Pelargonium ×hortorum

Pelargonium peltatum

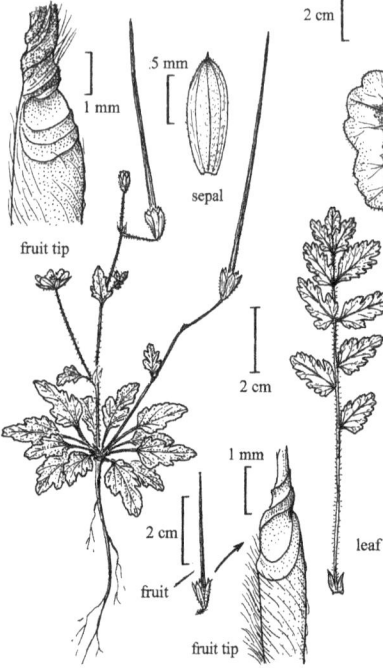

2 cm

HYDROPHYLLACEAE

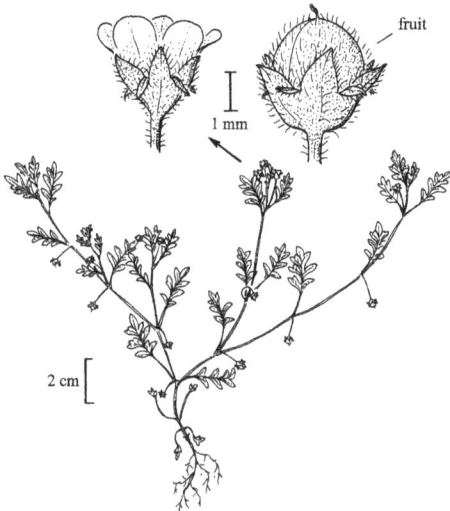

fruit

1 mm

1 cm

2 cm

2 cm

1 mm

Nemophila pedunculata

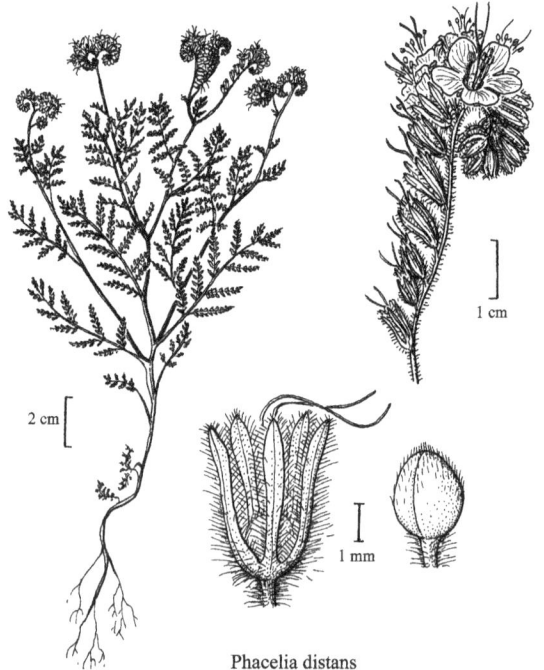

Phacelia distans

genera (e.g., *Hesperochiron, Nemophila,* and *Phacelia*) are cultivated as ornamentals. Some authors would include Hydrophyllaceae in Boraginaceae (Olmsted *et al.* 2000).

1. Flowers solitary, axillary; calyx lobes alternating with spreading to reflexed sepaloid appendages . . . .
. . . .**Nemophila**
1' Flowers in dense terminal coiled spikes; calyx without sepaloid appendages . . . .**Phacelia**

NEMOPHILA Nutt.

Ca. 11 species; temperate areas of w N America and se U.S. Some species (e.g., *N. maculata, N. menziesii*) are cultivated as ornamentals. (Greek: woodland-loving, referring to the habitat of some species)

N. pedunculata Benth. [*N. insularis* Eastw.] Plants to 8 cm tall. Mar-Apr. Scarce; canyon bottoms, moist flats, and slopes. Northern coastal flats and escarpment between Nicktown and airfield area; mesa near W and E Mesa canyons and at n edge of mesa w of airfield. San Miguel, Santa Rosa, and Santa Cruz islands; British Columbia to Baja CA, e to ID and NV.

PHACELIA Juss.

Ca. 150 species; N and S America. Some species (e.g., *P. tanacetifolia*) are cultivated as ornamentals. (Greek: cluster or bundle, referring to the dense inflorescences of some species)

P. distans Benth. [*P. cinerea* J.F.Macbr.] Plants to 3 dm tall. Apr. Rare; known from a single collection on the island and seen on "moist flats" in two localities by Blanche Trask in April 1901; not seen recently. All CA Channel islands; Los Coronados and Todos Santos islands; n CA (Tehama Co.) to Baja CA, e to NV and AZ. Plants from San Nicolas Island have been called *P. cinerea* J.F.Macbr., which was described as a perennial. However, large, sprawling plants of annual *P. distans* can be nearly 1 m across when growing on coastal flats of other CA Channel Islands and can appear to be perennials. Disregarding life cycle, it is very difficult to distinguish the type of *P. cinerea* from *P. distans.*

LAMIACEAE Lindl. [Labiatae Juss.]
MINT FAMILY

Ca. 252 genera and 6700 species; worldwide. Many species are cultivated for volatile oils, which are used in medicine and as flavorings for foods or teas. Several genera (e.g., *Leonotis, Monardella, Phlomis, Rosmarinus, Salvia,* and *Thymus*) are also cultivated as ornamentals.

MARRUBIUM L. HOREHOUND

Suffrutescent perennials. Ca. 30 species; temperate areas of Eurasia. Some species are cultivated for medicinal use or flavorings; some are toxic. (classical Latin name for horehound, based on Hebrew word meaning bitter)

*****M. vulgare** L. Plants to 13 dm tall. Mar-Jul. Scarce; canyon bottoms, slopes, and flats. North side in Celery Canyon and gullies on coastal flats nearby; mesa in upper portion of Mineral Canyon; s side in e fork of Cattail Canyon and upper e portion of Twin Rivers watershed. All CA Channel Islands except Anacapa and Santa Barbara; Todos Santos Island; AK to Baja CA, e to ME; Hawaii; native of Europe. First collected in "wash bottom" at unspecified location on island in June 1980. This taxon could be very invasive on the island and should be eliminated before it spreads. This species is used medicinally and to make horehound candy.

LOASACEAE Dumort.
STICK-LEAF FAMILY

Annual herbs. Ca. 14 genera and 260 species; temperate to subtropical areas of N and S America, Africa. Some taxa are cultivated for ornamental use.

MENTZELIA L. BLAZING STAR

Ca. 50 species; temperate to subtropical areas of w N America and S America. Seeds of several species were ground and eaten by Native Americans. (honoring Christian Mentzel, 1622-1701, German botanist)

M. affinis Greene HYDRA STICK-LEAF Plants to 15 cm tall. Mar-Apr. Scarce; sandy washes, open slopes, and ridgetops. Widely scattered locations on ridges of s escarpment between Grand Canyon and Dutch Harbor area. Santa Cruz, Santa Catalina, and San Clemente islands; central CA (Contra Costa Co.) to Baja CA, e to NV and AZ. On San Nicolas Island, this species is mostly restricted to soils derived from blue-grey siltstone and is often associated with *Eschscholzia ramosa.*

LYTHRACEAE J.St.-Hil.
LOOSESTRIFE FAMILY

Annual or perennial herbs. Ca. 27 genera and 600 species; temperate to tropical areas worldwide. Some genera (e.g., *Cuphea, Lagerstroemia,* and *Lythrum*) are cultivated as ornamentals, a dye is obtained from *Lawsonia* (henna), and some genera (e.g., *Heimia, Lagerstroemia*) yield hallucinogenic compounds.

LYTHRUM L. LOOSESTRIFE

Ca. 35 species; temperate areas worldwide. (Greek: blood, probably referring to the flower color or to the staining qualities of some species)

*L. hyssopifolia L. COMMON LOOSESTRIFE Stems to 5 dm long. Mar-Apr. Rare; moist flats and canyon bottoms. Mesa at wells area along Tufts Road, at Nicktown, and in gully on ne side of airfield runways. Santa Rosa, Santa Cruz, Santa Catalina, and San Clemente islands; WA to Baja CA, e (disjunctly) to ME; Australia; New Zealand; native of Europe. First collected in a gully on ne side of airfield in March 1984.

MALVACEAE Juss.
MALLOW FAMILY

Ca. 110 genera and 1800 species; temperate to tropical areas worldwide. Some species are cultivated for fiber, like cotton (*Gossypium*), or are used as ornamentals (e.g., *Alcea, Hibiscus, Lavatera,* and *Sidalcea*). Some species are food plants, such as okra (*Abelmoschus*).

1. Shrubs; petals 25-40 mm long . . . .**Lavatera**
1' Annuals; petals 4-5 mm long . . . .**Malva**

LAVATERA L. TREE-MALLOW

Shrubs. Ca. 25 species; temperate to subtropical areas, ± worldwide. Some species are cultivated as ornamentals, including several that are endemic to Pacific islands off Baja CA. Ray (1998) included *Lavatera* in the genus *Malva.* (honoring Johann Heinrich Lavater, 1611-1691, and possibly his brother Johann Jacob, 1594-1636, Swiss physicians and naturalists)

*L. assurgentiflora Kellogg subsp. **assurgentiflora** MALVA ROSA or NORTHERN ISLAND TREE-MALLOW Plants to 1.5 m tall. Mar-Jul. Rare; slopes and flats. North escarpment near W Mesa Canyon; mesa along Tufts Road (at Building 189) and behind Telemetry buildings on Harrigton Road (Building

LAMIACEAE

1 cm

inflorescence

2 mm

calyx

2 cm

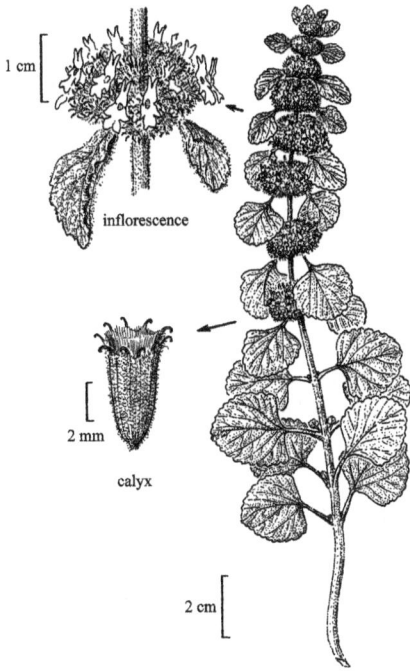

Marrubium vulgare

LOASACEAE

1 mm

stamen

2 cm

2 mm

flower

flower bract

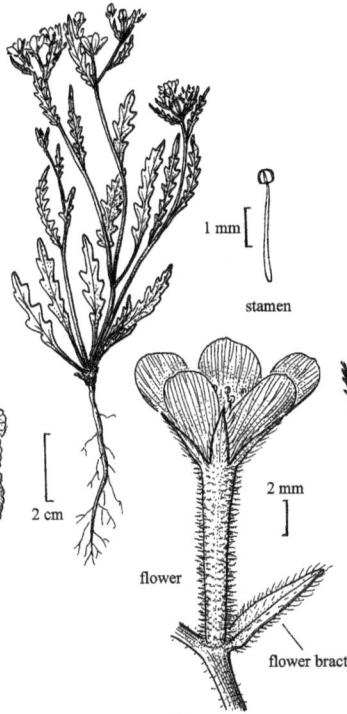

Mentzelia affinis

LYTHRACEAE

flower

2 mm

fruit

5 cm

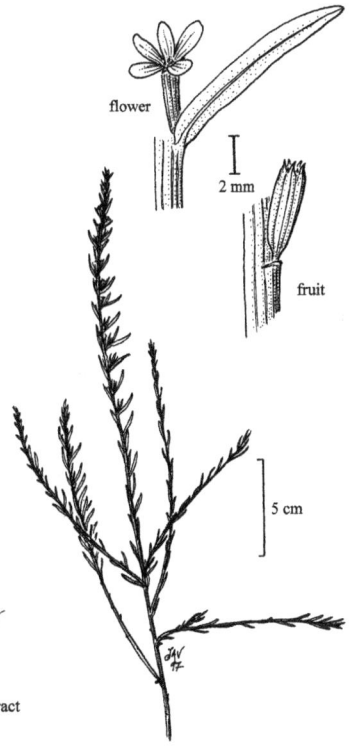

Lythrum hyssopifolia

MALVACEAE

2 cm

calyx

1 cm

bractlet

2 cm

fruit

5 mm

calyx

bractlet

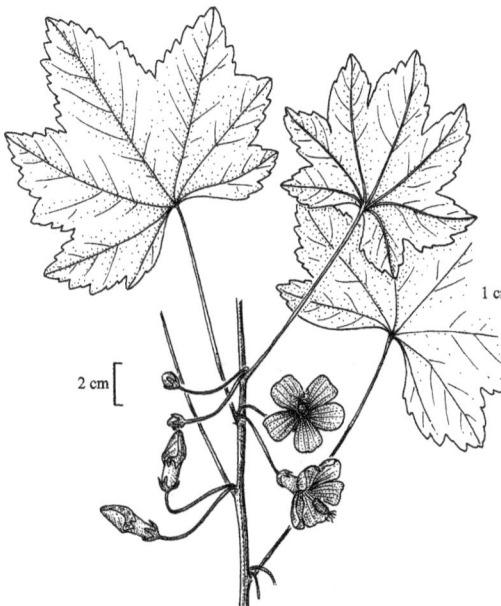

Lavatera assurgentiflora

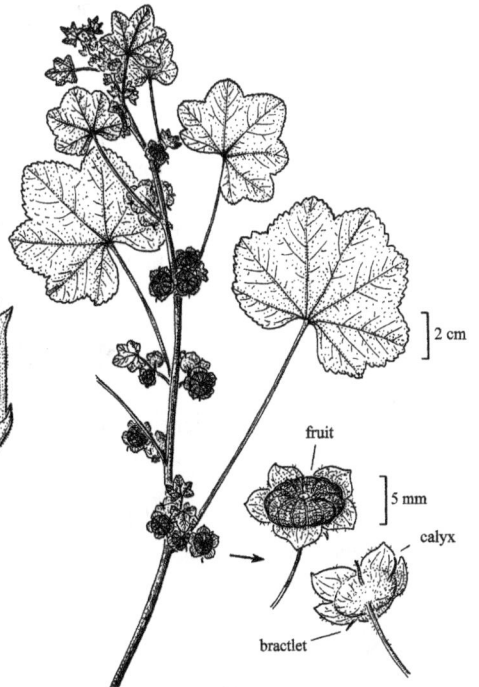

Malva parviflora

142 MYOPORACEAE – NYCTAGINACEAE

182). Native to San Miguel and Anacapa islands; apparently planted on Santa Rosa, Santa Cruz, Todos Santos, and Cedros islands; widely naturalized on CA mainland and in Baja CA. Subspecies *glabra* Philbrick occurs on Santa Catalina and San Clemente islands. First collected near buildings on Jackson Hill in April 1978. Although Nidever "found some high bushes, called by the natives malva real ..." on San Nicolas Island in the fall of 1852 (Dittman 1878, Ellison 1937), we have not found any early collections to document his observations. Some plants on the mesa were apparently introduced to San Nicolas Island from Santa Maria (J. Vanderwier, personal communication 1984). The plants growing on n escarpment near W Mesa Canyon need further study and may represent a native population.

MALVA L. MALLOW

Annual herbs. Ca. 40 species; temperate areas of Europe, Asia, and Africa. Some species are toxic to livestock due to selenium or nitrate accumulation. (Latin: mallow)

M. parviflora L. CHEESEWEED Plants to 17 dm tall. Mar-Jul. Common; disturbed sites, flats, and slopes. Widespread locations throughout much of island. All CA Channel Islands; all Baja CA islands except San Geronimo; WA to Baja CA, e to IA and disjunctly to MA and FL; Mexico; S America; Hawaii; Australia; New Zealand; native of Eurasia. First collected on "flat above stream" in April 1897. Fruits are edible raw; leaves are edible cooked and can be used to thicken stews. This species can be toxic to livestock.

MYOPORACEAE R.Br.
MYOPORUM FAMILY

Shrubs or trees. Ca. 3 genera and 235 species; Australia, Pacific islands, e Asia, s Africa, and West Indies. Many botanists would include this family in the Scrophulariaceae (figwort family). Some plants in this family yield edible fruit and some are harvested for timber.

MYOPORUM Forster f.

Ca. 28 species; Australia, New Zealand, Mauritius, e Asia, e Malesia, and Hawaii. Some species are cultivated for ornamental use and some are toxic to livestock. (Greek: closed pores, referring to translucent dots on leaves)

M. laetum Forster f. Plants to 10 m tall. Apr-Jul. Common; flats and canyon bottoms. Widespread locations on n side between Live-forever Canyon area and W Mesa Canyon area and on mesa. Santa Cruz and San Clemente islands; n CA (Sonoma Co.) to Baja CA; Hawaii; native of New Zealand. A detailed distribution map for this taxon can be found in Junak (2003b). First collected at Nicktown in July 1965. Leaves and fruits of this species can be toxic to livestock.

NYCTAGINACEAE Juss.
FOUR O'CLOCK FAMILY

Perennial herbs. Ca. 30 genera and 390 species; temperate to subtropical areas, ± worldwide. Plants in some genera (e.g., *Bougainvillea, Mirabilis*) are cultivated as ornamentals.

ABRONIA Juss. SAND-VERBENA

Ca. 20 species; w N America. Roots of some species were eaten by Native Americans. (Greek: graceful or delicate, referring to appearance of floral bracts in some species)

1. Perianth dark crimson to reddish purple; fruit with coarsely and reticulately veined, wing-like lobes
. . . .**A. maritima**
1' Perianth light rose to white; fruit with faintly net-veined, wing-like lobes . . . .**A. umbellata**

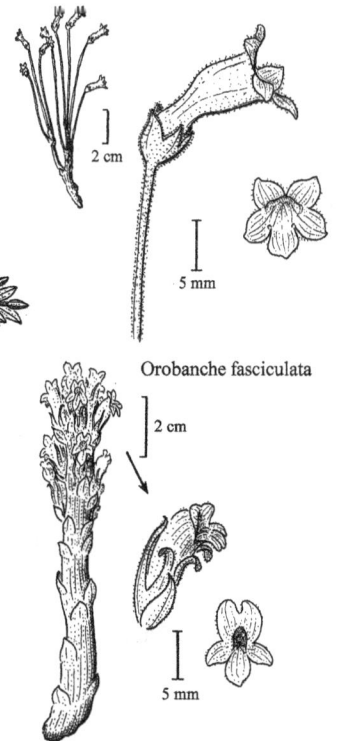

MYOPORACEAE

NYCTAGINACEAE

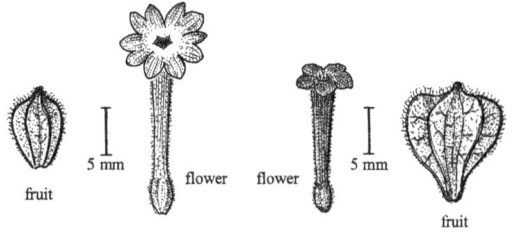

fruit

5 mm

flower

flower

5 mm

fruit

Abronia umbellata

Abronia maritima

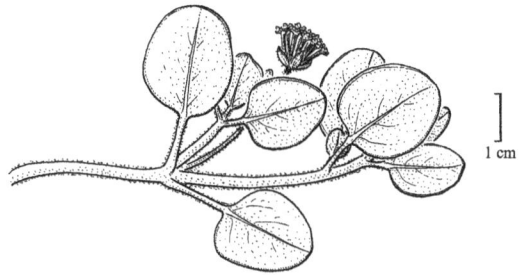

2 cm

2 mm

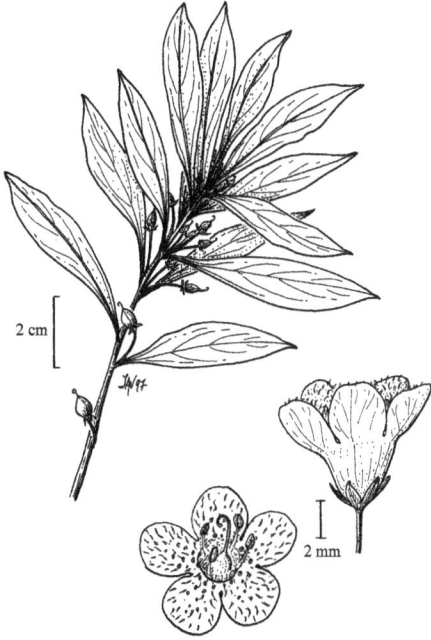

Myoporum laetum

1 cm

Abronia maritima

OROBANCHACEAE

ONAGRACEAE

2 cm

5 mm

2 cm

2 cm

5 mm

Orobanche fasciculata

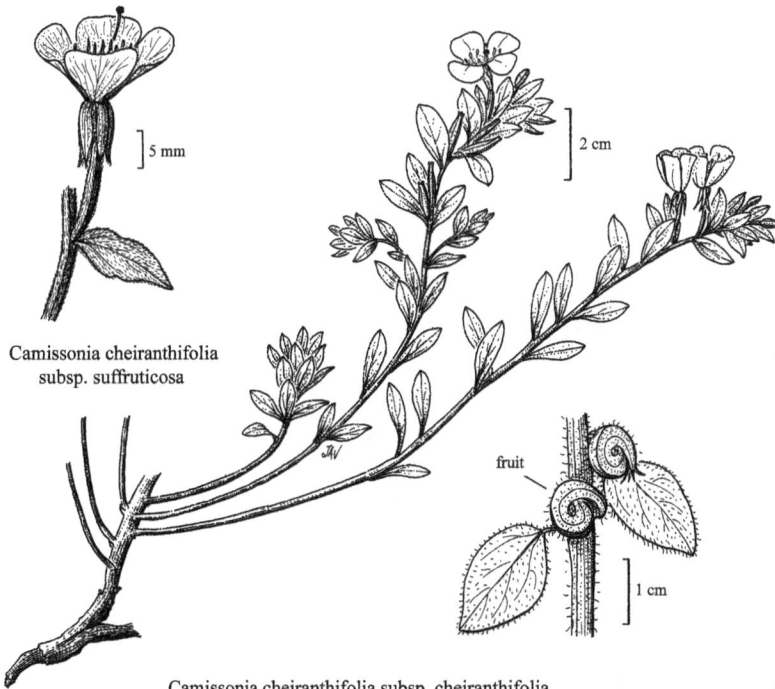

Camissonia cheiranthifolia
subsp. suffruticosa

2 cm

fruit

1 cm

5 mm

Camissonia cheiranthifolia subsp. cheiranthifolia

Orobanche parishii subsp. brachyloba

A. maritima S.Watson STICKY SAND-VERBENA Stems to 13 dm long. Apr-Jul. Common; sandy flats, stabilized and unstabilized dunes. Widespread locations, especially around perimeter of island. All CA Channel Islands except Santa Barbara; San Martin, Cedros, and Natividad islands; central CA (Monterey Co.) to Baja CA; Mexico. This taxon apparently hybridizes with *A. umbellata.*

A. umbellata Lam. var. **umbellata** BEACH SAND-VERBENA Stems to 1 m long. Mar-Aug. Common; sandy flats, stabilized and unstabilized dunes. Widespread locations throughout much of island. All CA Channel Islands except Anacapa and Santa Barbara; n CA (Sonoma Co.) to Baja CA. Variety *breviflora* (Standl.) L.A.Galloway occurs along the mainland coast of OR and n CA.

ONAGRACEAE Juss.
EVENING PRIMROSE FAMILY

Herbaceous or suffrutescent perennials. Ca. 18 genera and 650 species; temperate to subtropical areas worldwide, especially w N America. Plants in many genera (e.g., *Clarkia, Epilobium, Fuchsia,* and *Oenothera*) are cultivated as ornamentals.

CAMISSONIA Link

Ca. 62 species; temperate w N America, with 1 species in S America. (honoring Ludolf Karl Adelbert von Chamisso, 1781-1838, French-born German botanist and plant collector)

1. Suffrutescent perennials; petals mostly 10-20 mm long, usually with 1-2 red dots at base; anthers 2.2-3 mm long; style 13-23 mm long . . . .**C. cheiranthifolia** subsp. **suffruticosa**
1' Perennial herbs; petals mostly 6-10 mm long, rarely with 1-2 red dots at base; anthers 1-1.5 mm long; style 5-10 mm long . . . .**C. cheiranthifolia** subsp. **cheiranthifolia**

C. cheiranthifolia (Spreng.) Raim. subsp. **cheiranthifolia** BEACH-PRIMROSE Stems to 4 dm long. Mar-Jul. Common; sandy flats, slopes, canyon bottoms, and dunes. Widespread locations around perimeter of island and on mesa. All CA Channel Islands except Anacapa and Santa Catalina; OR to s CA (Ventura Co.).
C. cheiranthifolia (Spreng.) Raim. subsp. **suffruticosa** (S.Watson) Raven BEACH-PRIMROSE Stems to 5 dm long. Mar-Jul. Rare; unstabilized and partially stabilized sand dunes. North side in "open dunes one mile ese of Seal Beach"; nw end of mesa, on w side of canyon w of Tufts Road. San Martin Island; central CA (San Luis Obispo Co.) to Baja CA.

OROBANCHACEAE Vent.
BROOM-RAPE FAMILY

Annual or perennial herbs. Ca. 15 genera and 210 species; mostly in temperate areas of N America, Europe, and Asia.

OROBANCHE L. BROOM-RAPE

Ca. 150 species; temperate areas of N America, Europe, and Asia. Roots of several species were eaten by Native Americans. (Greek: vetch strangler, referring to parasitic habit)

1. Inflorescence open; pedicels 15-60 mm long . . . .**O. fasciculata**
1' Inflorescence dense, spike-like; pedicels 2-4.5 mm long . . . .**O. parishii** subsp. **brachyloba**

O. fasciculata Nutt. CLUSTERED BROOM-RAPE Plants to 12 cm tall. Apr-May. Rare; sandy flats. South side on coastal flats near Dutch Harbor and Daytona Beach. All CA Channel Islands except Santa Barbara; AK to Baja CA, e to MI and TX. On San Nicolas Island, this species is apparently parasitic on roots of *Eriogonum.* Entire plants were eaten by the Maidu Tribe.
O. parishii (Jeps.) Heckard subsp. **brachyloba** Heckard SHORT-LOBED BROOM-RAPE Plants to 15 cm tall. Apr-Jul. Common; eroded gullies, ridges, slopes, flats, sandy canyon bottoms, and washes. North

side near Live-forever and Mineral canyons and on coastal flats near Keyhole; s escarpment and coastal flats between w end of mesa and Sand Spit (especially in central portion of island). San Miguel, Santa Rosa, Santa Cruz, and Santa Catalina islands; s CA (San Diego Co.). Detailed distribution maps for this taxon can be found in Junak *et al.* (1996) and Junak (2003c). On San Nicolas Island, this taxon is apparently associated with *Isocoma menziesii*. Subspecies *parishii* occurs from s Sierra Nevada to Baja CA.

OXALIDACEAE R.Br.
OXALIS OR WOOD-SORREL FAMILY

Perennial herbs. Ca. 8 genera and 575 species; temperate to subtropical areas, ± worldwide.

OXALIS L. WOOD-SORREL

Ca. 480 species; temperate to subtropical areas, ± worldwide. Some species are cultivated as ornamentals or for edible tubers; other species are important weeds. Oxalates in foliage can be toxic to humans and livestock. (Greek: sour)

1. Leaflets 5-8 mm long; petioles 1.5-6 cm long; petals 4-8 mm long, light yellow . . . . . .**O. corniculata**
1' Leaflets 10-15 mm long; petioles 4.5-14 cm long; petals 18-25 mm long, bright yellow . . . .
. . . .**O. pes-caprae**

*O. corniculata L. CREEPING WOOD-SORREL Stems to 20 cm long. Apr-Jul. Rare; disturbed sites. Mesa at Nicktown and at airfield. Santa Rosa, Santa Cruz, Santa Catalina, and San Clemente islands; WA to Baja CA, e to ME; Hawaii; native of Europe. First collected at Nicktown in July 1979. This species has been used medicinally.
*O. pes-caprae L. BERMUDA-BUTTERCUP Plants to 30 cm tall. Feb-Mar. Rare; disturbed flats. Mesa at Nicktown, along Owens Road between Monroe Drive and Harrington Road, and s of airfield runways. Santa Cruz, Santa Catalina, and San Clemente islands; n CA (Humboldt Co.) to Baja CA, e to AZ and disjunctly to FL; S America; Australia; New Zealand; Europe; Asia; native of s Africa. First collected along Owens Road in March 1990. This species is toxic to sheep.

PAPAVERACEAE Juss.
POPPY FAMILY

Annual or perennial herbs. Ca. 23 genera and 230 species; temperate to tropical areas of N America, Europe, Asia, and S Africa. Taxa in some genera (e.g., *Dendromecon, Dicentra, Eschscholzia, Papaver,* and *Romneya*) are cultivated as ornamentals. *Papaver somniferum* L. is the natural source of morphine.

1. Sepals usually 2; petals usually 4, yellow to orange; leaves deeply lobed to dissected, cauline ones alternate
. . . .**Eschscholzia**
1' Sepals usually 3; petals usually 6, cream to yellow; leaves entire to minutely toothed, cauline ones (if present) opposite . . . .**Platystemon**

ESCHSCHOLZIA Cham.

Annual or perennial herbs. Ca. 12 species; w N America. *Eschscholzia californica* Cham., a widely cultivated species, is the state flower of California. (honoring Johann Friedrich von Eschscholtz, 1793-1831, Estonian surgeon and naturalist with Russian expeditions to Pacific Coast in 1816 and 1824)

1. Receptacle rim with outer collar, which is spreading to reflexed; petals 20-30 mm long , yellow to orange
. . . .**E. californica**

OXALIDACEAE

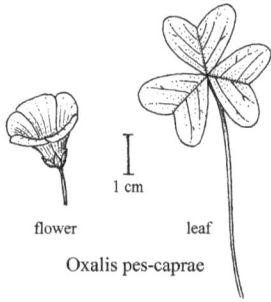

flower   leaf

Oxalis pes-caprae

1 cm

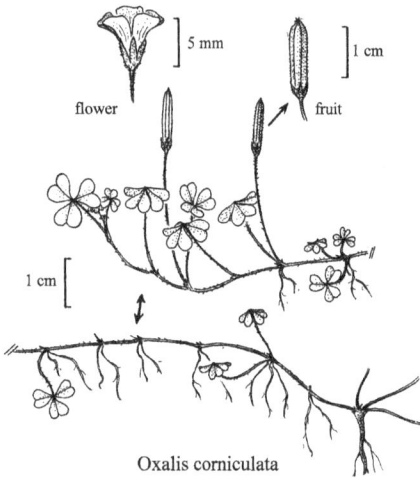

flower   fruit

5 mm   1 cm

1 cm

Oxalis corniculata

PAPAVERACEAE

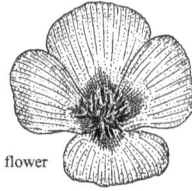

flower

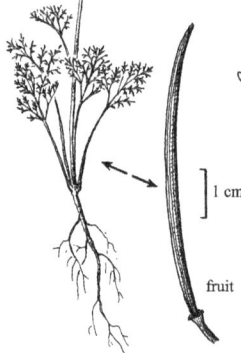

2 cm

Eschscholzia californica

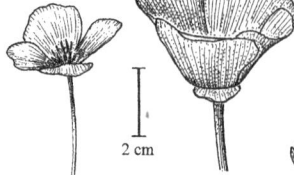

fruit

1 cm

Eschscholzia ramosa

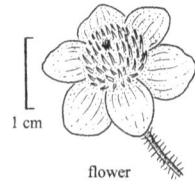

flower

1 cm

fruit

1 cm

2 cm

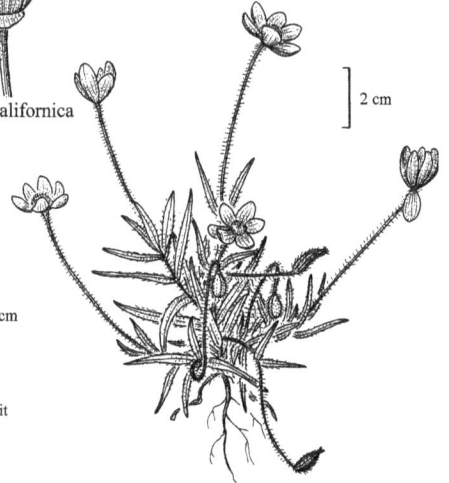

Platystemon californicus

PLANTAGINACEAE

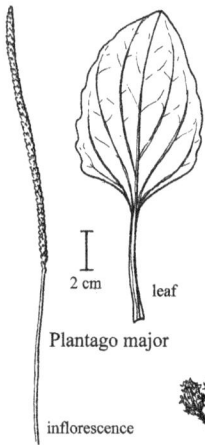

2 cm

leaf

Plantago major

inflorescence

1 mm

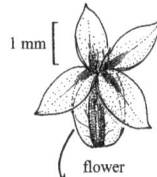

1 mm

flower

1 mm

bract

2 cm

Plantago ovata

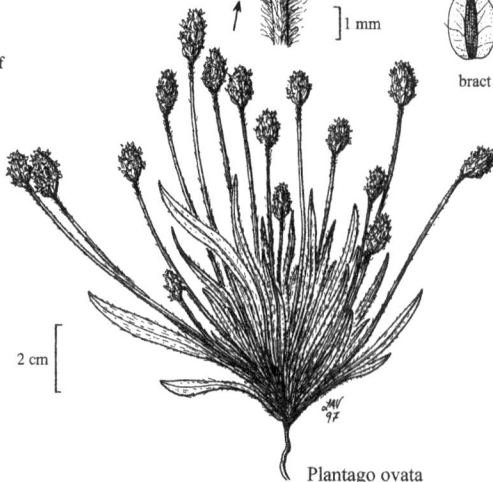

POLEMONIACEAE

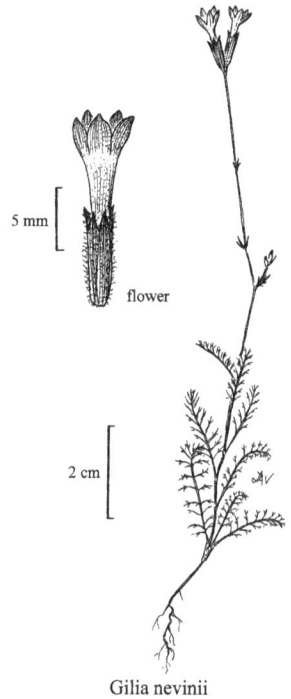

5 mm

flower

2 cm

Gilia nevinii

1' Receptacle without outer collar; petals 5-15 mm long, yellow with an orange dot near base . . . .
. . . .**E. ramosa**

*****E. californica** Cham. subsp. **californica** CALIFORNIA POPPY Plants to 12 cm tall. Apr-May. Rare; disturbed flats. Mesa at antenna farm n of Building 113 and at Nicktown. All CA Channel Islands except Anacapa and Santa Barbara (introduced on San Clemente); Los Coronados, Todos Santos, and Guadalupe islands (possibly introduced on Guadalupe); WA to Baja CA, e to NV; widely introduced in other parts of N America, Hawaii, and New Zealand. Although native to CA, this taxon was introduced to San Nicolas Island, where it was first collected at Nicktown in September 1988. Subspecies *mexicana* (Greene) C.Clark occurs in extreme e CA and ranges e to TX and Mexico.

E. ramosa (Greene) Greene ISLAND POPPY Plants to 15 cm tall. Apr-May. Scarce; dry ridgetops, slopes, and flats. Northeastern escarpment (a single population); widely scattered locations on s escarpment between e fork of Twin Rivers drainage and slopes above Sand Spit. ENDEMIC to all CA Channel Islands (except San Miguel and Anacapa) and all Baja CA islands (except San Geronimo). A detailed distribution map for this taxon can be found in Junak *et al.* (1995b). Most populations of this taxon have been found on sparsely-vegetated ridgetops or exposed slopes on the island, especially in low spots or saddles on ridges. Population sizes appear to be extremely variable from year to year; it is not seen on the island in extremely dry years (Junak 2003c). Distribution of this species on San Nicolas Island appears to be strongly correlated with a particular geologic substrate (a loosely-bedded blue or blue-grey siltstone) and also with earthquake faults (Junak *et al.* 1995b, Junak 2003c).

PLATYSTEMON Benth. CREAM CUPS

Annual herbs. One species. (Greek: wide stamen)

P. californicus Benth. [*P. c.* var. *ornithopus* (Greene) Munz] Plants to 10 cm tall. Feb-May. Occasional; sandy flats and stabilized dunes. North side between Red Eye Beach and vicinity of E Mesa Canyon; s side near Seal Beach, Dutch Harbor, and Daytona Beach. All CA Channel Islands except San Clemente; Guadalupe Island (possibly introduced on Guadalupe); OR to Baja CA, e to UT and AZ. This is a highly variable species with many locally-adapted, intergrading populations; plants from San Miguel, Santa Rosa, and San Nicolas islands that have been called var. *ornithopus* (Greene) Munz and those from Santa Barbara Island that have been called var. *ciliatus* Dunkle are doubtfully distinct. Leaves of this species were eaten by Native Americans.

PLANTAGINACEAE Juss.
PLANTAIN FAMILY

Annual, biennial, or perennial herbs. Ca. 3 genera and 275 species; mostly in temperate areas, ± worldwide.

PLANTAGO L. PLANTAIN

Ca. 270 species; ± worldwide. Some species (e.g., *Plantago afra* L.) are cultivated for use in laxatives. (Latin: footprint or sole of foot)

1. Perennial herbs with clustered fibrous roots; leaves widely elliptic to somewhat cordate . . . .**P. major**
1' Annual or biennial herbs with taproots; leaves linear to narrowly lanceolate
    2. Inflorescences erect, 2-20 cm long (including peduncle); leaves entire or with a few minute teeth
    . . . .**P. ovata**
    2' Inflorescences nodding, becoming erect or wavy in fruit, 5-40 cm long (including peduncle); leaves
      pinnately lobed . . . .**P. coronopus**

*****P. coronopus** L. CUT-LEAVED PLANTAIN Plants to 4 dm tall. Mar-May. Occasional; disturbed flats and roadsides. North side on coastal flats near S Spur Canyon; widely scattered locations on mesa between w end of Tufts Road and airfield. San Miguel, Anacapa, Santa Catalina, and San Clemente islands; WA to Baja CA, e (disjunctly) to TX and MA; Australia; New Zealand; Middle East; native of Europe. First collected on mesa near Jackson Hill in March 1999. This taxon has spread dramatically in recent years and should be eliminated as soon as possible. A detailed map of its distribution on the island can be found in Junak (2003b). Young leaves of this species are edible raw or cooked and have been used medicinally; seeds and seed capsules have been used to produce a laxative.

P. major L. COMMON PLANTAIN Plants to 6 dm tall. Sep. Rare; disturbed flats. Known only from a collection at unspecified location on island in 1985 and from a collection at Nicktown in September 1989. Santa Cruz and Santa Catalina islands; Cedros Island; WA to Baja CA, e to ME and FL; Hawaii; introduced ± worldwide; native of Eurasia. Fruits and leaves of this species have been used medicinally.

P. ovata Forssk. [*P. insularis* Eastw.] Plants to 2 dm tall. Mar-Aug. Common; sandy flats, slopes, swales, and canyon bottoms. Widely scattered locations throughout much of island, especially around perimeter. All CA Channel Islands; Guadalupe, San Benito, Cedros, and Natividad islands; central CA (San Joaquin Co.) to Baja CA, e to UT and AZ and disjunctly to TX; Europe. Possibly introduced on San Nicolas Island. Seed capsules of this species are used to produce a laxative similar to that from *Plantago afra* L. ("psyllium").

POLEMONIACEAE Juss.
PHLOX FAMILY

Annual herbs. Ca. 20 genera and 290 species; mostly in temperate areas of N and S America, Europe, and Asia. Taxa in some genera (e.g., *Cantua, Collomia, Ipomopsis,* and *Phlox*) are cultivated as ornamentals.

GILIA Ruiz & Pav.

Ca. 40 species; w N America and S America. Some species (e.g., *Gilia tricolor*) are cultivated as ornamentals. (honoring Filippo Luigi Gil, 1756-1821, Italian naturalist, clergyman, and director of Vatican Observatory)

G. nevinii A.Gray ISLAND GILIA Plants to 3 cm tall. Apr. Rare; sandy sites. Known only from collections of very depauperate plants made in April 1897 and April 1901; these were found on "sand cliffs above a brackish stream" in 1897 and at an unspecified locality in 1901. ENDEMIC to Santa Rosa, Santa Cruz, Anacapa, Santa Barbara, San Nicolas, Santa Catalina, San Clemente, and Guadalupe islands.

POLYGONACEAE Juss.
KNOTWEED OR BUCKWHEAT FAMILY

Annual herbs, perennial herbs, or shrubs. Ca. 46 genera and 1100 species; worldwide, especially in n temperate areas. Taxa in some genera are cultivated for ornamental use (e.g., *Antigonon, Coccoloba, Eriogonum, Muehlenbeckia,* and *Polygonum*) or for food (e.g., *Fagopyrum, Rheum*).

1. Stipules present, fused into hyaline sheath surrounding stem
    2. Perianth parts 5, petaloid, greenish with white to pink margins, erect . . . .**Polygonum**
    2' Perianth parts 6, sepaloid, green, outer 3 spreading, inner 3 ± erect . . . .**Rumex**
1' Stipules absent
    3. Perennial herbs or shrubs with ± erect stems; leaves basal or cauline, alternate; flowers in dense, often capitate, clusters . . . .**Eriogonum**
    3' Annuals with prostrate stems; leaves cauline, opposite; flowers axillary . . . .**Pterostegia**

ERIOGONUM Michx. WILD BUCKWHEAT

Perennial herbs or shrubs. Ca. 240 species; N America. Several taxa, including some endemic to the CA Channel Islands (e.g., *Eriogonum arborescens, E. giganteum* var. *giganteum,* and *E. grande* var. *rubescens*),

POLYGONACEAE

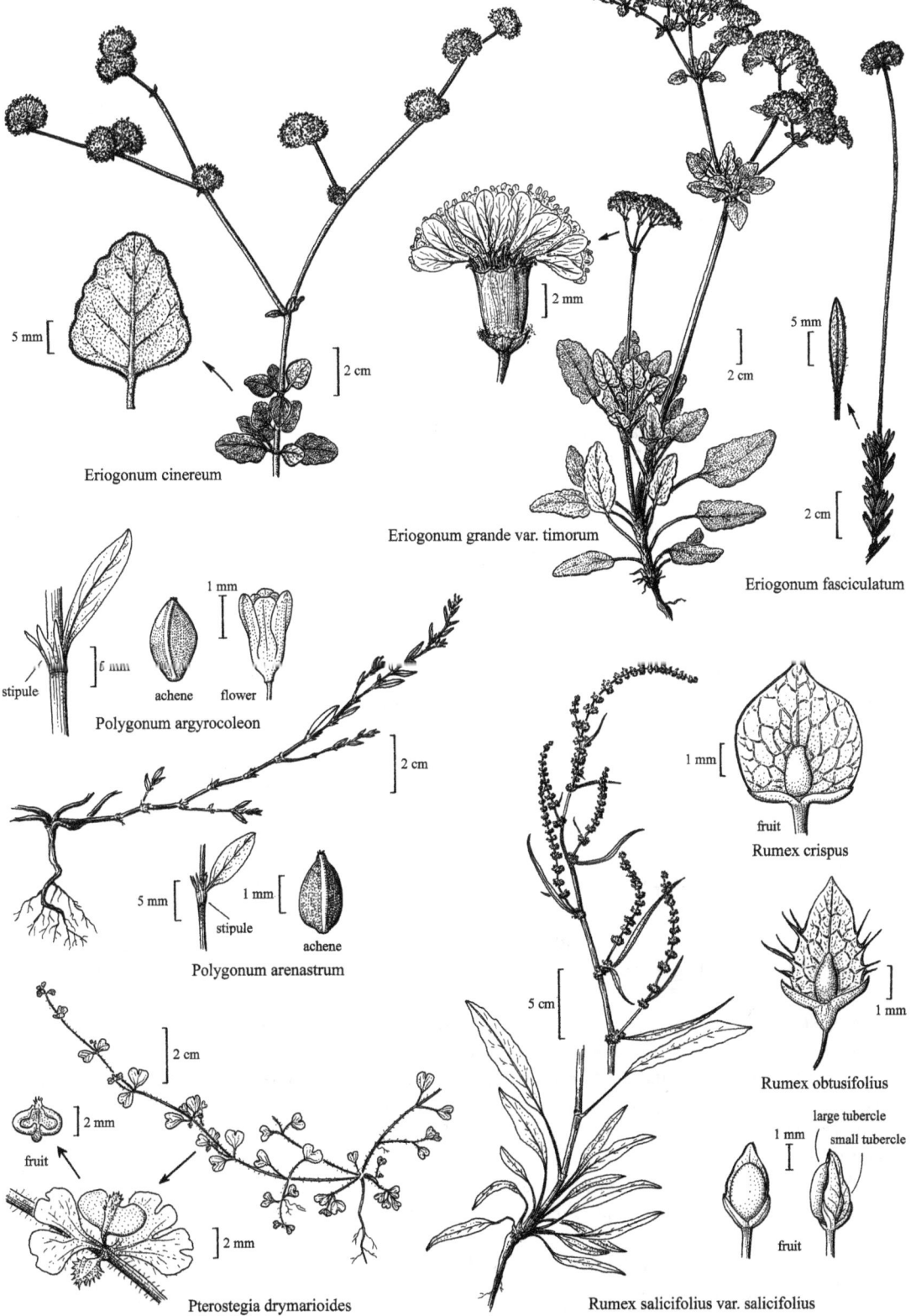

Eriogonum cinereum

5 mm

2 cm

Eriogonum grande var. timorum

2 mm

2 cm

5 mm

2 cm

Eriogonum fasciculatum

stipule

achene

flower

1 mm

5 mm

Polygonum argyrocoleon

2 cm

5 mm

stipule

1 mm

achene

Polygonum arenastrum

1 mm

fruit

Rumex crispus

1 mm

Rumex obtusifolius

2 cm

2 mm

fruit

2 mm

Pterostegia drymarioides

5 cm

Rumex salicifolius var. salicifolius

large tubercle

small tubercle

1 mm

fruit

are cultivated as ornamentals. Roots, leaves, and young stems of some species were eaten by Native Americans. (Greek: woolly joint or knee, referring to hairy nodes of type species)

1. Plants essentially herbaceous, only woody at base; leaves in dense clusters near base .....**E. grande**
1' Plants definitely shrubby; leaves well-distributed along stems
    2. Leaves clustered, ± linear; leaf blades 7-20 mm long, 2-4 mm wide ....**E. fasciculatum**
    2' Leaves borne singly, ovate to oblong-ovate, leaf blades 15-30 mm long, 8-15 mm wide ....
    ....**E. cinereum**

***E. cinereum** Benth. ASHY-LEAF BUCKWHEAT Plants to 8 dm tall. Feb-Aug. Rare; flats and slopes. South side on s escarpment near Jackson Hill (below Building 112). Santa Rosa Island; s CA (Summerland area to Palos Verdes peninsula). First noted near Building 112 in August 1984, but already well-established at that time. Although native to CA, this taxon was introduced to the island, has naturalized, and is spreading. This taxon, along with *E. fasciculatum,* should be removed from San Nicolas Island immediately. *Eriogonum cinereum* and *E. fasciculatum* are in the same subgenus as the endemic *E. grande* var. *timorum* (Reveal 1989) and may pose a threat to its genetic integrity if hybridization occurs.
***E. fasciculatum** Benth. var. **polifolium** (A.DC.) Torr. & A.Gray EASTERN MOJAVE BUCKWHEAT Plants to 1 m tall. Feb-Aug. Rare; flats and slopes. Mesa near Jackson Hill (near Building 113); s side on escarpment below Building 112. Central CA (San Joaquin Co.) to Baja CA; e to UT and AZ. First noted near Building 112 in August 1984, but already well-established at that time. Although native to CA, this taxon was introduced to the island, has naturalized, and is spreading. This taxon, as well as *E. cinereum,* should be removed from San Nicolas Island immediately. Three other varieties of *E. fasciculatum* occur on CA mainland.
E. grande Greene var. **timorum** Reveal SAN NICOLAS ISLAND BUCKWHEAT Plants to 6 dm tall. Mar-Jun. Abundant; dry slopes and canyon walls. North side near Corral Harbor and Sand Spit; extreme se end of mesa; s side near Grand Canyon and Twin Rivers drainage and on escarpment and coastal flats between e side of Dutch Harbor and Sand Spit. ENDEMIC to San Nicolas Island. Detailed distribution maps for this taxon can be found in Junak et al. (1995b) and Junak (2003c). Two other varieties are endemic to other CA Channel Islands; another is endemic to nw Baja CA mainland and Todos Santos Island.

POLYGONUM L. KNOTWEED or SMARTWEED
Annual herbs. Ca. 300 species; worldwide, especially in temperate areas of N Hemisphere. Dried rhizomes of some species have been used medicinally; leaves and stems of several species are cooked for food. (Greek: many joints or knees, referring to swollen nodes of some species)

1. Stems prostrate to decumbent; perianth 1.5-2 mm long; achenes minutely striate, dull ..**P. arenastrum**
1' Stems erect to ascending; perianth 2-3 mm long; achenes smooth, shiny ....**P. argyrocoleon**

***P. arenastrum** Boreau [*P. aviculare* L. misapplied] COMMON KNOTWEED Stems to 3 dm long. May-Jun. Scarce; disturbed flats. Scattered locations on mesa, especially near airfield. All CA Channel Islands except San Miguel and Santa Barbara; Cedros Island; AK to Baja CA, e to ME and FL; Hawaii; native of Europe. First confirmed collection was made at airfield in June 1987.
***P. argyrocoleon** Steud. ex Kunze PERSIAN KNOTWEED Plants to 8 dm tall. May-Nov. Scarce; disturbed flats. North side near Thousand Springs; on mesa at old borrow pit along Monroe Drive. Santa Cruz, Santa Catalina, and San Clemente islands; n CA (Yolo Co.) to Baja CA, e to TX and disjunctly to MA and FL; Hawaii; native of Asia. First collected at Thousand Springs in August 1969.

PTEROSTEGIA Fisch. & C.A.Mey.
Annual herbs. One species. (Greek: winged cover, referring to involucre surrounding the pistillate flowers)

P. drymarioides Fisch. & C.A.Mey. FAIRY MIST Stems to 4 dm long. Mar-May. Occasional; moist flats and slopes, often under shrubs. North side on ne coastal flats and on ne escarpment; mesa in upper reaches of E Mesa Canyon and along n edge of mesa w of airfield. All CA Channel Islands; Los Coronados, Todos Santos, San Martin, Guadalupe, and Cedros islands; OR to Baja CA, e to NM. Population sizes of this taxon appear to be extremely variable from year to year; it is not seen on the island in extremely dry years.

RUMEX L. DOCK or SORREL

Perennial herbs. Ca. 200 species; mostly in temperate areas of N America, Europe, and Asia. Several species have edible parts, medicinal uses, or practical uses (e.g., stain removal, leather tanning). Most species contain soluble oxalates and are toxic to livestock (especially sheep) if eaten in large quantities. (ancient Latin name for *R. acetosella* L.)

1. Margins of inner perianth parts toothed in fruit . . . .**R. obtusifolius**
1' Margins of inner perianth parts entire in fruit
    2. Inner perianth parts 2-3 mm long in fruit, each with a tubercle, one of which is much wider than the other two . . . .**R. salicifolius**
    2' Inner perianth parts 4-5 mm long in fruit, without tubercles . . . .**R. crispus**

*R. crispus L. subsp. **crispus** CURLY DOCK Plants to 1 m tall. Mar-Jun. Occasional; disturbed sites and wet areas. Widely scattered locations in canyons on n escarpment and on mesa. All CA Channel Islands except Santa Barbara; AK to Baja CA, e to ME and FL; Hawaii; native of Europe. First collected at Jackson Hill in April 1966. Young leaves can be boiled and eaten; seeds can be ground into flour; roots have been used medicinally.
*R. obtusifolius L. BITTER DOCK Plants to 13 dm tall. Apr-Jun. Rare; moist areas. Mesa on n side of airfield runways. Alaska to Baja CA, e to ME and FL; Hawaii; native of Europe. First collected on mesa in June 1987. Young leaves can be boiled and eaten.
R. salicifolius Weinm. var. **salicifolius** WILLOW DOCK Plants to 5 dm tall. Apr-Jun. Scarce; moist canyon bottoms, dune swales, and seeps. North side between Tender Beach and Corral Harbor (especially in Tule Creek area) and in lower portion of W Mesa Canyon; s side at Army Springs. All CA Channel Islands except Anacapa and Santa Barbara; n CA (Del Norte Co.) to Baja CA. Six other varieties are known from mainland CA and elsewhere in N America. Seeds were eaten by the Maidu and Miwok tribes.

PORTULACACEAE Juss.
PURSLANE FAMILY

Annual herbs. Ca. 20 genera and 400 species; temperate areas of the Americas, Australia, and s Africa. Taxa in some genera (e.g., *Calandrinia, Lewisia,* and *Portulaca*) are cultivated as ornamentals.

1. Cauline leaves 2, fused and perfoliate, sessile . . . .**Claytonia**
1' Cauline leaves several, free, alternate, usually petiolate . . . .**Calandrinia**

CALANDRINIA Kunth

Ca. 14 species; temperate areas in w N America and w S America. (honoring Jean Louis Calandrini, 1703-1758, Swiss botanist and professor of mathematics and philosophy)

C. ciliata (Ruiz & Pav.) DC. [*C. c.* var. *menziesii* (Hook.) J.F.Macbr.] RED-MAIDS Stems to 15 cm long. Mar-May. Rare; flats and canyon slopes. Northeastern portion of mesa, n and w of airfield. All CA Channel Islands; Todos Santos, San Martin, and Guadalupe islands; British Columbia to Baja CA, e to NM and Sonora, Mexico; Central America (Guatemala); S America. Seeds, roots, and leaves were eaten by Native Americans.

CLAYTONIA L.

Ca. 28 species; N America and e Asia. (honoring John Clayton, 1694-1773, colonial American botanist)

1. Most basal leaves linear to narrowly oblanceolate, blades gradually tapered to bases, blade length more than 3 times width . . . . **C. parviflora** subsp. **parviflora**
1' Most basal leaves elliptic to ± cordate, blade bases rounded to cordate, blade length less than 3 times width . . . .**C. perfoliata** subsp. **mexicana**

C. parviflora Hook. subsp. **parviflora** [*C. perfoliata* var. *parviflora* (Hook.) Torr.] Plants to 12 cm tall. Feb-Apr. Scarce; moist slopes and banks. North side in canyons of n escarpment between Corral Harbor area and vicinity of "L" Canyon; n edge of mesa n of airfield. San Miguel, Santa Cruz, Santa Barbara, and Santa Catalina islands; British Columbia to Baja CA, e to MT and AZ. Three other subspecies occur on CA mainland.

C. perfoliata Willd. subsp. **mexicana** (Rydb.) J.M.Miller & K.L.Chambers MINER'S LETTUCE Plants to 2 dm tall. Feb-Apr. Occasional; moist slopes, shaded hillsides, and canyon bottoms. North side in canyons and on hillsides of n escarpment between Tender Beach and airfield; n edge of mesa between Corral Harbor and airfield and on moist flats e of Nicktown. All CA Channel Islands; Los Coronados, Todos Santos, San Martin, Guadalupe, and Cedros islands; n CA (Humboldt Co.) to Baja CA, e to NM; Mexico; Central America (Guatemala). Can be abundant in wet years. Entire plant is edible.

PRIMULACEAE Vent.
PRIMROSE FAMILY

Annual or perennial herbs. Ca. 21 genera and 825 species; ± worldwide, especially in temperate areas of N Hemisphere. Taxa in several genera (e.g., *Cyclamen, Primula*) are cultivated as ornamentals.

1. Leaves in basal rosettes; inflorescence an umbel . . . .**Primula**
1' Leaves cauline, opposite; flowers solitary, axillary . . . .**Anagallis**

ANAGALLIS L. PIMPERNEL

Annual herbs. Ca. 28 species; temperate areas of S America, Europe, and Africa. Some species are cultivated as ornamentals. This genus is probably best placed in the Myrsinaceae (Manns and Anderberg 2005). (ancient Greek name for pimpernel)

*****A. arvensis** L. SCARLET PIMPERNEL Stems to 3 dm long. Feb-Apr (Jun). Scarce; disturbed flats. Mesa in vicinity of Nicktown and at airfield; s side on s escarpment at antenna near Building 176. All CA Channel Islands; Guadalupe Island; WA to Baja CA, e to ME and FL; Hawaii; native of Europe. First collected at airfield in June 1987. A cleansing soap can be made from flowers and foliage, but entire plant is toxic to humans if ingested. This species has been used medicinally.

PRIMULA L. PRIMROSE

Perennial herbs. Ca. 445 species; mostly in temperate areas of N Hemisphere. Roots, stems, and leaves of some species (e.g., *P. hendersonii*) were roasted and eaten by Native Americans. Species formerly in *Dodecatheon* are now considered to be part of the genus *Primula* (Mast and Reveal 2007). (Latin: diminutive of first, referring to early flowers)

P. clevelandii (Greene) A.R.Mast & Reveal subsp. **insulare** (H.J.Thomps.) A.R.Mast & Reveal [*Dodecatheon c.* Greene subsp. *i.* (Greene) H.J.Thomps.] SHOOTING STAR Plants to 5 dm tall. Feb-Mar (Jun). Scarce; moist slopes. Widely scattered locations on coastal flats and ne escarpment between Nicktown area and Sand Spit. All CA Channel Islands except Santa Barbara; Guadalupe Island; central CA (Monterey Co.) to s CA (San Diego Co.). On San Nicolas Island, this taxon is known only from isolated individuals

PORTULACACEAE

Calandrinia ciliata

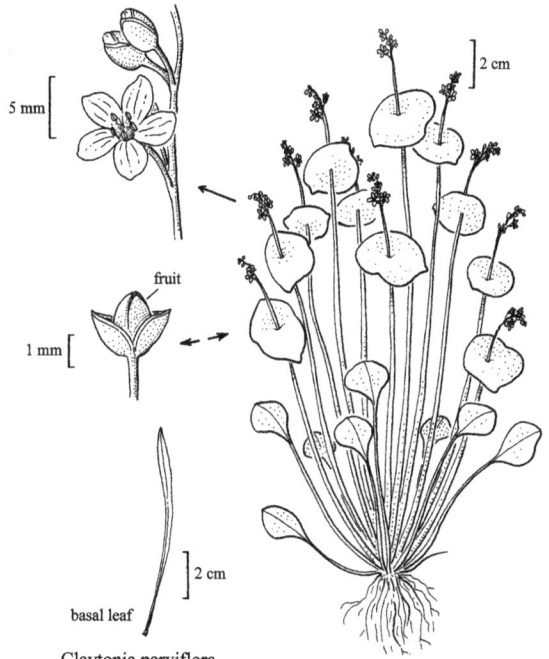

Claytonia parviflora
subsp. parviflora

Claytonia perfoliata subsp. mexicana

PRIMULACEAE

RESEDACEAE

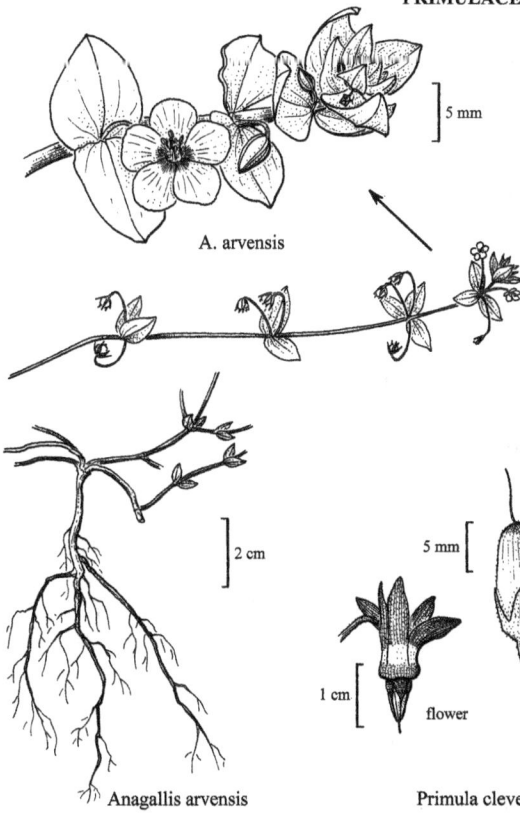

A. arvensis

Anagallis arvensis

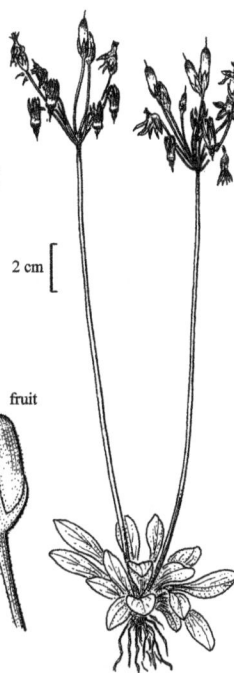

Primula clevelandii subsp. insulare

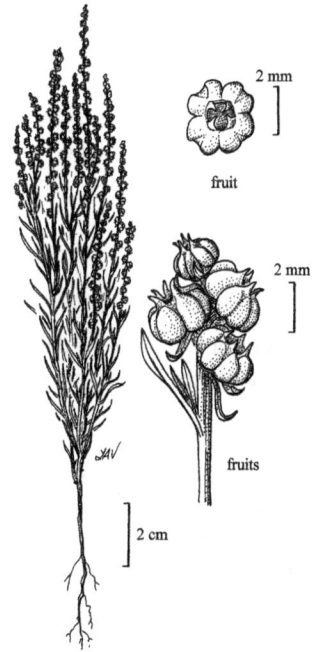

Oligomeris linifolia

or very small populations. Three other subspecies occur on the CA mainland.

RESEDACEAE Gray
MIGNONETTE FAMILY

Annual herbs. Ca. 6 genera and 80 species; temperate areas of N America, Europe, Africa, and Asia. Several taxa in the genus *Reseda* are cultivated as ornamentals.

OLIGOMERIS Chambess. OLIGOMERIS

Ca. 3 species; dry temperate areas of N Hemisphere and in s Africa. (Greek: small parts)

O. linifolia (M.Vahl) J.F.Macbr. Plants to 3 dm tall. Apr-Jun (Aug). Occasional; barren ridgetops, dry eroded slopes, saline flats, and sand dunes. Widely scattered locations around much of island's perimeter; along s rim of mesa s of airfield. All CA Channel Islands; all Baja CA islands except San Martin and San Geronimo; central CA (Santa Barbara Co.) to Baja CA, e to TX; Eurasia.

RUBIACEAE Juss.
MADDER OR COFFEE FAMILY

Annual herbs. Ca. 630 genera and 10,200 species; worldwide, especially in temperate to tropical areas. Many taxa are cultivated for economic products (e.g., coffee, quinine) or for ornamental use (e.g., *Bouvardia, Gardenia*).

GALIUM L. BEDSTRAW or CLEAVERS

Ca. 300 species; worldwide, especially temperate areas. Cheese rennet (*Galium verum* L.) contains a milk-curdling enzyme which has been used in cheese production; the species also yields a yellow dye which was used for coloring cheese and butter. (Greek: milk)

G. aparine L. CLEAVERS or GOOSE-GRASS Stems to 13 dm long. Mar-Apr. Occasional; flats, slopes, and canyon bottoms. Widely scattered locations on ne escarpment and on mesa. All CA Channel Islands; Los Coronados, San Martin, Guadalupe, and Cedros islands; AK to Baja CA, e to ME and FL; ± worldwide. This species may have been introduced to San Nicolas Island. It is often found in *Opuntia* patches and was first collected on the island "about *Opuntia*" in April 1901. Fruits can be roasted and used as coffee substitute.

SALICACEAE Mirbel
WILLOW OR POPLAR FAMILY

Shrubs or trees. Two genera and ca. 435 species; mostly in N Hemisphere temperate areas. Some species are cultivated as ornamentals or used for wood.

SALIX L. WILLOW

Ca. 400 species; ± worldwide, mostly in arctic to temperate areas. Salicin and related compounds used in pharmaceutical products (especially aspirin) were formerly extracted from the bark of some species. Ancient Asian records indicate that willow bark has been used to reduce pain for at least 2400 years. (ancient Latin name)

1. Leaves linear to linear-lanceolate, 2-11 mm wide, subsessile, both surfaces gray-tomentose or lanate . . . .**S. exigua**
1' Leaves lanceolate to elliptic or oblanceolate, 10-30 mm wide, petiolate, upper surfaces pubescent, lower surfaces glabrous, glaucous, or tomentose . . . .**S. lasiolepis**

RUBIACEAE

SALICACEAE

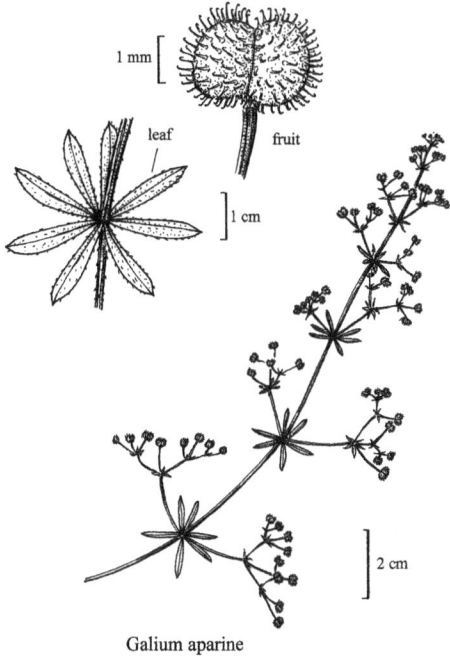

Galium aparine

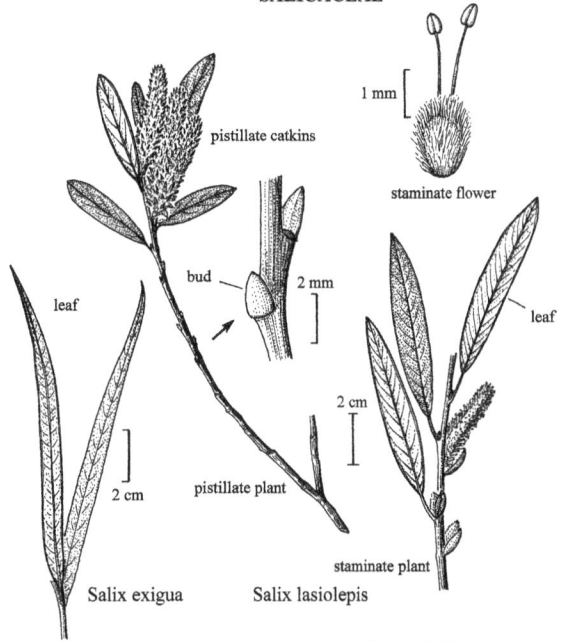

Salix exigua   Salix lasiolepis

SCROPHULARIACEAE

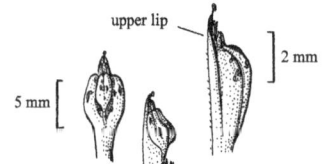

flower

SAURURACEAE

SAXIFRAGACEAE

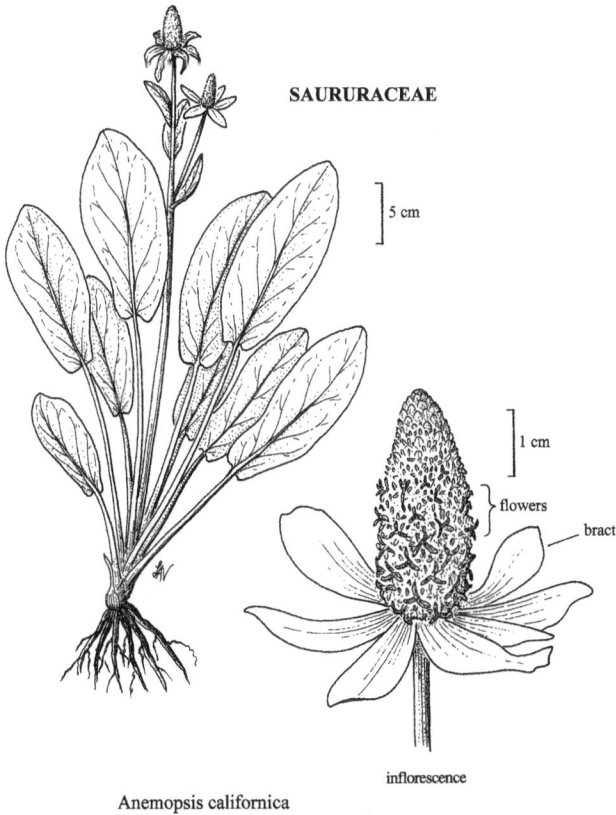

Anemopsis californica   inflorescence

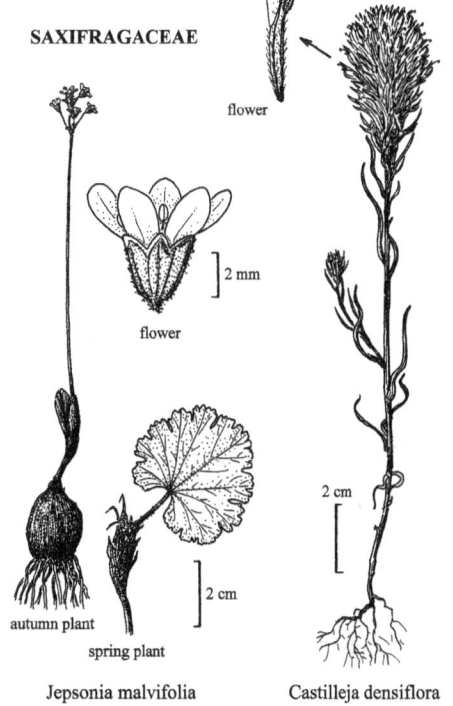

Jepsonia malvifolia   Castilleja densiflora

S. exigua Nutt. [*S. hindsiana* Benth.] SAND-BAR WILLOW Plants to 15 dm tall. Mar-May. Rare; canyon bottom. S side in lower portion of Grand Canyon. Santa Cruz and Santa Catalina islands; WA to Baja CA, e to TX. Bark and stems were used medicinally by Native Americans.

S. lasiolepis Benth. ARROYO WILLOW Plants to 6 m tall. Mar-May. Scarce; moist sites and canyon bottoms. Mesa at wells area on Tufts Road and in upper portion of W Mesa Canyon near Nicktown; s side at Army Springs. All CA Channel Islands except Santa Barbara and San Clemente; WA to Baja CA, e to ID and TX. Bark and leaves were used medicinally by Native Americans.

SAURURACEAE E.Meyer
LIZARD'S-TAIL FAMILY

Perennial herbs. Ca. 4 genera and 6 species; N America and e Asia, mostly in temperate areas. Some species are cultivated for ornamental use.

ANEMOPSIS Hook. & Arn. YERBA MANSA

One species. (Greek: anemone-like, referring to the inflorescence)

A. californica (Nutt.) Hook. & Arn. Plants to 3 dm tall. Apr-Aug. Rare; moist seeps on coastal bluffs. North side at Red Eye Beach. Santa Cruz (introduced), Santa Catalina, and San Clemente islands; Cedros Island; OR to Baja CA, e to TX; Mexico. Bark and roots were used medicinally by Native Americans.

SAXIFRAGACEAE Juss.
SAXIFRAGE FAMILY

Perennial herbs. Ca. 35 genera and 660 species; arctic, alpine, and temperate areas ± worldwide, mostly in N Hemisphere. Taxa in several genera (e.g., *Bergenia, Heuchera, Saxifraga, Tellima,* and *Tolmiea*) are cultivated as ornamentals.

JEPSONIA Small

Three species; California and n Baja California. (for Willis Linn Jepson, 1867-1946, CA botanist)

J. malvifolia (Greene) Small ISLAND JEPSONIA Plants to 3 dm tall. Oct-Dec. Occasional; clay slopes and canyon walls. Northeastern escarpment between Live-forever Canyon and Sand Spit area. ENDEMIC to Santa Rosa, Santa Cruz, San Nicolas, Santa Catalina, San Clemente, and Guadalupe islands.

SCROPHULARIACEAE Juss.
FIGWORT FAMILY

Annual herbs. Ca. 200 genera and 3000 species; mostly in temperate areas, ± worldwide. Taxa in some genera (e.g., *Antirrhinum, Gambelia, Hebe, Mimulus,* and *Penstemon*) are cultivated for ornamental uses; others are cultivated for medicinal uses (e.g., *Digitalis*). Many authors (e.g., Olmstead 2002) would divide the Scrophulariaceae into several families and place *Castilleja* in the Orobanchaceae.

CASTILLEJA L.f. PAINTBRUSH or OWL'S-CLOVER

Ca. 200 species; mostly w N America. Seeds of several species were eaten by Native Americans; entire plants of some species were also used medicinally. (honoring Domingo Castillejo, 1744-1793, botanist and instructor of botany at Cadiz, Spain)

C. densiflora (Benth.) T.I.Chuang & Heckard [*Orthocarpus d.* Benth.] OWL'S-CLOVER Plants to 4 dm tall. Mar-May. Occasional; flats. North side on ne coastal flats near W and E Mesa canyons; n edge of mesa between Nicktown and airfield area. San Miguel, Santa Rosa, and Santa Cruz islands; n CA

(Mendocino Co.) to Baja CA.

SOLANACEAE Juss.
NIGHTSHADE OR TOMATO FAMILY

Annual herbs, perennial herbs, suffrutescent perennials, or shrubs. Ca. 94 genera and 2950 species; temperate to tropical areas worldwide. Several species are widely cultivated for food (e.g., peppers, potatoes, tomatillos, and tomatoes) or as ornamentals (e.g., *Brugmansia, Petunia*). Many species are toxic, especially tobacco and the deadly nightshades.

1. Shrubs, with spiny, woody twigs; leaves clustered at stem nodes . . . . **Lycium**
1' Annuals or perennial herbs, stems sometimes woody near base; leaves not clustered
    2. Corollas ± salverform; anthers not connivent; fruit a capsule . . . . **Nicotiana**
    2' Corollas ± rotate; anthers connivent; fruit a berry . . . . **Solanum**

LYCIUM L. BOX-THORN

Shrubs. Ca. 100 species; warm temperate areas of N America and Europe. Raw or dried fruits of several species were eaten by Native Americans; dried roots also were baked and eaten by some tribes. (Latin: Lycia, ancient country of Asia Minor)

1. Seeds 2; leaf ± round in cross-section . . . . **L. californicum**
1' Seeds several-many; leaf flat to elliptic in cross-section
    2. Stamens exserted, attached near middle of corolla tube . . . . **L. brevipes**
    2' Stamens barely exserted, attached at top of corolla tube . . . . **L. verrucosum**

L. brevipes Benth. DESERT BOX-THORN Plants to 1 m tall. Apr. Rare; canyon walls. Known only from middle portion of Celery Canyon. San Clemente Island; Todos Santos, San Martin, San Geronimo, and Cedros islands; s CA to Baja CA, e to AZ.
L. californicum Nutt. CALIFORNIA BOX-THORN Plants to 1 m tall. Mar-Apr. Abundant; eroded slopes, canyon rims, slopes, and flats. North side, primarily between Live-forever canyon area and Sand Spit, with a few scattered plants near Thousand Springs and Corral Harbor; widely scattered locations on e half of mesa, especially near the edges; s side near Daytona Beach and Sand Spit, with isolated plants in Twin Rivers drainage and between Dutch Harbor and Daytona Beach. All CA Channel Islands except Santa Rosa; Los Coronados, Todos Santos, San Martin, Guadalupe, San Benito, and Natividad islands; s CA (Palos Verde peninsula, Los Angeles Co.) to Baja CA. This taxon is often associated with *Opuntia* patches. A detailed distribution map for this taxon can be found in Junak (2003a).
L. verrucosum Eastw. SAN NICOLAS ISLAND BOXTHORN Plants to ca. 24 dm tall. Apr. Rare; "arroya cliffs". This species is known only from a single collection made in April 1897 and is now presumed to be extinct. ENDEMIC to San Nicolas Island, where it reportedly "grew in several localities on arroya cliffs, with its branches hanging over the arroyas ..." (Eastwood 1898). This taxon needs further study and may be an aberrant form of *L. brevipes* (Hitchcock 1932, Chiang-Cabrera 1981).

NICOTIANA L. TOBACCO

Shrubs. Ca. 67 species; N and S America, Pacific islands, Australia, and s Africa. The natural source of nicotine; some species are also grown as ornamentals or for insecticide production. Although leaves of several species were smoked and used medicinally by Native Americans, all species contain alkaloids and can be toxic to humans. (honoring Jean Nicot, 1530-1600, French ambassador to Lisbon who introduced tobacco into France)

*N. glauca Graham TREE TOBACCO Plants to 2 m tall. Apr-May. Rare; disturbed flats and roadsides. North side on coastal flats near Rock Jetty and near Sand Spit; mesa at old landfill site near Nicktown; sw end of island at S.L.A.M. missile target site. Santa Cruz, Santa Catalina, and San Clemente islands; Guadalupe Island; n CA coast ranges to Mexico, e to Sierra Nevada foothills; native of Argentina. First collected on island at "rocket launch site at western edge of mesa" in November 1971. This species contains the alkaloid anabasine and is extremely toxic to humans and livestock.

SOLANUM L. NIGHTSHADE

Annual herbs, perennial herbs, or suffrutescent perennials. Ca. 1700 species; ± worldwide, especially tropical America. Some species are cultivated for food (e.g., eggplant, pepinos, potatoes, and tomatoes) or as ornamentals. Foliage and unripe fruits of most species are toxic to humans, especially children. (Latin: quieting, referring to narcotic properties of some species)

1. Leaves compound; corollas yellow . . . .*S. lycopersicum*
1' Leaves simple; corollas white to purplish
    2. Corollas 3-6 mm wide; anthers generally 1.4-2.2 mm long . . . .*S. americanum*
    2' Corollas 8-15 mm wide; anthers generally 2.5-4 mm long . . . .*S. douglasii*

*S. americanum Mill. [*S. nodiflorum* Jacq.] WHITE NIGHTSHADE Plants to 1 m tall. Mar-Sep. Occasional; moist sites, seeps, and flats. North side at Red Eye Beach, Thousand Springs, Tule Creek, and on ne coastal flats near Rock Jetty and Sand Spit; mesa in W Mesa Canyon near Nicktown and at airfield; s side at Army Springs. Santa Rosa, Anacapa, and San Clemente islands; Los Coronados, Todos Santos, San Martin, and Guadalupe islands; WA to Baja CA, e to FL; Hawaii; Mexico; S America. Although this taxon may be native to N America (which is questionable), it was probably introduced to San Nicolas Island. First collected on island at Tule Creek and at Nicktown in April 1961.

S. douglasii Dunal DOUGLAS' NIGHTSHADE Plants to 1 m tall. Mar-May. Rare; flats and swales. North escarpment near Nicktown; nw portion of mesa near Tufts Road. All CA Channel Islands; Guadalupe Island; n CA (Humboldt Co.) to Baja CA, e to MS. Leaves and fruits were used medicinally by Native Americans.

*S. lycopersicum L. var. **lycopersicum** [*Lycopersicon esculentum* Mill.] TOMATO Stems to 3 m long. Feb-Apr. Rare; flats. Known from a single collection in Nicktown in February 1990. San Miguel, Santa Cruz, Anacapa, Santa Barbara, and San Clemente islands; Los Coronados, San Geronimo, and Cedros islands; OR to Baja CA, e to AZ and disjunctly to ME and FL; Hawaii; native of S America. Ripe fruits are edible; unripe fruits, stems, and leaves are toxic to humans and livestock.

TAMARICACEAE Link
TAMARISK FAMILY

Trees. Ca. 4 genera and 78 species; warm temperate areas of Europe, Asia, and n Africa.

TAMARIX L. TAMARISK

Ca. 54 species; Europe, Asia, and n Africa. Some species are cultivated as ornamentals or for wind breaks but are invasive weeds, especially along stream banks and irrigation canals. Galls on some species are rich in tannins and are used for tanning leather; "manna" of the Bedouin is produced as a result of punctures in stems made by scale insects. (early Latin name)

*T. ramosissima Lebed. [*T. pentandra* Pall.] SALTCEDAR Plants to 5 m tall. Apr-May (Oct). Scarce; moist sites and canyon bottoms. North side near Mineral Canyon; mesa in gully west of upper Tule Creek (e of Building 110), Nicktown (planted there), and at old borrow pit along Monroe Drive. San Miguel, Santa Rosa, Santa Cruz, Santa Catalina, and San Clemente islands; Cedros Island; n CA (Shasta Co.) disjunctly to Baja CA, e to MS and disjunctly to VA; native of e Asia. First collected w of intersection of Tufts and Shannon roads in October 1983. These trees pose a threat to native vegetation in mesic sites (Neill 1985) and should not be allowed to spread on the island.

SOLANACEAE

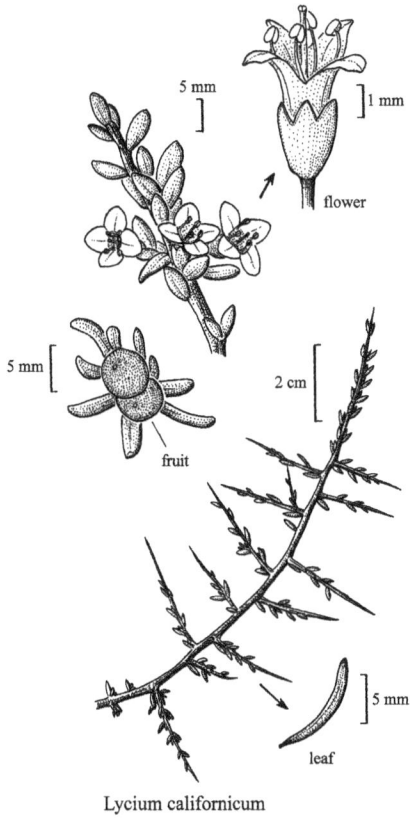

Lycium californicum

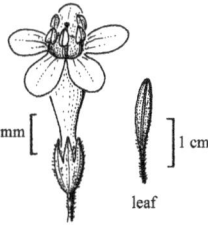

Lycium brevipes

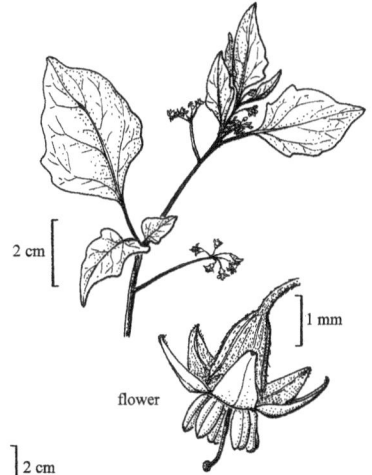

Solanum americanum

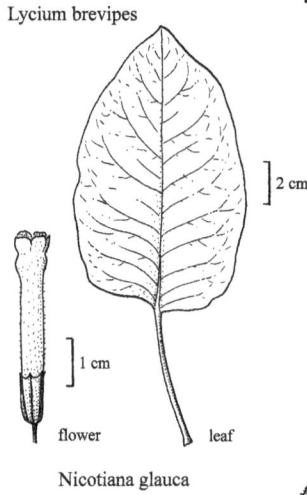

Nicotiana glauca

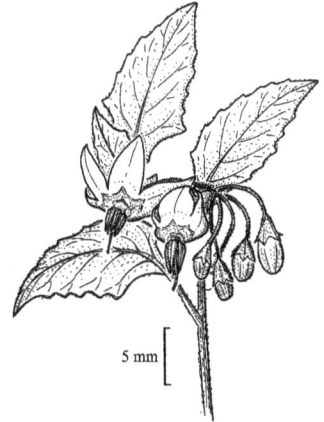

Solanum douglasii

TAMARICACEAE

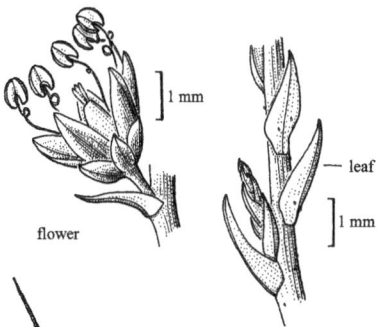

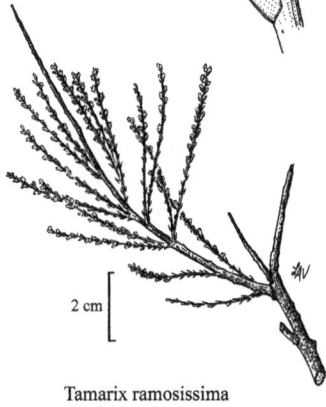

Tamarix ramosissima

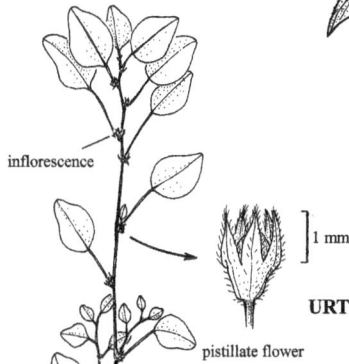

URTICACEAE

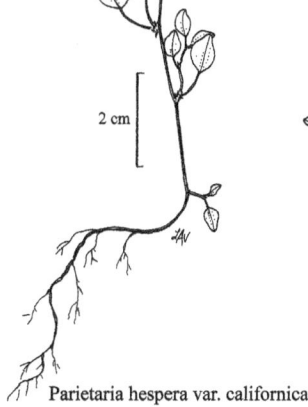

Parietaria hespera var. californica

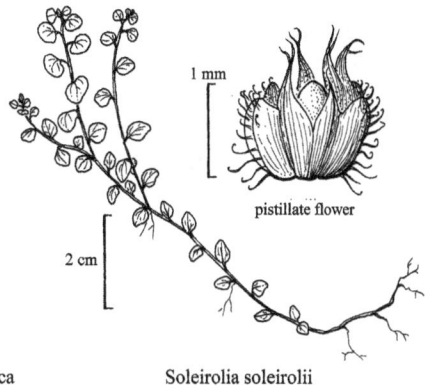

Soleirolia soleirolii

URTICACEAE Juss.
NETTLE FAMILY

Annual or perennial herbs. Ca. 48 genera and 1050 species; temperate to tropical areas ± worldwide. Taxa in some genera (e.g., *Pilea, Soleirolia*) are cultivated as ornamentals; taxa in some genera are grown for fiber (e.g., *Boehmeria, Urtica*).

1. Flowers solitary; stems rooting at nodes . . . .**Soleirolia**
1' Flowers in cymes; stems not rooting at nodes . . . .**Parietaria**

PARIETARIA L. PELLITORY

Annual herbs. Ca. 10 species; ± worldwide. (Latin: of walls, referring to habitat of some species)

P. hespera B.D.Hinton var. **californica** B.D.Hinton WESTERN PELLITORY Stems to 7 dm long. Mar-Jun. Scarce; shaded slopes and flats. Widely scattered locations on ne coastal flats and escarpment between E Mesa Canyon and Rock Jetty area; s escarpment near Twin Towers (Building 186) and in second canyon e of Towers Canyon. All CA Channel Islands; Los Coronados, Todos Santos, San Martin, Guadalupe, and Cedros islands; central CA (Monterey Co.) to Baja CA. On San Nicolas Island, this taxon is usually on slopes shaded by *Coreopsis gigantea* or in *Opuntia* patches. Variety *hespera* occurs throughout much of sw N America.

SOLEIROLIA Gaudich. BABY'S TEARS

Perennial herbs. One species. (honoring Captain Joseph Francois Soleirol, 1781-1863, botanical collector in Corsica)

*S. soleirolii (Req.) Dandy [*Helxine s.* Req.] MOTHER-OF-THOUSANDS Stems to 1 dm long. Mar-May. Rare; moist cliff face. Known only from Thousand Springs. Northern CA (Lake Co.) to s CA (San Diego Co.); native of Corsica and Sardinia. First collected at Thousand Springs in April 1961.

VERBENACEAE J.St.-Hil.
VERBENA OR VERVAIN FAMILY

Suffrutescent perennials. Ca. 41 genera and 950 species; worldwide, mostly in temperate to tropical areas. Taxa in some genera (e.g., *Clerodendron, Lantana,* and *Verbena*) are cultivated as ornamentals; others are grown for wood. Recent evaluations of Lamiales would move many genera in Verbenaceae to Lamiaceae (mint family).

VERBENA L.

Ca. 200 species; temperate to tropical areas of N and S America, Europe, and n Africa. Several species have been used medicinally; some are cultivated as ornamentals. (Latin name for leaves and shoots of laurel, myrtle, and other plants used in religious ceremonies and also in medicine)

*V. lasiostachys Link var. **lasiostachys** WESTERN VERBENA Stems to 7 dm long. Apr-Jun. Rare; sandy flats. Known only from mesa (at wells area along Tufts Road). Santa Catalina and San Clemente islands; OR to Baja CA. Although native to mainland CA, this taxon was probably introduced to San Nicolas Island. First collected on mesa in May 1985. Variety *scabrida* Moldenke occurs on the Northern Channel Islands and CA mainland.

VERBENACEAE

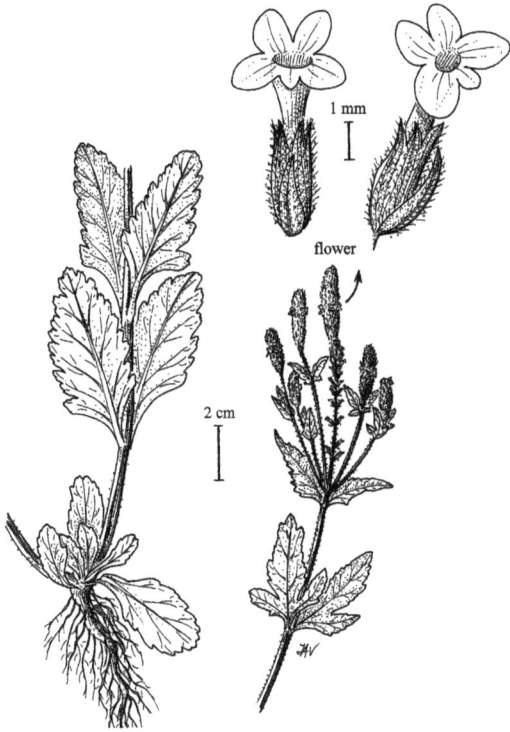

Verbena lasiostachys var. scabrida

ALLIACEAE

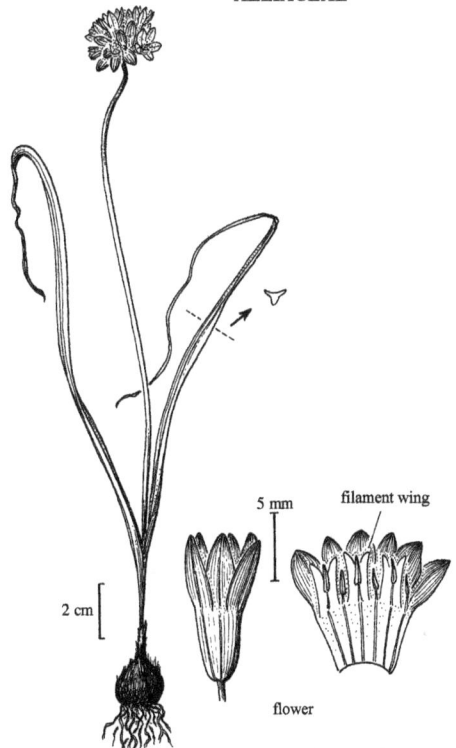

Dichelostemma capitatum subsp. capitatum

CYPERACEAE

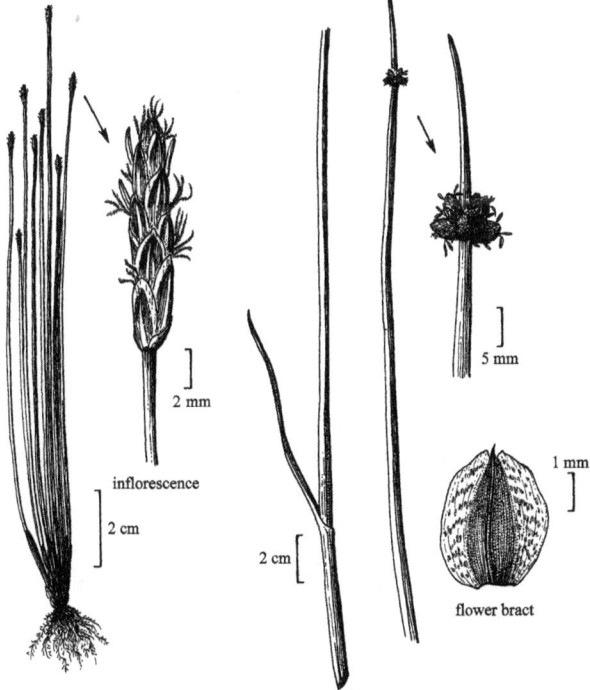

Eleocharis macrostachya

Schoenoplectus americanus

JUNCACEAE

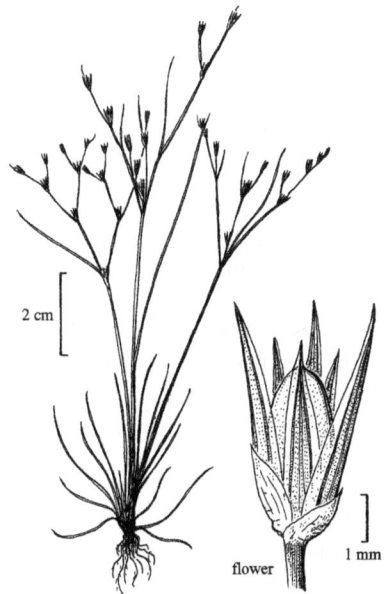

Juncus bufonius

MONOCOTYLEDONOUS ANGIOSPERMS (MONOCOTS)

ALLIACEAE J.Agardh
ONION FAMILY

Perennial herbs. Ca. 30 genera and 850 species; ± worldwide, mostly in temperate areas. This family includes many genera formerly placed in Amaryllidaceae, which differ primarily by having an inferior ovary. Taxa in some genera (e.g., *Allium*) are cultivated for food or as ornamentals (e.g., *Ipheion, Nothoscordum*).

DICHELOSTEMMA Kunth

Ca. 5 species; w N America. (Greek: toothed crown, referring to stamen appendages)

D. capitatum (Benth.) A.W.Wood subsp. **capitatum** [*Brodiaea pulchella* (Salisb.) Greene, *D. p.* (Salisb.) A.A.Heller] BLUE DICKS Plants to 8 dm tall. Mar-Jun. Common; canyon bottoms, moist slopes, flats, and ridge tops. Widely scattered locations on n side between Nicktown and Sand Spit area, on ne and se edges of mesa, and on s escarpment between Twin Rivers drainage and Sand Spit. All CA Channel Islands; all Baja CA islands except San Geronimo and Natividad; OR to Baja CA, e to UT and NM; n Mexico. Bulbs were eaten raw or cooked by Native Americans. Subspecies *pauciflorum* (Torrey) Keator occurs in CA deserts and sw N America.

CYPERACEAE Juss.
SEDGE FAMILY

Annual or perennial herbs. Ca. 98 genera and 4350 species; temperate to tropical areas worldwide. Some taxa are used for thatching, basketry, or paper-making.

1. Summit of achene with tubercle of different color and texture; involucral leaves none . . . .**Eleocharis**
1' Summit of achene beaked, tubercle none; involucral leaf present, appearing to be a continuation of the stem
. . . .**Schoenoplectus**

ELEOCHARIS R.Br. SPIKERUSH

Perennial herbs. Ca. 120 species; worldwide. Some species have edible corms (e.g., Chinese water-chestnut); leaves of some species are used to make matting or rustic clothing. (Greek: marsh grace, referring to typical habitat)

E. macrostachya Britton PALE SPIKERUSH Plants to 1 m tall. Mar-Jul. Scarce; vernal ponds and moist flats. Mesa n and w of airfield runways, especially at old borrow pit along Monroe Drive. Santa Rosa, Santa Catalina, and San Clemente islands; AK to Baja CA, e to GA.

SCHOENOPLECTUS (Reichb.) Palla NAKED-STEMMED BULRUSH

Perennial herbs. Ca. 77 species; worldwide. Seeds, pollen, and rhizomes of several species were eaten by Native Americans; culms were used to construct water bottles, dwellings, and other items. (Greek: twisted or woven rush or reed, referring to use of culms in functional items)

S. americanus (Pers.) Schinz & R.Keller [*Scirpus americanus* Pers., *S. olneyi* A.Gray] THREE-SQUARE Plants to 2 m tall. Mar-Jul. Rare; moist canyon bottom. North side in Tule Creek. Santa Rosa Island; AK to Baja CA, e to ME and FL; Mexico; Central and S America.

JUNCACEAE Juss.
RUSH FAMILY

Annual herbs. Ca. 7 genera and 430 species; worldwide, mostly in arctic and cold temperate areas. Some taxa are cultivated as ornamentals; some are used for producing matting.

JUNCUS L. RUSH

Ca. 300 species; worldwide, mostly in temperate areas of N Hemisphere. Stems of some species were roasted, boiled, or eaten raw by Native Americans; stems of several species were used in basketry. (Latin: to join or bind, referring to use of stems)

J. bufonius L. TOAD RUSH Plants to 3 dm tall. Mar-Apr. Scarce; vernal ponds, moist flats, and swales. North side in Tule Creek; widely scattered locations on mesa (such as wells area along Tufts Road, vernal ponds between Nicktown area and airfield, and old borrow pit along Monroe Drive). All CA Channel Islands except Santa Barbara; Guadalupe Island; AK to Baja CA, e to ME and FL; Hawaii; ± worldwide.

POACEAE Juss. [Gramineae Juss.]
GRASS FAMILY

Annual or perennial herbs. Ca. 668 genera and 9500 species; worldwide. A family of great economic importance, providing food (e.g., corn, rice, and wheat), ornamental plants, noxious weeds, and species used for building, thatching, and weaving. Caryopses of many native taxa were eaten by Native Americans.

1. Spikelets enclosed within upper leaf sheaths, 1-3; plants matted, stems stoloniferous; leaves strongly over lapping, sheaths concealing internodes . . . .**Pennisetum**
1' Spikelets usually many, 1-few only on short plants, inflorescences well-developed, conspicuous, usually not concealed by leaf sheaths; leaves not usually strongly overlapping
   2. Plants usually taller than 1.5 m
      3. Leaf blade margins entire, smooth; plants and spikelets bisexual . . . .**Arundo**
      3' Leaf blade margins serrulate; plants dioecious, spikelets unisexual . . . .**Cortaderia**
   2' Plants usually shorter than 1.5 m
      4. Most spikelets sessile, sometimes sunken in inflorescence axis; inflorescence either solitary and spikate or paniculate and composed of spicate branches; spikelets often in 2 rows on opposite sides of axis or borne along 1 side of inflorescence axis . . . .**GROUP 1**
      4' Most spikelets short- to long-pedicellate; inflorescence usually racemose or paniculate, open (if appearing spicate then spikelets short-pedicellate); spikelets evenly distributed around inflorescence axis, solitary or in small clusters, but not in 2 rows or along 1 side of axis
         5. Most or all spikelets with 1 fertile floret, sometimes with sterile florets much reduced in size or different from fertile ones . . . .**GROUP 2**
         5' Most or all spikelets with 2-many fertile florets . . . .**GROUP 3**

**GROUP 1.   Most fertile spikelets sessile or sunken in inflorescence axis,
often in 2 rows on opposite sides of axis or borne along 1 side of inflorescence axis.**

1. Inflorescence composed of 2-7 spicate branches, each with many spikelets; spikelets evenly distributed around inflorescence axis or borne along 1 side of axis . . . .**Cynodon**
1' Inflorescence solitary
   2. Spikelets on 1 side of inflorescence axis; inflorescence axis conspicuously flattened and thickened
   . . . .**Stenotaphrum**
   2' Spikelets usually in 2 rows on opposite sides of axis or solitary; inflorescence axis not conspicuously flattened and thickened
      3. Spikelets in clusters of 2-3 at each node . . . .**Hordeum**

3' Spikelets solitary at each node

    4. Spikelets sunken in inflorescence axis; each spikelet with 1 bisexual floret; plants of coastal, often moist and saline, habitats . . . .**Parapholis**

    4' Spikelets sessile to subsessile, not usually sunken in inflorescence axis; each spikelet with 2-many bisexual florets; plants of interior, often dry habitats

        5. Spikelets in 2 rows on opposite sides of axis

            6. Flat surface of compressed spikelet facing inflorescence axis; inflorescence axis often breaking apart in fruit; each spikelet with 2-5 florets . . . .**Triticum**

            6' Edge of compressed spikelet facing inflorescence axis; inflorescence axis not breaking apart in fruit; each spikelet with 8-14 florets . . . .**Lolium**

        5' Spikelets not in 2 rows

            7. Spikelets densely clustered; plants strongly rhizomatous; leaves in 2 conspicuous ranks; lemmas awnless . . . .**Distichlis**

            7' Spikelets solitary; annuals or perennial herbs without extensive rhizomes; leaves not conspicuously ranked; lemmas awned

                8. Lemmas apices acute, awned from tip; palea margins conspicuously stiff-ciliate . . . .**Brachypodium**

                8' Lemma apices bidentate, awned from between teeth; palea margins glabrous to ciliolate, but not stiff-ciliate . . . .**Bromus**

GROUP 2. Spikelets pedicellate, evenly distributed around inflorescence axis, solitary or in small clusters; most or all spikelets with 1 fertile floret.

1. Spikelets in distinct clusters that fall together; clusters composed of 1 fertile spikelet subtended or surrounded by 1-5 sterile spikelets

    2. Spikelets spreading to nodding; fertile spikelet with lemmas awned from bidentate apices; plants 10-40 cm tall . . . .**Lamarckia**

    2' Spikelets erect; fertile spikelet awnless, lemma apices acute to acuminate; plants 30-100 cm tall . . . .**Phalaris paradoxa**

1' Spikelets not in distinct clusters; spikelets all alike, disarticulating either below glumes or between glumes and fertile floret

    3. Spikelets disarticulating below glumes, falling entire from their pedicels . . . .**Polypogon**

    3' Spikelets usually disarticulating above glumes, floret falling from spikelet

        4. Lemma of fertile floret awnless

            5. Floret 1 per spikelet . . . .**Ammophila**

            5' Florets 2-7 per spikelet

                6. Inflorescence spicate, dense, cylindrical; lemma of fertile floret puberulent . . . .**Phalaris**

                6' Inflorescence usually paniculate, open; lemma of fertile floret glabrous to minutely scabrous . . . .**Melica**

        4' Lemma of fertile floret awned dorsally near base, near middle, or from apex

            7. Awn less than 5 mm long; lemmas ± thin, often translucent . . . .**Piptatherum**

            7' Awn 10-110 mm long; lemmas thick, becoming hard in fruit

                8. Inflorescence open, branches spreading to nodding; lemma slightly constricted near apex; floret callus sharply acute . . . .**Nassella**

                8' Inflorescence dense, branches appressed; lemma tapered to apex; floret callus obtuse . . . .**Achnatherum**

GROUP 3. Spikelets pedicellate, evenly distributed around inflorescence axis, solitary or in small clusters; most or all spikelets with 2-many fertile florets.

1. One or both glumes usually as long as or longer than lowest floret, sometimes as long as spikelet; lemmas dorsally awned or awnless
    2. Lemmas awned from near middle, awns 15-40 mm long; glumes 15-30 mm long; lemmas 10-25 mm long . . . .**Avena**
    2' Lemmas acute to short-awned from apex, awns less than 1.5 mm long; glumes 2-12 mm long; lemmas 2-12 mm long, acute to short-awned, awn less than 1.5 mm long
        3. Perennial herbs 20-70 dm tall; leaf blades 2-5 cm wide; lemmas villous . . . .**Arundo**
        3' Annuals less than 2 dm tall; leaf blades less than 10 cm wide; lemmas glabrous or ciliate . . . .**Schismus**
1' Glumes usually shorter than lowest floret; lemmas awned from apices or awnless
    4. Stems usually matted, with scaly rhizomes; inflorescences unisexual, staminate ones usually exceeding leaves, pistillate ones equal to or shorter than uppermost leaves . . . .**Distichlis**
    4' Stems not matted, rhizomes if present, not scaly; inflorescences bisexual
        5. Lemma apices rounded to acute, awnless . . . .**Poa**
        5' Lemmas awned from tip, awns 0.5-55 mm long (sometimes absent in *Festuca*)
            6. Spikelets strongly compressed; lemmas dorsally keeled . . . .**Bromus**
            6' Spikelets not strongly compressed, lemmas dorsally rounded
                7. Lemmas lanceolate to narrowly ovate, apices bidentate, awned from between teeth . . . .**Bromus**
                7' Lemmas narrowly lanceolate to attenuate, tapered to awned tip
                    8. Perennial herbs, with short rhizomes; awns 0.5-1.5 mm long, sometimes absent . . . .**Festuca**
                    8' Annuals; awns 2-12 mm long
                        9. Ligules puberulent; palea margins conspicuously stiffly ciliate; spikelets 1-3 per inflorescence . . . .**Brachypodium**
                        9' Ligules glabrous to minutely scabrous; palea margins glabrous to ciliolate, but not stiffly ciliate; spikelets usually many per inflorescence . . . . .**Vulpia**

ACHNATHERUM P.Beauv. NEEDLEGRASS
Perennial herbs. Ca. 75 species; temperate areas worldwide. (Greek: awned scale, referring to lemma)

A. diegoense (Swallen) Barkworth [*Stipa d.* Swallen] SAN DIEGO NEEDLEGRASS Plants to 14 dm tall. Feb-May. Abundant; canyon walls and moist slopes. North side in canyons of n escarpment between Live-forever Canyon area and Sand Spit; s side in canyons of s escarpment between Twin Rivers drainage and Sand Spit. San Miguel, Santa Rosa, Santa Cruz, and Anacapa islands; Los Coronados and Todos Santos islands; s CA (San Diego Co.) to Baja CA. Detailed distribution maps for this taxon can be found in Junak *et al.* (1995b) and Junak (2003c).

AMMOPHILA Host BEACHGRASS
Perennial herbs. Two species; coastal areas of e N America and Europe. (Greek: sand-loving)

*****A. arenaria** (L.) Link EUROPEAN BEACHGRASS Plants to 15 dm tall. May-Jun. Scarce; sand dunes. Mesa near Shannon Road n of Tufts Road; s side near Army Springs. WA to s CA (Ventura Co.); PA (historical occurrence); Hawaii; native of n and w Europe. First collected near head of Tule Canyon in April 1961, where it was intentionally introduced for dune stabilization. This aggressive grass poses a threat to native dune vegetation (Sweet 1985) and should be eliminated on San Nicolas Island if possible. This species has been planted for dune stabilization and is also used for making baskets and brooms.

ARUNDO L.

Perennial herbs. Three species; warm temperate to subtropical areas of Europe and Asia. (Latin: reed)

*A. donax L. GIANT REED Plants to 45 dm tall. Rarely flowers. Scarce; seeps, ravines, canyon bottoms, flats, and beaches. Mesa near wells area along Tufts Road, in Tule Creek, at Nicktown, near fire station at intersection of Owens Road and Monroe Drive, n of airfield, and at old borrow pit on Monroe Drive; s side at w end of Daytona Beach. San Miguel, Santa Rosa, Santa Cruz, and Santa Catalina islands; Cedros Island; n CA (Butte Co.) to Baja CA, e to FL; Hawaii; New Zealand; S America (Argentina and Chile); native of India. First collected on s side of island "near road e of Dutch Harbor light" in July 1965. Eradication efforts were initiated at Tule Creek in 2000 (Junak 2003b). This species was introduced to CA mainland by Mission padres for roof construction. Stems have also been used for making musical instruments and for constructing houses, lattice-work, screens, and other useful items (e.g., walking sticks and fishing poles).

AVENA L. OATS

Annual herbs. Ca. 29 species; temperate and cold areas of Europe, n Africa, and central Asia. Some species are cultivated for grain and hay. (Latin: oats)

1. Lemmas pubescent throughout, apical teeth setaceous, 2-5 mm long . . . .**A. barbata**
1' Lemmas subglabrous to glabrous over upper half, apical teeth acute, 0.5-1 mm long . . . .**A. fatua**

*A. barbata Link SLENDER WILD OATS Plants to 7 dm tall. Mar-Jun. Common; flats and slopes. Widespread locations throughout much of island. All CA Channel Islands; Los Coronados, Todos Santos, Guadalupe, and Cedros islands; OR to Baja CA, e to MT and NM; native of Europe and central Asia. First reported for island by Eastwood (1898).
*A. fatua L. WILD OATS Plants to 8 dm tall. Apr-Jun. Common; flats and slopes. Widespread locations throughout much of island, but on s side it is found primarily in canyon bottoms and in Opuntia patches. All CA Channel Islands; Los Coronados, Todos Santos, San Martin, Guadalupe, and Cedros islands; AK to Baja CA, e to ME; temperate areas ± worldwide; native of Europe and central Asia. First collected at "one locality about *Opuntia*" in April 1897.

BRACHYPODIUM P.Beauv.

Annual herbs. Ca. 18 species; Mexico and Eurasia.

*B. distachyon (L.) P.Beauv. [Trachynia distachya (L.) Link] PURPLE FALSE-BROME Plants to 3 dm tall. Mar-May. Rare; disturbed flats. Known only from one site on mesa (parking area between Buildings 176 and 273). Santa Rosa, Santa Cruz, Santa Catalina, and San Clemente islands; n CA (Tehama Co.) to Baja CA, e (disjunctly) to TX and NJ; Hawaii; Australia; native of s Europe. First collected on island in March 2000.

BROMUS L. BROME

Annual or perennial herbs. Ca. 150 species; temperate areas worldwide. Some species are grown for hay, but others have long, rough awns that can injure grazing animals. (ancient Greek name meaning oats or food)

1. Spikelets strongly compressed; glumes and lemmas dorsally keeled
    2. Upper glume ± equal in length to lower lemma (excluding awn); lemma margins puberulent below
        middle; annuals . . . .**B. arizonicus**
    2' Upper glume usually shorter than lower lemma (excluding awn); lemmas glabrous to puberulent
        or minutely scabrous throughout
        3. Lemma awns 1-3 mm long; glumes and lemmas usually minutely scabrous . . .**B. catharticus**

POACEAE: Achnatherum-Avena

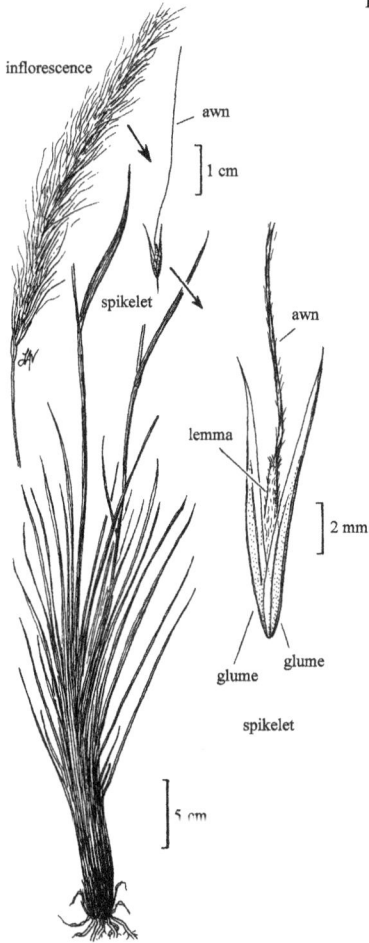

inflorescence

awn

1 cm

spikelet

awn

lemma

2 mm

glume glume

spikelet

5 cm

Achnatherum diegoense

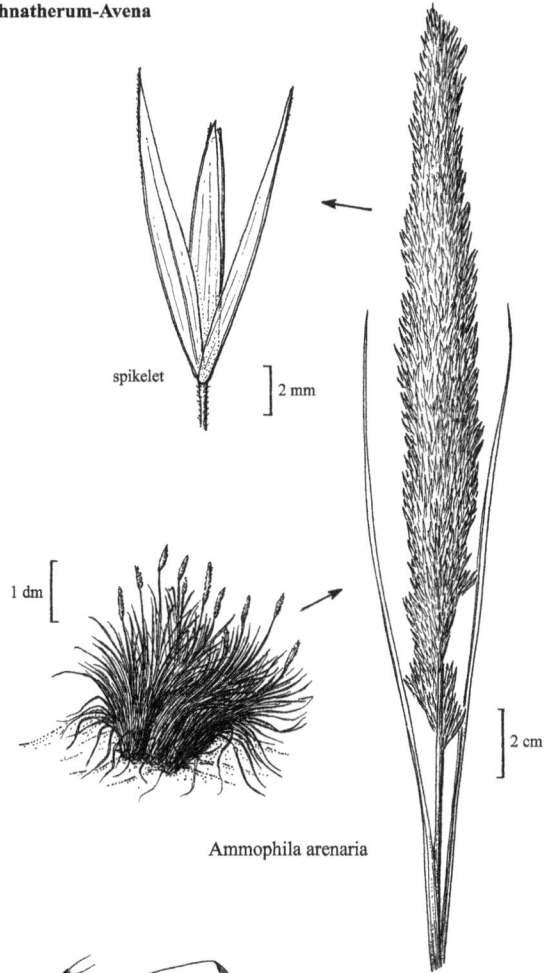

spikelet

2 mm

1 dm

Ammophila arenaria

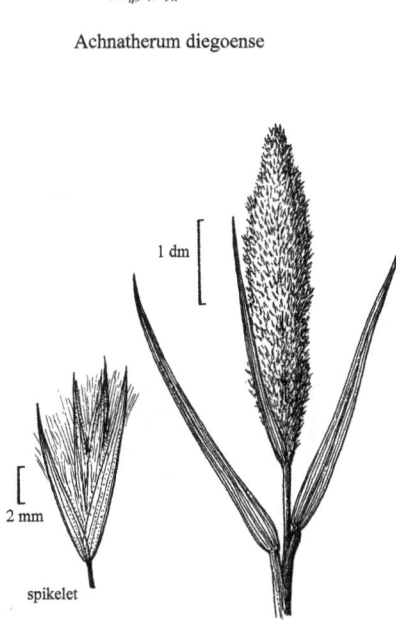

2 cm

1 dm

2 mm

spikelet

Arundo donax

2 cm

5 mm

spikelet

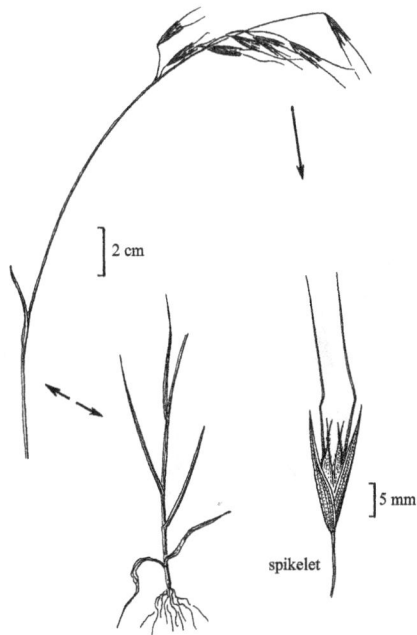

Avena barbata

5 mm

spikelet

Avena fatua

POACEAE: Brachypodium-Bromus

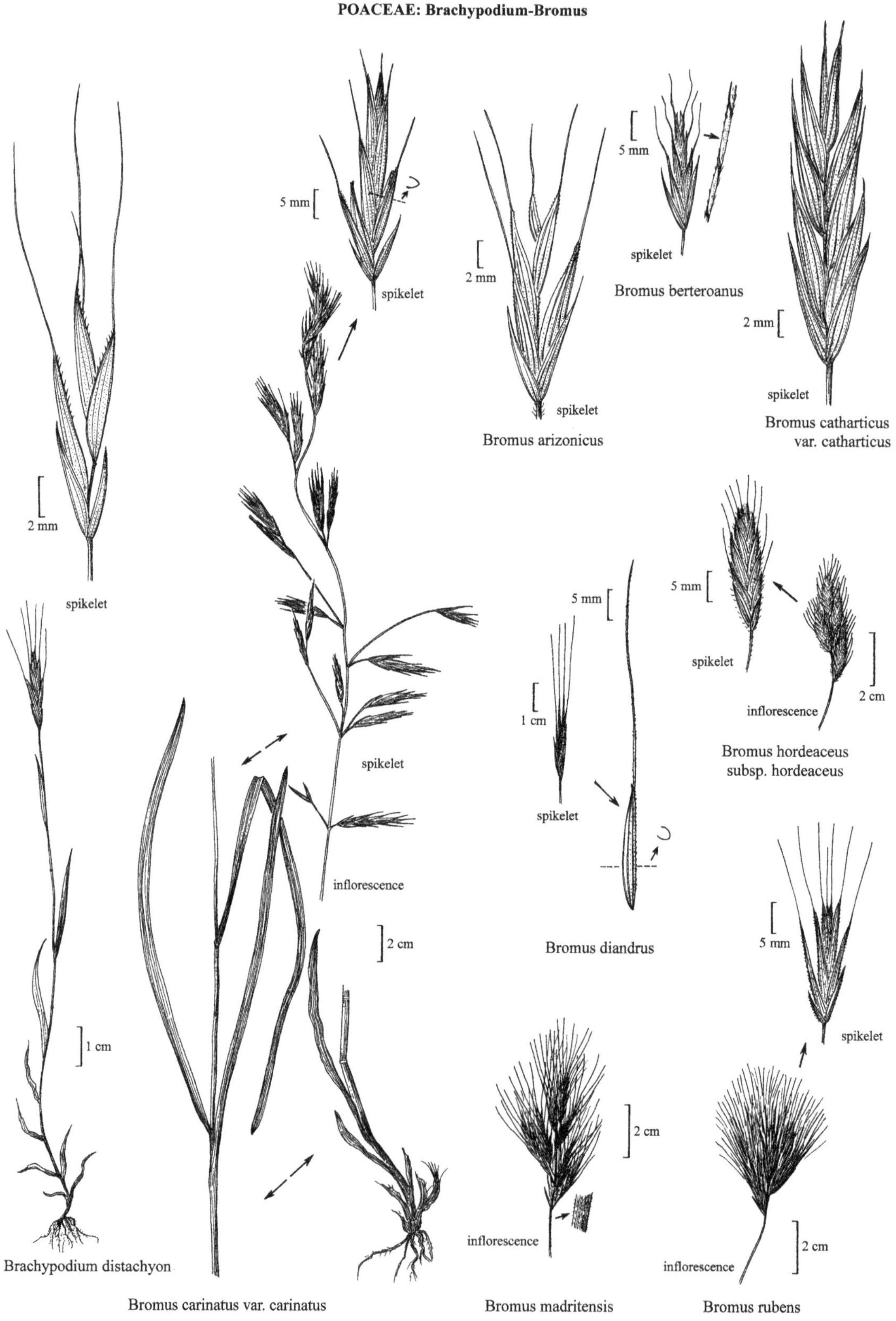

5 mm

spikelet

5 mm

spikelet

Bromus berteroanus

2 mm

spikelet

Bromus arizonicus

2 mm

spikelet

Bromus catharticus
var. catharticus

2 mm

spikelet

5 mm

spikelet

inflorescence

2 cm

Bromus hordeaceus
subsp. hordeaceus

5 mm

1 cm

spikelet

spikelet

Bromus diandrus

spikelet

2 mm

spikelet

inflorescence

2 cm

1 cm

Brachypodium distachyon

Bromus carinatus var. carinatus

inflorescence

2 cm

Bromus madritensis

5 mm

spikelet

inflorescence

2 cm

Bromus rubens

3' Lemma awns 4-15 mm long; glumes and lemmas glabrous to puberulent
    4. Inflorescence open, lower branches widely spreading, usually longer than spikelets, pedicels 5-24 mm long; most spikelets not overlapping ....**B. carinatus**
    4' Inflorescence ± contracted, lower branches ascending, usually shorter than spikelets, pedicels 2-8 mm long; most spikelets overlapping ....**B. maritimus**
1' Spikelets terete to slightly compressed; glumes and lemmas dorsally rounded
    5. Lemma awns bent, distal part divergent ....**B. berteroanus**
    5' Lemma awns straight, ascending to erect, not divergent
        6. Inflorescence open, lower branches spreading to drooping ....**B. diandrus**
        6' Inflorescence somewhat open to dense, branches ascending to erect
            7. Lemma awns 4-9 mm long, apical teeth 0.5-1.5 mm long; spikelets greenish to tan ....**B. hordeaceus**
            7' Lemma awns 12-25 mm long, apical teeth 2-3.5 mm long; spikelets becoming reddish or purplish
                8. Stems and leaf sheaths usually glabrous; spikelets not strongly overlapping, bases of upper ones usually visible ....**B. madritensis**
                8' Stems and leaf sheaths usually puberulent; spikelets strongly overlapping, bases of middle and upper ones obscured by lower spikelets ....**B. rubens**

B. arizonicus (Shear) Stebbins ARIZONA BROME Plants to 7 dm tall. Mar-Apr. Scarce; flats. North side on ne coastal flats; ne end of mesa n of airfield. All CA Channel Islands; n CA (Yolo Co.) to s CA (San Diego Co.), e to AZ.

*B. berteroanus Colla [*B. trinii* Desv.] CHILEAN CHESS Plants to 6 dm tall. Apr. Rare; flats. Known only from collections in "one locality, about *Opuntia*" in April 1897; not seen recently. All CA Channel Islands; San Martin, Guadalupe, and Cedros islands; British Columbia; n CA to Baja CA, e to MT, UT, and AZ; believed to be native of Chile. Probably introduced to San Nicolas Island.

B. carinatus Hook. & Arn. var. **carinatus** CALIFORNIA BROME Plants to 8 dm tall. Apr-May. Rare; flats and slopes. Lower slopes of sw escarpment. All CA Channel Islands except Santa Barbara; Los Coronados, Todos Santos, and San Martin islands; AK to Baja CA, e to CO and TX. Variety *marginatus* (Nees) Barkworth & Anderton occurs in w N America.

*B. catharticus Vahl var. **catharticus** [*B. willdenovii* Kunth] RESCUE GRASS Plants to 1 m tall. Mar-Jul (Nov). Occasional; disturbed flats and roadsides. Northeastern coastal flats n of airfield; mesa near Building 176, w end of Tufts Road, wells area along Tufts Road, Building 127 (Radar Row), Nicktown, and airfield. All CA Channel Islands except Anacapa and Santa Barbara; OR to Baja CA, e to NY and FL; Hawaii; Europe; native of S America. First collected at airfield parking area in June 1978.

*B. diandrus Roth [*B. rigidus* Roth] RIPGUT BROME Plants to 7 dm tall. Feb-May. Abundant; flats, slopes, and canyon bottoms. Widespread locations throughout much of island; on s side, it is found primarily in canyon bottoms and in *Opuntia* patches. All CA Channel Islands; Los Coronados, Todos Santos, and Guadalupe islands; British Columbia to S America; native of Europe. First collected from unspecified location on island in April 1940.

*B. hordeaceus L. subsp. **hordeaceus** [The name *B. mollis* L. has been misapplied to this species] SOFT CHESS Plants to 8 dm tall. Feb-May. Abundant; flats and slopes. Widely scattered locations throughout much of island, especially on ne side of island and on mesa. All CA Channel Islands; Los Coronados, Todos Santos, San Martin, and Guadalupe islands; AK to Baja CA, e to ME; Hawaii; native of Europe. First collected at Corral Harbor in July 1939. Three other subspecies occur in N America.

*B. madritensis L. MADRID BROME. Plants to 6 dm tall. Mar-May. Rare; flats and slopes. North side of island. Santa Rosa, Santa Cruz, Santa Catalina, and San Clemente islands; OR to Baja CA, e to AZ; native of Europe. First collected on escarpment above NavFac buildings in April 1989.

B. maritimus (Piper) C. Hitchc. [*B. carinatus* Hook. & Arn. var. *m.* (Piper) C. Hitchc.] MARITIME BROME Plants to 8 dm tall. Apr-May. Rare; flats and slopes. Northeastern side of island. All CA Channel Islands except Santa Barbara and San Clemente; OR to s CA (Los Angeles Co.).

*B. rubens L. [*B. madritensis* L. subsp. *r.* (L.) Husn.] RED BROME Plants to 5 dm tall. Feb-May. Abundant; flats, slopes, and swales. Widespread locations throughout much of island. All CA Channel Islands; Los Coronados, Todos Santos, San Martin, Guadalupe, and Cedros islands; WA to Baja CA, e to TX; native of Europe. Already widespread on island in April 1966, when it was collected in sand dunes at w end of island, in small valley overlooking Tender Beach, and at Tule Creek.

CORTADERIA Stapf

Perennial herbs. Ca. 25 species; temperate to subtropical areas of S America, New Guinea, and New Zealand. Some species are cultivated as ornamentals or for their showy inflorescences. (Spanish for cutting, referring to sharp leaf margins)

*C. selloana (Schult. & Schult.f.) Asch. & Graebn. PAMPAS GRASS Plants to 13 dm tall. Mar-May. Scarce; disturbed flats and sand dunes. North side of island in Tule Creek; mesa near airfield; s side in sand dunes at w end of island. San Miguel, Santa Cruz, and Santa Catalina islands; OR to Baja CA, e (disjunctly) to NJ; Australia; native of central S America. First collected at airfield in April 1990. Eradication efforts on San Nicolas Island were initiated in 1990, starting with the removal of plants at the w end (Junak 2003b). This is an extremely aggressive grass (Cowan 1976, Kerbavaz 1985) and should not be allowed to spread on San Nicolas Island.

CYNODON Rich.

Perennial herbs. Ca. 9 species; warm temperate to tropical areas of Eurasia and Africa. Several species are cultivated for lawns, pastures, or forage. (Greek: dog tooth, referring to hardened scales on rhizomes)

*C. dactylon (L.) Pers. var. **dactylon** BERMUDA GRASS Plants to 4 dm tall. Apr-Aug (Dec). Common; disturbed flats and canyon bottoms. Scattered locations on n side between Corral Harbor area and Coast Guard Beach area; mesa near wells area along Tufts Road, at Nicktown, and near airfield. All CA Channel Islands; Los Coronados and Cedros islands; WA to Baja CA, e to NH and FL; Hawaii; native of Africa. First collected "above a brackish stream" at an elevation of ca. 1000 feet in April 1897. A detailed distribution map for this taxon can be found in Junak (2003b). Variety *aridus* J.R.Harlan & deWet occurs in AZ. This taxon can be extremely invasive and was observed covering a large *Opuntia* patch in "L" Canyon (Junak 2003b). It is sometimes grown for erosion control; leaves (which are reportedly rich in Vitamin C) have been used for tea.

DISTICHLIS Raf. ALKALI GRASS

Perennial herbs. Ca. 5 species; N and S America, Australia. (Greek: two-ranked, referring to leaf arrangement)

D. spicata (L.) Greene SALTGRASS Plants to 5 dm tall. Apr-Jul. Occasional; low, saline, and moist sites, especially in canyon bottoms and near beaches. Widely scattered locations around perimeter of island. All CA Channel Islands except Santa Barbara; Los Coronados, San Martin, and Cedros islands; British Columbia to Baja CA, e to ME and FL; Hawaii; Australia. Native Americans obtained salt from burned plants; this species has also been used medicinally.

FESTUCA L. FESCUE

Perennial herbs. Ca. 450 species; worldwide, especially in temperate areas. A number of species are important lawn and pasture grasses. (ancient Latin name)

*F. arundinacea Schreb. [*Schedonorus a.* (Schreb.) Dumort., *S. phoenix* (Scop.) Holub] TALL FESCUE Plants to 14 dm tall. Mar-May. Scarce; moist flats and canyon bottoms. North side near NavFac buildings, in "L" Canyon, and in Jetty Canyon; mesa at wells area along Tufts Road, in Nicktown, and at airfield area. All CA Channel Islands except Anacapa and Santa Barbara; AK to Baja CA, e to ME and FL;

POACEAE: Cynodon-Hordeum

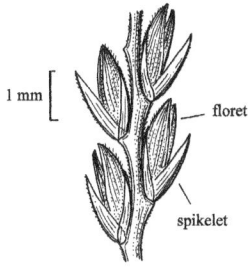

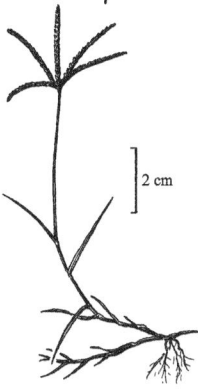

1 mm

floret

spikelet

2 cm

Cynodon dactylon

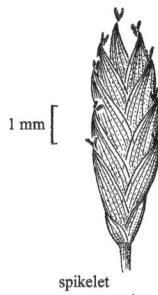

1 mm

spikelet

2 cm

Distichlis spicata

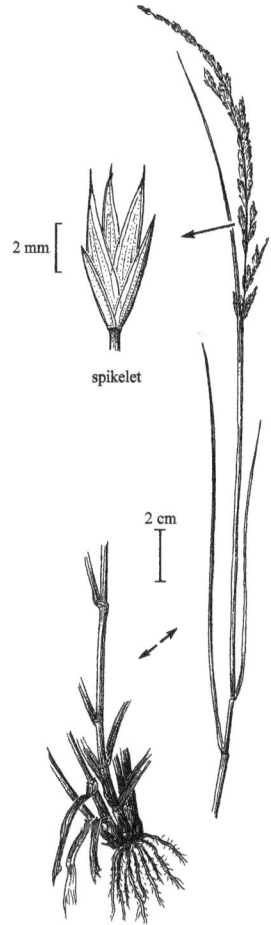

2 mm

spikelet

2 cm

Festuca arundinacea

2 mm

2 cm

Hordeum brachyantherum
subsp. californicum

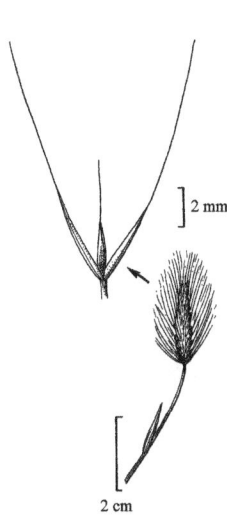

2 mm

2 cm

Hordeum marinum
subsp. gussoneanum

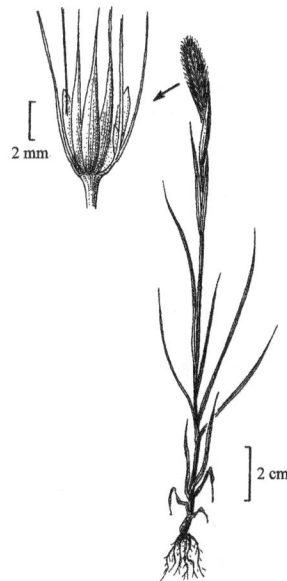

2 mm

2 cm

Hordeum intercedens

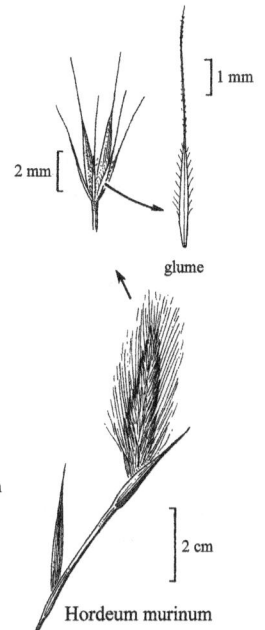

1 mm

2 mm

glume

2 cm

Hordeum murinum

Hawaii; native of Eurasia. First seen on San Nicolas Island in May 1985, this grass can be extremely invasive and should be removed; it has spread rapidly on San Miguel Island. This species is frequently infected with endophytic fungi, which produce ergot alkaloids that are toxic to livestock.

HORDEUM L. BARLEY

Annual or perennial herbs. Ca. 32 species; temperate areas of N America, Europe, and Asia. (ancient Latin name for barley)

1. All 3 spikelets alike, sessile, bisexual, lemma awns 50-150 mm long . . . .**H. vulgare**
1' Central spikelet bisexual, producing fruits, sessile, lemma awns 4.5-30 mm long; lateral spikelets staminate or sterile, lemmas usually reduced, smaller than those of central spikelet
  2. Glumes conspicuously ciliate near bases
    3. Central floret (excluding awn) slightly shorter than lateral florets; internodes of inflorescence axis 1-2 mm long with 5-8 spikelet clusters per cm of axis . . . .**H. murinum** subsp. **glaucum**
    3' Central floret (excluding awn) slightly shorter than lateral florets; internodes of inflorescence axis 2-3 mm long, with 3-5 spikelet clusters per cm of axis . .**H. murinum** subsp. **leporinum**
  2' Glumes not ciliate near bases, glabrous to minutely scabrous
    4. Perennial herb; spikelets purple-tinged . . . .**H. brachyantherum** subsp. **californicum**
    4' Annual; spikelets greenish to tan, not purple-tinged
      5. Central spikelet glumes slightly wider near bases, 8-15 mm long, lemma awn 6-12 mm long; lateral spikelet lemmas with awns 3-6 mm long . . . .**H. intercedens**
      5' Central spikelet glumes linear, 12-22 mm long, lemma awn 8-23 mm long; lateral spikelet lemmas acute or with awns less than 1.5 mm long . . .**H. marinum** subsp. **gussoneanum**

H. brachyantherum Nevski subsp. **californicum** (Covas & Stebbins) Bothmer, N.Jacobsen & Seberg [*H. c.* Covas & Stebbins] CALIFORNIA BARLEY Stems to 8 dm long. Apr-Jun. Occasional; flats, gullies, bluffs, sand dunes, and canyon rims. North side near shoreline w of Thousand Springs and on n coastal flats e of W Mesa Canyon; mesa in Wells area on Tufts Road, on n edge se of Corral Harbor, and s of Nicktown; s side in dunes ese of Seal Beach and in canyons of s escarpment. All CA Channel Islands except Santa Barbara and San Clemente; n CA (Humboldt Co.) to s CA (San Diego Co.). Subspecies *brachyantherum* is widespread in w N America.
H. intercedens Nevski [The name *Hordeum pusillum* Nutt. has been misapplied to this species] VERNAL BARLEY or LITTLE BARLEY Plants to 4 dm tall. Mar-May. Occasional; flats and small depressions. North side on coastal flats near Sand Spit; mesa near Jackson Hill and in e and se portions; s side near Daytona Beach. All CA Channel Islands; central CA (San Benito Co.) to Baja CA. A detailed distribution map for this taxon can be found in Junak *et al.* (1996). Population sizes appear to be extremely variable from year to year on San Nicolas Island. Many CA mainland populations of *H. intercedens* have been extirpated, so preservation of insular populations is extremely important for continued survival of this grass.
*H. marinum Hudson subsp. **gussoneanum** (Parl.) Thell. [*H. geniculatum* All.] MEDITERRANEAN BARLEY Plants to 3 dm tall. Apr-May. Scarce; disturbed flats. Mesa near Jackson Hill and near airfield. Santa Rosa, Santa Cruz, Santa Catalina, and San Clemente islands; WA to Baja CA, e to MT and AZ and disjunctly to MA; native of Europe. First collected at airfield in May 1999.
*H. murinum L. subsp. **glaucum** (Steud.) Tzvelev FOXTAIL Plants to 4 dm tall. Mar-Jun. Common; flats and slopes. Widely scattered locations throughout much of island. All CA Channel Islands; Los Coronados, Todos Santos, San Martin, Guadalupe, and Cedros islands; British Columbia to Baja CA, e to TX; native of Europe. First collected "in fertile spots and in sand" in April 1897.
*H. murinum L. subsp. **leporinum** (Link) Arcang. FOXTAIL Plants to 5 dm tall. Mar-Jun. Common; flats and slopes. Widely scattered locations throughout much of island. All CA Channel Islands; Los Coronados, Todos Santos, San Martin, and Guadalupe islands; British Columbia to Baja CA, e to TX and disjunctly to ME; Hawaii; native of Europe. First collected at unspecified location on island in March 1932.

*H. vulgare L. var. **vulgare** CULTIVATED BARLEY Plants to 7 dm tall. Mar-Apr. Rare; disturbed flats. Known from single collection near Nicktown on mesa (at old landfill site) in April 1992. Santa Rosa, Santa Cruz, Santa Catalina, and San Clemente islands; cultivated throughout N America and elsewhere; native of Eurasia.

LAMARCKIA Moench GOLDENTOP
Annual herbs. One species. (honoring Jean Baptiste Lamarck, 1744-1829, French botanist)

*L. aurea (L.) Moench GOLDENTOP Plants to 2 dm tall. Feb-May (-Jul). Occasional; throughout island but especially in eastern portion. All CA Channel Islands; Los Coronados, Todos Santos, San Martin, and Guadalupe islands; n CA (Butte Co.) to Baja CA, e to AZ; Hawaii; native of Europe. First collected at u   n   -specified location on island in April 1940.

LOLIUM L. RYEGRASS
Annual or perennial herbs. Ca. 5 species; temperate areas of Europe, n Africa, and Asia. Some species are cultivated as forage grasses. (ancient Latin name for ryegrass)

1. Florets 10-22 per spikelet; lemmas usually with awns to ca. 15 mm long; annuals or short-lived perennials
. . . .**L. multiflorum**
1' Florets 2-10 per spikelet; lemmas without awns or with awns to ca. 8 mm long; long-lived perennials
. . . .**L. perenne**

*L. multiflorum Lam. ANNUAL RYEGRASS or ITALIAN RYE Plants to 6 dm tall. Apr-Jul. Common; flats, shallow swales, and canyon bottoms. Widespread locations throughout much of island, especially on mesa near airfield. All CA Channel Islands except Anacapa; throughout N America; native of Europe. First collected near Nicktown in April 1961.
*L. perenne L. PERENNIAL RYEGRASS or ENGLISH RYE Plants to 5 dm tall. Apr-Jul. Common; flats, shallow swales, and canyon bottoms. Widespread locations throughout much of island. Santa Cruz, Santa Catalina, and San Clemente islands; throughout N America; native of Europe. First collected at airfield in July 1965. This species can be toxic to livestock.

MELICA L. MELIC
Perennial herbs. Ca. 80 species; temperate areas ± worldwide (except Australia). Some species are cultivated as ornamentals. (ancient Greek name for some grass)

M. imperfecta Trin. COAST RANGE MELIC or SMALL-FLOWERED MELIC Plants to 1 m tall. Mar-Jun.   Scarce; slopes, swales, and gullies. North escarpment below Nicktown and in canyon near Airfield Grade; shallow gullies on mesa n of airfield. All CA Channel Islands; Los Coronados, Todos Santos, San Martin, Guadalupe, and Cedros islands; central CA (Contra Costa Co.) to Baja CA, e to s NV.

NASSELLA (Trin.) E.Desv.
Perennial herbs. Ca. 116 species; temperate areas of w N and S America. (Latin: *nassa,* a basket with a narrow neck that was used for catching fish, referring to shape of lemma)

1. Terminal segment of awn straight, ± stiff . . . .**N. pulchra**
1' Terminal segment of awn flexuous, soft
    2. Glumes 5-12 mm long; lemmas (excluding awns) 4-6.5 mm long; awns 20-40 mm long . . . .
. . . .**N. lepida**
    2' Glumes 13-22 mm long; lemmas (excluding awns) 7-12 mm long; awns 50-110 mm long . . . . . .
. . . .**N. cernua**

POACEAE: Lamarkia-Nassella

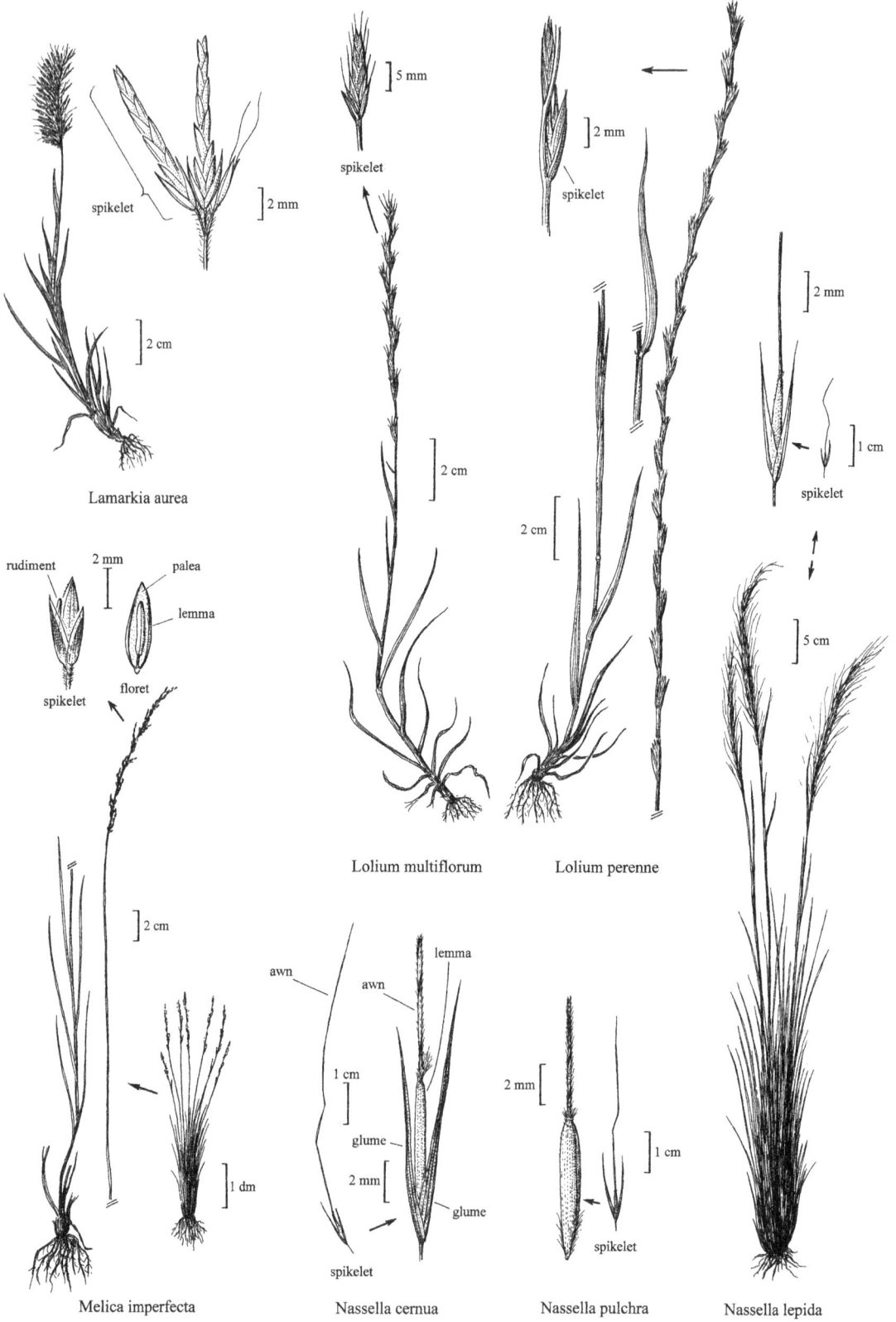

Lamarkia aurea

Lolium multiflorum

Lolium perenne

Melica imperfecta

Nassella cernua

Nassella pulchra

Nassella lepida

N. cernua (Stebbins & Love) Barkworth [*Stipa c.* Stebbins & Love] NODDING NEEDLEGRASS Plants to 12 dm tall. Mar-May. Occasional; slopes, ridges, gullies, and canyon walls. Northeastern coastal flats n of airfield; central and e portions of mesa; canyons and slopes on s escarpment between Cattail Canyon area and Sand Spit. All CA Channel Islands except Santa Barbara; n CA (Tehama Co.) to Baja CA. Detailed distribution maps for *N. cernua, N. lepida,* and *N. pulchra* on San Nicolas Island can be found in Junak *et al.* (1996).

N. lepida (Hitchc.) Barkworth [*Stipa l.* Hitchc.] FOOTHILL NEEDLEGRASS or SMALL-FLOWERED NEEDLEGRASS Plants to 14 dm tall. Mar-May. Occasional; flats, gullies, ridges, swales, and dry canyon bottoms. Northeastern coastal flats and canyons on n escarpment between Nicktown and Sand Spit; se edge of mesa near Twin Towers (Building 186); canyons on s escarpment between Twin Towers and Sand Spit. All CA Channel Islands; Los Coronados, Guadalupe, and Cedros islands; n CA (Humboldt Co.) to Baja CA. On San Nicolas Island, *N. lepida* and *N. pulchra* are usually found growing in *Opuntia* patches or associated with stands of *Coreopsis gigantea.*

N. pulchra (Hitchc.) Barkworth [*Stipa p.* Hitchc.] PURPLE NEEDLEGRASS. Plants to 14 dm tall. Feb-May. Occasional; flats, slopes, and gullies. Northeastern escarpment near "L" Canyon and S Spur Canyon and on coastal flats near S Spur Canyon; on eroded clay flats near Jackson Hill, w of upper end of Mineral Canyon, in and near W Mesa and S Spur canyons, and near Peak 606 at se end of mesa. All CA Channel Islands; Los Coronados Islands; n CA (Humboldt Co.) to Baja CA.

PARAPHOLIS C.E.Hubb.

Annual herbs. Ca. 6 species; temperate areas from w Europe to India. (Greek: beside scale, referring to the 2 adjacent glumes)

*P. incurva (L.) C.E. Hubb. SICKLE GRASS Plants to 4 dm tall. Mar-Jun. Common; bluff tops, saline flats, slopes, sandy canyon bottoms, seeps, and vernal ponds. Widely scattered locations throughout much of island. All CA Channel Islands; OR to Baja CA, e (disjunctly) to NJ; native of Europe. First collected at unspecified location on island in April 1940.

PENNISETUM Rich. ex Pers.

Perennial herbs. Ca. 80 species; warm temperate to tropical areas of Europe, Asia, and Africa. Several species are cultivated for ornamental use or for edible grain. (Latin: feather bristle, referring to involucre of plumose bristles surrounding spikelets of some species)

*P. clandestinum Chiov. KIKUYU GRASS Stems to 8 dm long. Apr-Jun. Scarce; disturbed flats and canyon bottoms. North side on ne coastal flats just w of Rock Jetty; disturbed sites on mesa, including area around fire station at intersection of Owens Road and Monroe Drive, at head of canyon just e of Nicktown, and on flats s of Bldg 312. San Miguel, Santa Rosa, Santa Cruz, Anacapa, and Santa Catalina islands; coastal CA to n Baja CA; native of Africa. First collected in canyon just e of Nicktown in June 1969. This is an aggressive grass and should be eradicated on the island; it is a problem on Santa Cruz Island.

PHALARIS L. CANARY GRASS

Annual or perennial herbs. Ca. 22 species; temperate areas of N America, Europe, and Asia. A few species are cultivated for animal fodder and one species (*P. canariensis* L.) is cultivated for bird seed. (ancient Greek name for a grass with shiny spikelets)

1. Spikelets deciduous in clusters of 6-7, each cluster composed of 1 fertile floret surrounded by 5-6 sterile, reduced spikelets . . . .**P. paradoxa**
1' Spikelets all alike, solitary, not deciduous
    2. Perennial herbs, 8-15 dm tall; keel of glumes barely winged if at all; lemmas of sterile florets . . . unequal or sterile floret 1 . . . .**P. aquatica**

POACEAE: Parapholis-Piptatherum

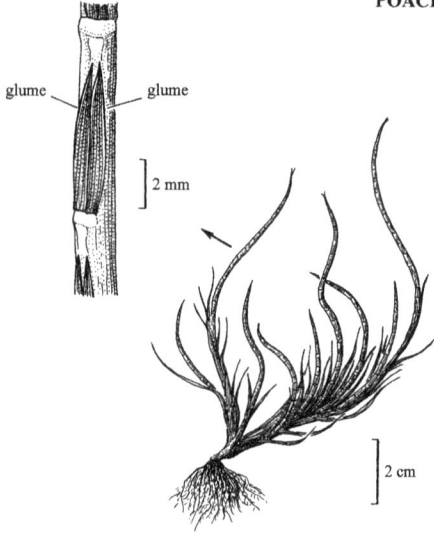

glume   glume

2 mm

2 cm

Parapholis incurva

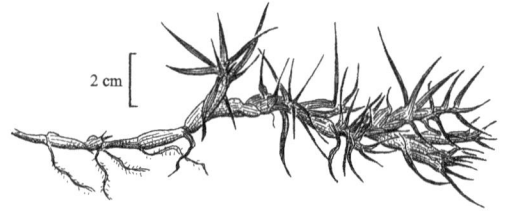

2 cm

Pennisetum clandestinum

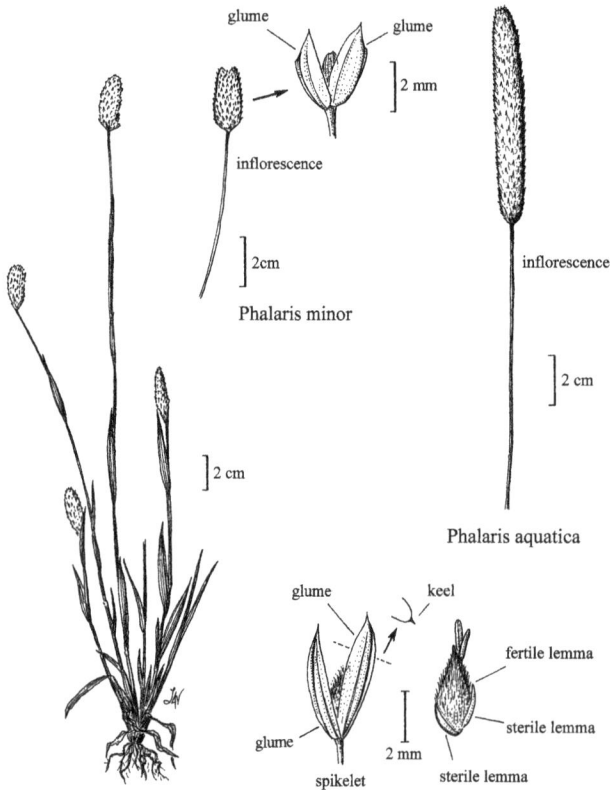

glume   glume

2 mm

inflorescence

2cm

Phalaris minor

inflorescence

2 cm

Phalaris aquatica

2 cm

1 mm   glume

glume

spikelet

2 mm

Piptatherum miliaceum

2 cm

glume   keel

fertile lemma

sterile lemma

glume

2 mm   sterile lemma

spikelet

Phalaris caroliniana

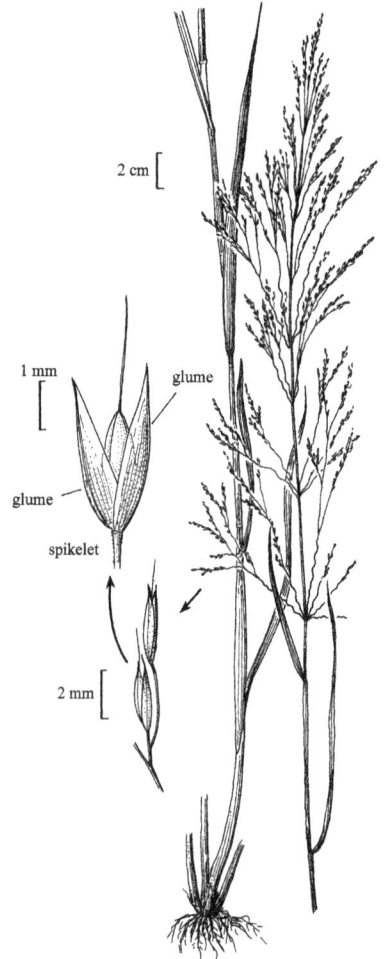

2' Annuals, 2-9 dm tall; glumes dorsally winged; lemmas of sterile florets equal or sterile floret 1 in
   P. minor
     3. Glumes not notched near apex; sterile florets 2 . . . .**P. caroliniana**
     3' Glumes notched near apex; sterile floret 1 . . . .**P. minor**

*P. aquatica L. [*P. tuberosa* L. var. *stenoptera* (Hackel) Hitchc.] HARDING GRASS Plants to 15 dm tall. Mar-May. Rare; disturbed flats, canyons, and roadsides. North side in Celery Canyon, on ne coastal flats at base of Airfield Grade, and in upper portion of "L" Canyon; mesa along Beach Road at w end of airfield runways. Santa Cruz and Santa Catalina islands; OR to Baja CA, e to MT; Hawaii; introduced ± worldwide; native of Europe. First collected on ne coastal flats in March 1991. This is an aggressive grass and should be eliminated before it can spread. It is spreading rapidly on Santa Cruz and Santa Catalina islands.
*P. caroliniana Walter CAROLINA CANARY GRASS Plants to 6 dm tall. Apr. Rare; flats. Collected on a "fertile flat near the sea" in April 1897; not seen recently. Santa Cruz, Santa Barbara, Santa Catalina, and San Clemente islands; OR to Baja CA, e to FL; Australia; Europe; believed to be native of e N America. Possibly introduced on San Nicolas Island.
*P. minor Retz. MEDITERRANEAN CANARY GRASS Plants to 9 dm tall. Apr-Jun. Occasional; flats, swales, arroyos, and canyons. Widely scattered locations, especially on mesa. All CA Channel Islands; Todos Santos, Guadalupe, and Natividad islands; OR to Baja CA, e to FL; Hawaii; native of Europe. First collected near Dutch Harbor in April 1966. Upper parts of plants, especially new growth, can be toxic to cattle.
*P. paradoxa L. HOODED CANARY GRASS Plants to 7 dm tall. May-Jun. Rare; disturbed flats. Mesa n of airfield. Santa Cruz, Anacapa, Santa Catalina, and San Clemente islands; WA to Baja CA, e to AZ and (disjunctly) to NJ; Hawaii; introduced ± worldwide, especially in harbor areas; native of Europe. First collected on mesa in June 2000.

PIPTATHERUM P.Beauv.

Perennial herbs. Ca. 30 species, arid temperate to subtropical areas of Europe, Asia, and Africa. (Greek: falling awn)

*P. miliaceum (L.) Cosson subsp. **miliaceum** [*Oryzopsis m.* (L.) Asch. & Schweinf.] RICE GRASS or SMILO GRASS Plants to 17 dm tall. May-Nov. Occasional; disturbed flats, slopes, and sand dunes. North side near Corral Harbor and near mouth of Celery Canyon; mesa along Tufts Road , along Jackson Highway at sw end of mesa, near Jackson Hill, and near airfield; s side near Army Springs, Dizon's Ravine, and below sw end of mesa. All CA Channel Islands except Santa Barbara; n CA (Tehama Co.) to Baja CA, e to AZ with a disjunct occurrence in MD; native of Eurasia. First collected at wells area on Tufts Road in November 1989. A detailed distribution map for this taxon can be found in Junak (2003b). This is an invasive grass and should be eliminated before it spreads further. It has spread rapidly on San Clemente and Santa Cruz islands.

POA L. BLUEGRASS

Annual or perennial herbs. Ca. 500 species; worldwide, primarily in temperate and boreal areas. Several species (e.g., *P. pratensis* L.) are cultivated for lawns and pastures; several species are important native forage grasses in w N America. (ancient Greek name)

1. Annual herbs; spikelets compressed, 3-5 mm long; lemmas dorsally keeled or ridged . . . . . .**P. annua**
1' Perennials; spikelets subterete, 5-10 mm long; lemmas dorsally rounded . .**P. secunda** subsp. **secunda**

*P. annua L. ANNUAL BLUEGRASS Plants to 2 dm tall. Mar-Apr. Rare; on mesa along access road north of airfield runways; Nicktown. All CA Channel Islands; Guadalupe Island; AK to Baja CA, e to ME and FL; Hawaii; introduced ± worldwide; native of Eurasia. First collected on n side of airfield runways in April 1989.

P. secunda J.Presl subsp. **secunda** [*P. scabrella* (Thurb.) Vasey] PACIFIC BLUEGRASS Plants to 15 dm tall. Mar-Apr. Scarce; flats, slopes, and canyons. Northeastern escarpment between "L" Canyon and Rock Jetty; ne side of mesa in E Mesa Canyon. All CA Channel Islands except Santa Barbara; Todos Santos and Guadalupe islands; AK to Baja CA, e to NM; S America (disjunct population). Subspecies *juncifolia* (Scribn.) Soreng occurs in w N America.

POLYPOGON Desf. BEARD GRASS

Annual herbs. Ca. 18 species; warm temperate areas of Europe, Asia, Africa, and S America. Some species are cultivated as ornamentals. (Greek: much bearded, referring to bristly inflorescences of some species)

*P. monspeliensis (L.) Desf. RABBITSFOOT GRASS Plants to 1 m tall. Apr-Aug. Occasional; seeps, moist swales, and canyon bottoms. North side at w end of Red Eye Beach and in canyons of ne escarpment; scattered locations on mesa; s side at Army Springs and Cattail Canyon. All CA Channel Islands; Todos Santos, Guadalupe, and Cedros islands; AK to Baja CA, e to ME and FL; Hawaii; introduced ± worldwide; native of s Europe and Turkey. First collected in "sand-swept arroyos" and "by the sides of brackish water courses" in April 1897.

SCHISMUS P.Beauv. MEDITERRANEAN GRASS

Annual herbs. Ca. 5 species; Africa, and Asia. (Greek: split, referring to notched lemma)

*S. arabicus Nees ARABIAN SCHISMUS Plants to 1 dm tall. Apr-Jun. Rare; disturbed flats, open ridges, and slopes. Mesa at Nicktown; s side on ridge at Sand Spit. Santa Cruz, Santa Barbara, Santa Catalina, and San Clemente islands; central CA (Stanislaus Co.) to Baja CA, e to NM; native of sw Asia. First collected at Nicktown in June 2000. This invasive grass should be removed from San Nicolas Island as soon as possible.

STENOTAPHRUM Trin.

Perennial herbs. Ca. 7 species; primarily in tropical areas around Indian Ocean rim, with three species endemic to Madagascar. (Greek: narrow trench, referring to spikelet scars on axis of inflorescence)

*S. secundatum (Walter) Kuntze SAINT AUGUSTINE GRASS Stems to 6 dm long. May-Nov. Rare; disturbed flats. Mesa at fire station at intersection of Owens Road and Monroe Drive and near airfield. Central CA (Marin Co.) to s CA (San Diego Co.), e to FL; Hawaii; Africa; may be native of se United States. First collected near air terminal building in May 1986. This grass has been spreading on San Nicolas Island in recent years and should be controlled.

TRITICUM L.

Annual herbs. Ca. 25 species; w and central Asia. This genus was first cultivated in w Asia at least 9000 years ago and is now the world's most important food crop. (Latin name for wheat)

*T. aestivum L. COMMON WHEAT Plants to 1 m tall. Apr-May. Rare; canyon bottoms. Northeastern coastal flats in "L" Canyon and N Spur Canyon. Santa Catalina and San Clemente islands; San Martin, Guadalupe, and Cedros islands; cultivated throughout N America and elsewhere; originated in Asia. First collected on ne coastal flats in "L" Canyon in May 1995.

VULPIA C.C.Gmel.

Annual herbs. Ca. 30 species; temperate areas of Europe and w N America. (honoring Johann Samuel Vulpius, 1760-1846, German chemist, pharmacist, and botanist)

1. Lower glumes less than ½ the length of upper glumes . . . . **V. myuros**
1' Lower glumes ½ or more the length of upper glumes . . . . **V. octoflora**

POACEAE: Poa-Vulpia

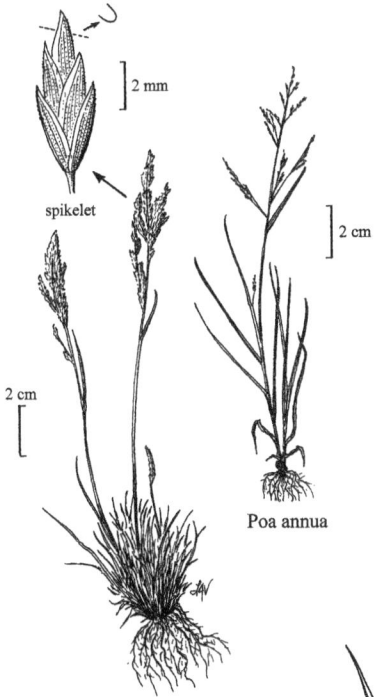

2 mm

spikelet

2 cm

2 cm

Poa annua

Poa secunda subsp. secunda

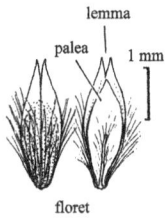

lemma

palea

1 mm

floret

Schismus arabicus

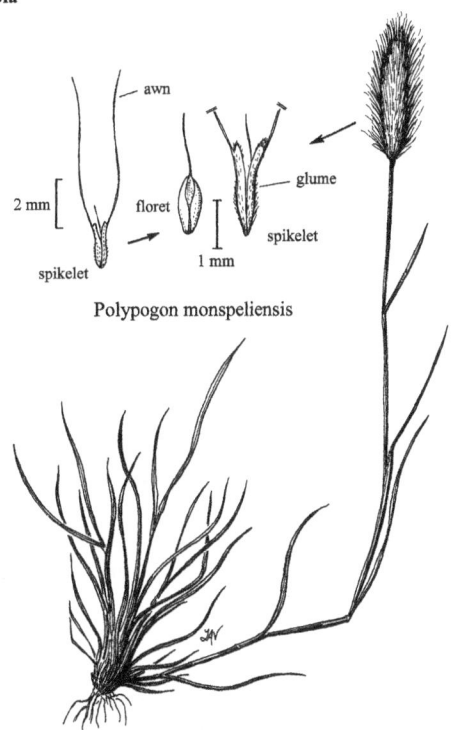

awn

2 mm

floret

spikelet

1 mm

glume

spikelet

Polypogon monspeliensis

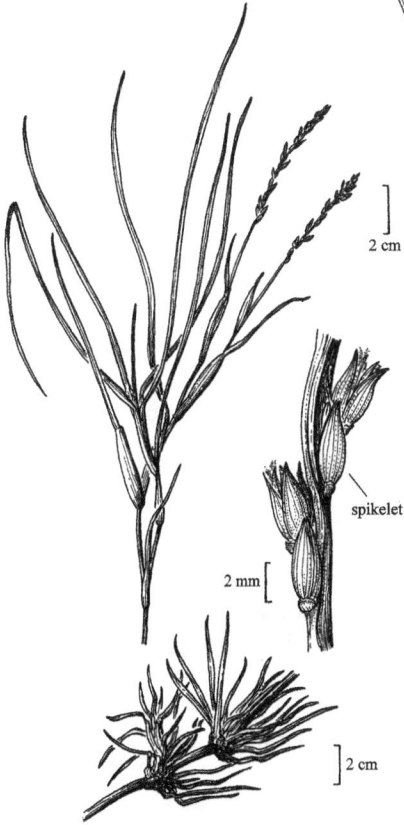

Polypogon monspeliensis

2 cm

2 cm

2 cm

2 mm

spikelet

Stenotaphrum secundatum

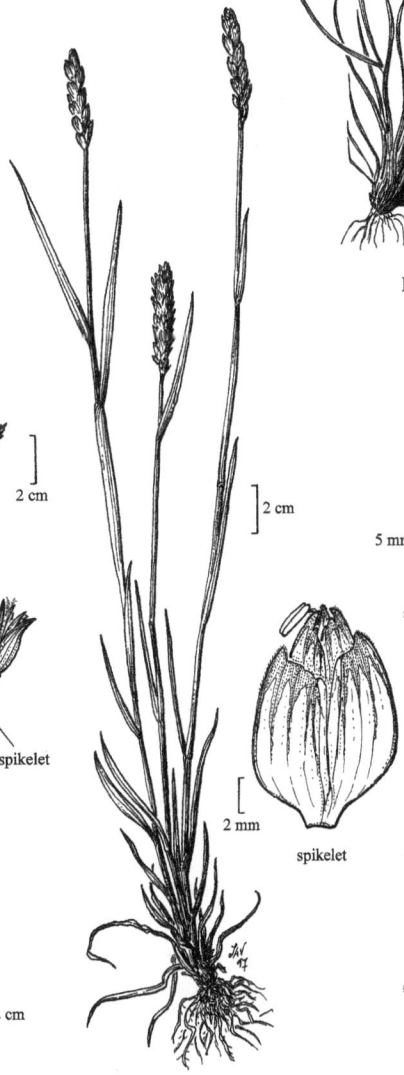

2 cm

2 mm

spikelet

Triticum aestivum

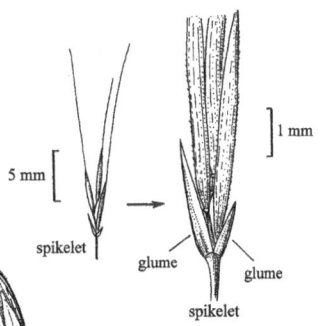

5 mm

spikelet

1 mm

glume

glume

spikelet

Vulpia myuros var. hirsuta

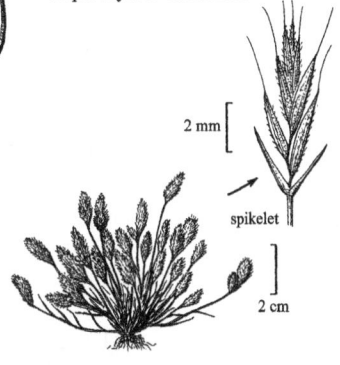

2 mm

spikelet

2 cm

Vulpia octoflora var. hirtella

*V. myuros (L.) C.C.Gmel. [*Festuca megalura* Nutt., *V. m.* (Nutt.) Rydb.] FOXTAIL FESCUE Plants to 7 dm tall. Mar-Apr (Oct). Common; flats and slopes. Widespread locations throughout much of island, especially on n coastal flats and on mesa. All CA Channel Islands; Los Coronados, Todos Santos, Guadalupe, and Cedros islands; AK to Baja CA, e to ME and FL; Hawaii; native of Europe and n Africa. First collected at Nicktown in June 1969.

V. octoflora (Walter) Rybd. var. **hirtella** (Piper) Henrard SIX-WEEKS FESCUE Plants to 5 dm tall. Mar-May. Occasional; flats, slopes, and sandy areas. North side on ne coastal flats and in canyons and on slopes of ne escarpment; n portion of mesa between Corral Harbor area and airfield; slopes and coastal flats at se end of island. All CA Channel Islands; San Martin, Guadalupe, and Cedros islands; WA to Baja CA, e to TX and disjunctly to FL. Abundance and distribution of this taxon appears to vary dramatically from year to year. Variety *octoflora* occurs on adjacent CA mainland.

RUPPIACEAE Hutchinson
DITCH-GRASS FAMILY

Annual herbs. One genus and ca. 10 species; nearly worldwide.

RUPPIA L. DITCH-GRASS

Ca. 10 species; nearly worldwide. (for Heinrich Bernard Ruppius, 1688-1719, German botanist)

R. maritima L. Stems to 3 dm long. Apr-Oct. Rare; freshwater pond. Known from a single collection at Thousand Springs in August 1969; not seen recently (habitat on island has been destroyed). Santa Rosa, Santa Cruz, Santa Catalina, and San Clemente islands; AK to Baja CA, e (disjunctly) to ME and FL; Bermuda; West Indies; Mexico; Central and S America; Africa; Australia; Eurasia.

TYPHACEAE Juss.
CATTAIL FAMILY

Perennial herbs. One genus and ca. 12 species; boreal to tropical areas worldwide.

TYPHA L. CATTAIL

Ca. 12 species; worldwide, mostly in temperate areas. Flower buds, pollen, seeds, and rhizomes were eaten by Native Americans; flowers and rhizomes also were used medicinally. (ancient Greek name)

1. Staminate segment of spike separated from pistillate segment by naked axis; pistillate segment 15-25 wide in fruit; pistillate pedicel subtended by minute, linear to narrowly spatulate bract (often obscured by villous pubescence) . . . .**T. domingensis**

1' Staminate segment of spike confluent with pistillate segment; pistillate segment 25-35 mm wide in fruit; pistillate pedicel bractless . . . .**T. latifolia**

T. domingensis Pers. NARROWLEAF CATTAIL or SOUTHERN CATTAIL Plants to 2 m tall. May-Jul. Occasional; seeps, moist sites, and canyon bottoms. Widely scattered locations in canyons and moist areas around perimeter of island (as in Celery Canyon, W Mesa Canyon, and in many wet canyons on s side); mesa at borrow pit along Monroe Drive. All CA Channel Islands except Anacapa and Santa Barbara; Cedros Island; to Baja CA, e to FL; Australia; New Zealand; West Indies; Mexico; Central and S America; Eurasia; Africa. According to annotations by S. Galen Smith at SBBG, some plants on San Nicolas Island appear to be hybrids with *T. angustifolia* or *T. latifolia*.

T. latifolia L. BROADLEAF CATTAIL Plants to 2 m tall. Apr. Scarce; seeps, moist sites, and canyon bottoms. North side in Tule Canyon and W Mesa Canyon; mesa n of airfield; s side at Army Springs and in Cattail Canyon. Santa Cruz, Santa Catalina, and San Clemente islands; AK to Baja CA, e to ME and

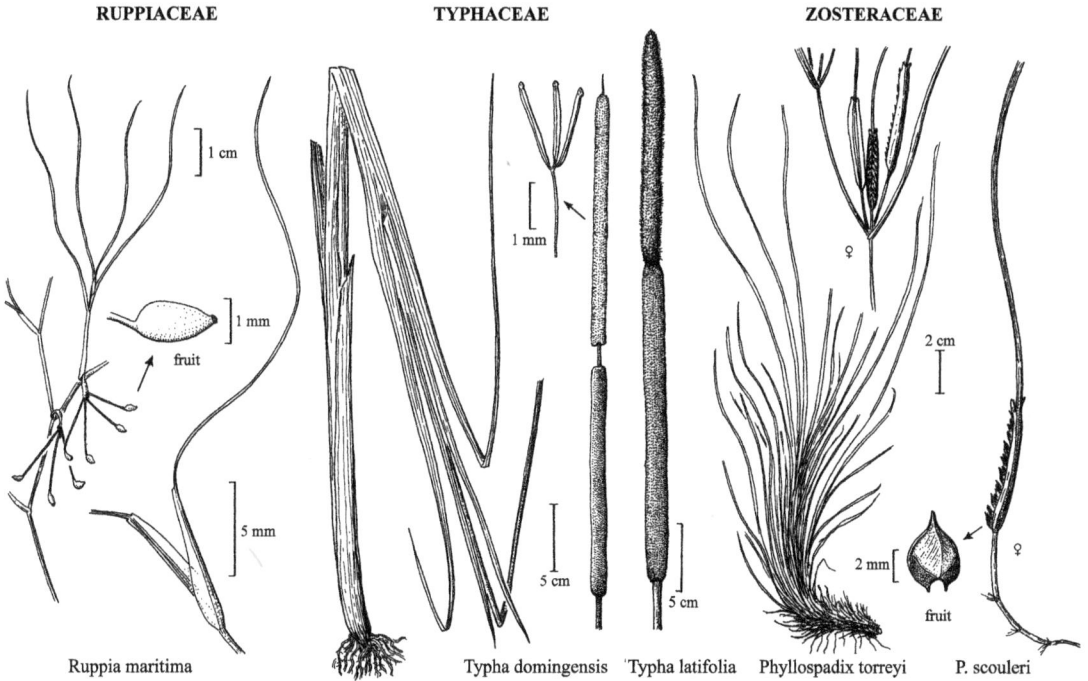

RUPPIACEAE

TYPHACEAE

ZOSTERACEAE

1 cm

1 mm

fruit

5 mm

Ruppia maritima

1 mm

5 cm

5 cm

Typha domingensis   Typha latifolia

2 cm

2 mm

fruit

Phyllospadix torreyi   P. scouleri

FL; Australia (introduced in Tasmania); Mexico; Central and S America; Eurasia; Africa.

ZOSTERACEAE Dumort.
EEL-GRASS FAMILY

Perennial marine herbs. Three genera and ca. 18 species; cold to temperate seacoasts ± worldwide. Some species are an important source of food for marine animals and were used by Native Americans.

1. Plants with short, thick rhizomes, occurring in tidal zone, often on wave-swept rocks; leaves tufted, 1-5 mm wide, blades thick to subterete; inflorescence subtended by leaf-like bract . . . . .**Phyllospadix**
1' Plants with long, slender rhizomes, usually occurring in deep waters below tidal zone, often found stranded on beaches; leaves cauline, well-separated, 2-18 mm wide, blades thin; inflorescence enclosed by sheath-like leaf base . . . .**Zostera**

PHYLLOSPADIX Hook. SURF-GRASS

Five species; n Pacific coast. Insular distributions of both species found in CA need further study. Leaves of both CA species were woven into mats, water bottles, and other items by Native Americans. (Greek: leaf-like inflorescence)

1. Leaves thin, flat, 1.5-4 mm wide; pistillate inflorescences near base, usually 1 per stem . . .**P. scouleri**
1' Leaves thick, cylindrical to elliptic in cross-section, 0.5-2 mm wide; pistillate inflorescences cauline, 3-7 per stem . . . .**P. torreyi**

P. scouleri Hook. SCOULER'S SURF-GRASS Plants to 1 m tall. Apr-May. Common; shallow reefs, rocky shorelines, and tidal pools. Ocean waters around perimeter of island, especially at nw end. All CA Channel Islands except San Miguel and Santa Rosa; all Baja CA islands except Todos Santos and Guadalupe; AK to Baja CA.

P. torreyi S. Watson TORREY'S SURF-GRASS Plants to 7 dm tall. May-Jul. Occasional; shallow reefs, rocky shorelines, and tidal pools. Ocean waters around perimeter of island, especially at nw end. All CA Channel Islands; Todos Santos, San Martin, and Guadalupe islands; OR to Baja CA.

ZOSTERA L. EEL-GRASS

Perennial marine herbs. Ca. 12 species; cold to temperate marine waters, ± worldwide. (Greek: belt, referring to ribbon-like leaves)

Z. pacifica S.Watson [*Z. latifolia* Morong] Leaves to 1.5 m long. Aug. Rare; subtidal sandy flats. North side of island ca. 0.5 miles nw of Rock Jetty. Santa Rosa, Santa Cruz, Anacapa, and Santa Catalina islands; British Columbia to Baja CA. See Coyer *et al.* (in press) for more information on distribution along our coastline.

LITERATURE CITED

Anonymous. 1891. Miles of human bones: discoveries of a party who went to San Nicolas. Santa Barbara Morning Press, 30 August 1891.

Anonymous. 1898. San Diego party returns from San Nicolas Island. Los Angeles Times, 25 September 1898, p. A15.

Anonymous. 1900. San Nicolas explorers back. Los Angeles Times, 12 April 1900, p. I15.

Anonymous. 1901. Successful botanists. Los Angeles Times, 15 May 1901, p. 15.

Anonymous. 1902a. Avalon brevities. Los Angeles Times, 25 April 1902, p. A7.

Anonymous. 1902b. Rough run to San Nicholas. Los Angeles Times, 6 May 1902, p. A7.

Anonymous. 1902c. Santa Catalina brevities. Los Angeles Times, 4 June 1902, p. A5.

Anonymous. 1902d. San Nicolas afire. Los Angeles Times, 25 June 1902, p. A7.

Anonymous. 1902e. San Nicolas to be exploited for oil. Los Angeles Times, 17 April 1902, p. A1.

Anonymous. 1902f. Prospectors have San Nicolas Island. Los Angeles Times, 6 May 1902, p. 11.

Anonymous. 1904. Wild man story again to the fore. Santa Barbara Morning Press, 17 September 1904.

Anonymous. 1909. Indian relics shipped north: Mrs. Blanche Trask donates them to museum. Los Angeles Times, 22 December 1909, p. II 10.

Anonymous. 1917. Island tax report. Oxnard Courier, 20 April 1917.

Anonymous. 1924a. Rafts used to save starving sheep on isle: twenty-five hundred head transferred to San Nicholas. Santa Barbara Morning Press, 24 October 1924.

Anonymous. 1924b. Sheep on island prove problem: feeders unable to ship to market. Los Angeles Times, 27 October 1924, p. A8.

Anonymous. 1925. West coast isle, peril spot which yachtsmen shun. Helena Daily Independent, 11 December 1925, p. 7.

Anonymous. 1938. San Diego scientists return; tell of sea disaster. San Diego Union, 8 May 1938, p. I 13.

Anonymous. 1949. Sea lion herds bask on island. Los Angeles Times, 25 April 1949, pp. A1-A2.

Anonymous. 1965. Navy to fertilize base on island from the air. Santa Barbara News-Press, 21 January 1965.

Anonymous. 1970. Then came the nurserymen. Ventura County Star-Free Press, 22 March 1970.

Anonymous. 1971. Navy barber leaves barren island legacy. Santa Barbara News-Press, 7 May 1971.

Bean, J. and K. Staubel. 1972. *Temalpakh:* Cahuilla Indian knowledge and usage of plants. Malki Museum Press, Banning, CA. 225 pp.

Beard, F. 1994. San Nicholas Island: Agee and Elliott 1929-1943. Unpublished manuscript on file at Santa Cruz Island Foundation, Santa Barbara, CA. 79 pp.

Beauchamp, R. 1987. San Clemente Island: remodeling the museum. Pp. 575-579 IN: T. Elias (ed.) Conservation and management of rare and endangered plants. California Native Plant Society, Sacramento, CA.

Beauchamp, R. 1997. The plants of the Southern Channel Islands: opportunities for horticultural exploitation. Pp. 47-61 IN: B. O'Brien, L. Fuentes, and L. Newcombe (eds.) Out of the wild and into the garden I: a symposium of California's horticulturally significant plants. Occasional Publication 1. Rancho Santa Ana Botanic Garden, Claremont, CA.

Bowers, S. 1889. Nineteen days on San Nicholas Island. II. Geology. Ventura Vidette, 13 November 1889.

Bowers, S. 1890. San Nicolas Island. Pp. 57-61 IN: California State Mining Bureau. Ninth Annual Report of the State Mineralogist, for the year ending December 1, 1889. California State Printing Office, Sacramento, CA.

Brooks, R. 1933. Bill of sale for San Nicolas Island sheep operation from Robert L. Brooks to Roy E. Agee *et al.* Copy on file at Santa Cruz Island Foundation, Santa Barbara, CA (Brooks binder). 2 pp.

Brummitt, R. and C. Powell (eds.) 1992. Authors of plant names. Royal Botanic Gardens, Kew, England. 732 pp.

Bryan, B. 1970. Archaeological explorations on San Nicolas Island. Southwest Museum Paper 22. Southwest Museum, Los Angeles, CA. 160 pp.

Burnham, W., F. Kunkel, W. Hofmann, and W. Peterson. 1963. Hydrogeologic reconnaissance of San Nicolas Island, California. U.S. Geological Survey Water Supply Paper 1539-O. U.S. Government Printing Office, Washington, D.C. 43 pp.

Cantelow, E. and H. Cantelow. 1957. Biographical notes on persons in whose honor Alice Eastwood named native plants. Leaflets of Western Botany 8: 83-101.

Carlquist, S. 1974. Island biology. Columbia University Press, New York, NY. 660 pp.

Carroll, M., L. Laughrin, and A. Bromfield. 1993. Fire on the California Islands: does it play a role in chaparral and closed-cone pine forest habitats? Pp. 73-88 IN: F. Hochberg (ed.) Third California Islands symposium: recent advances in research on the California Islands. Santa Barbara Museum of Natural History, Santa Barbara, CA.

Chess, K., W. Halvorson, and K. McEachern. 1996. San Nicolas Island Vegetation Monitoring Report 1993-1996. Technical Report 56. U.S. Geological Survey, Cooperative Park Studies Unit, University of Arizona, Tucson, AZ. 46 pp.

Chiang-Cabrera, F. 1981. A taxonomic study of the North American species of *Lycium* (Solanaceae). Ph.D. Dissertation, University of Texas, Austin, TX. 287 pp.

Clark, C. 1977. Edible and useful plants of California. University of California Press, Berkeley, CA. 280 pp.

Clark, R. and W. Halvorson. 1989. Status of the endangered and rare plants on Santa Barbara Island, Channel Islands National Park: final report. Unpublished contract study for Endangered Plant Program, California Department of Fish and Game. Channel Islands National Park, Ventura, CA. 56 pp.

Clark, R., W. Halvorson, A. Sawdo, and K. Danielsen. 1990. Plant communities of Santa Rosa Island, Channel Islands National Park. Technical Report 42. Cooperative National Park Resources Study Unit, University of California, Davis, CA. 93 pp.

Coan, E. 1982. James Graham Cooper: pioneer western naturalist. University Press of Idaho, Moscow, ID. 253 pp.

Cockerell, T. 1937. The botany of the California Islands. Torreya 37: 117-123.

Cockerell, T. 1939. Natural history of Santa Catalina Island. Scientific Monthly 48: 308-318.

Cowan, B. 1976. The menace of pampas grass. Fremontia 4 (2): 14-16.

Coyer, J., K. Miller, J. Engle, J. Veldsink, A. Cabello-Pasini, W. Stam, and J. Olsen. In press. Eelgrass meadows in the California Channel Islands and adjacent coast reveal a mosaic of two species, evidence for introgression and variable clonality. Annals of Botany.

Crocker, T. 1933. The Templeton Crocker Expedition of the California Academy of Sciences, 1932. No. 2. Introductory statement. Proceedings of the California Academy of Sciences, 4th Series, 21 (2): 3-9.

Dahlgren, R., H. Clifford, and P. Yeo. 1985. The families of the monocotyledons: structure, evolution, and taxonomy. Springer-Verlag, Berlin. 520 pp.

Daily, M. and C. Stanton. 1983. Historical highlights of Santa Cruz Island. La Reata 5: 14-19.

Daily, M. 1987. California's Channel Islands: 1001 questions answered. McNally and Loftin, Publishers, Santa Barbara, CA. 284 pp.

D'Antonio, C., W. Halvorson, and D. Fenn. 1992. Restoration of denuded areas and iceplant areas on Santa Barbara Island, Channel Islands National Park. Technical Report NPS/WRUC/NRTR-92/46. Cooperative National Park Studies Unit, University of California, Davis, CA. 90 pp.

Davis, W. and S. Junak. 1993. Hybridization between *Malacothrix polycephala* and *M. incana* (Asteraceae) on San Nicolas Island, California. Pp. 89-95 IN: F. Hochberg (ed.) Third California Islands symposium: recent advances in research on the California Islands. Santa Barbara Museum of Natural History, Santa Barbara, CA.

Davis, W. 1997. The systematics of annual species of *Malacothrix* (Asteraceae: Lactuceae) endemic to the California Islands. Madroño 44 (3): 223-244.

De Violini, R. 1974. Climatic handbook for Point Mugu and San Nicolas Island, Part I, surface data. Technical Publication PMR-TP-74-1. Pacific Missile Range, Point Mugu, CA. 139 pp.

Dittman, C. 1878. Narrative of a seafaring life on the coast of California. Unpublished manuscript,

Bancroft Library, University of California, Berkeley, CA. 10 pp.

Drost, C. and G. Fellers. 1991. Density cycles in an island population of deer mice, *Peromyscus maniculatus.* Oikos 60: 351-364.

Dulka, K., M. Dalke, and L. Barker. 1993. San Nicolas Island and Santa Cruz Island site manual. Naval Air Warfare Center Weapons Division, Point Mugu, CA.

Dunkle, M. 1939. Botany. Pp. 33-34 IN: D. Meadows. Progress report of the Los Angeles Museum-Channel Islands Biological Survey: 4th expedition. Unpublished manuscript on file at Santa Barbara Botanic Garden, Santa Barbara, CA.

Dunkle, M. 1950. Plant ecology of the Channel Islands of California. Allan Hancock Pacific Expeditions 13 (3): 247-386. University of Southern California Press, Los Angeles, CA.

Eastwood, A. 1898. Studies in the herbarium and the field II. Notes on the plants of San Nicolas Island. Proceedings of the California Academy of Sciences, 3rd Series, Botany 1 (3): 89-120, 140.

Eastwood, A. 1941. The islands of southern California and a list of the recorded plants. Leaflets of Western Botany 3 (2): 27-36, 3 (3): 54-78.

Elliott, J. 1897. Indenture between J.V. Elliott and Peter Cazes, 15 November 1897. Ventura County Official Records, Deed Book 54, pp. 562-563.

Ellison, W. (ed.) 1937. The life and adventures of George Nidever [1802-1883]. University of California Press, Berkeley, CA. 128 pp.

Fellers, G. and C. Drost. 1991. Ecology of the island night lizard, Xantusia riversiana, on Santa Barbara Island, California. Herpetological Monographs 5: 28-78.

Flora of North America Editorial Committee. 1993. Flora of North America, north of Mexico. Volume 2: Pteridophytes and Gymosperms. Oxford University Press, New York, NY. 475 pp.

Foreman, R. 1967. Observations on the flora and ecology of San Nicolas Island. Report USNRDL-TR-67-8. U.S. Naval Radiological Defense Laboratory, San Francisco, CA. 79 pp.

Forney, S. 1879a. Topographic Survey Map T-1523. U.S. Coast and Geodetic Survey, Washington, D.C.

Forney, S. 1879b. Report of the completion of the survey of San Nicolas Island, Santa Barbara Channel, coast of California, July, August, and September, 1879. Manuscript submitted to Charles Patterson, Superintendent of U.S. Coast Survey, Washington, D.C., 17 December 1879. U.S. National Archives, Record Group 23, U.S. Coast Survey, Superintendent's File (1866-1910), Box 384 (Assistants 1879).

Foster, L. 1992. Living on San Nicolas Island 1938-1939. Unpublished manuscript on file at Santa Barbara Botanic Garden, Santa Barbara, CA. 38 pp.

Gherini, P. 1966. Island rancho. Noticias 12: 14-20.

Gonderman, R. 1966. Comments on fire-resistant plants. LASCA Leaves 16 (3): 64-67.

Greenwell, W. 1858. U.S. Coast Survey. Section X. Descriptions of signals. Secondary triangulation of the islands of San Miguel and San Nicolas, Santa Barbara Channel. U.S. National Archives, Record Group 23, U.S. Coast Survey. 19 pp.

Halsey, A. 1872. Indenture between Abraham Halsey and Agnes M., Alice E., and Maggie A. Hamilton, 22 January 1872. Ventura County Official Records, Deeds Book E, pp. 401-402.

Halvorson, W., R. Clark, and C. Soiseth. 1992. Rare plants of Anacapa, Santa Barbara, and San Miguel in Channel Islands National Park. Technical Report NPS/WRUC/NRTR-92/47. Cooperative National Park Studies Unit, University of California, Davis, CA. 134 pp.

Halvorson, W., S. Junak, C. Schwemm, and T. Keeney. 1996. Plant communities of San Nicolas Island, California. Technical Report 55. National Biological Service, Cooperative Park Studies Unit, University of Arizona, Tucson, AZ. 47 pp.

Hamilton, A. 1872. Indenture between Agnes M., Alice E., and Maggie A. Hamilton and the Pacific Wool Growing Company, 3 September 1872. Ventura County Official Records, Deeds Book E, pp. 403-405.

Handbury, T. 1902. Letter to Lighthouse Board, Washington, D.C., 1 July 1902. U.S. National Archives, Record Group 26, Lighthouse Board Correspondence (1901-1910), File 2998 (San Nicolas Island).

Hickman, J. (ed.) 1993. The Jepson Manual: higher plants of California. University of California Press, Berkeley, CA. 1400 pp.

Hitchcock, C. 1932. A monographic study of the genus *Lycium* of the Western Hemisphere. Annals of the

Missouri Botanical Garden 19: 179-374.

Hillyard, D. 1985. Status reports on invasive weeds: artichoke thistle. Fremontia 12 (4): 21-22.

Holder, C. 1899. The wind-swept island of San Nicolas. Scientific American 81 (15): 233-234.

Holder, C. 1910. The Channel Islands of California. A.C. McClurg and Company, Chicago, IL. 397 pp.

Holmgren, P., N. Holmgren, and L. Barnett (eds.) 1990. Index Herbariorum. Part 1: The herbaria of the world. Regnum Vegetabile 120: 1-693.

Howell, J. 1932. Field notes for 12-13 March 1932. Unpublished manuscript on file in Botany Department Library at California Academy of Sciences, San Francisco, CA.

Howell, J. 1933. Field notes on the manzanitas of Santa Cruz Island. Leaflets of Western Botany 1 (7): 63-64.

Howell, J. 1935. The Templeton Crocker Expedition of the California Academy of Sciences, 1932. No. 22. The vascular plants of San Nicolas Island, California. Proceedings of the California Academy of Sciences, 4th Series, 21 (22): 277-284.

Howell, J. 1941. The closed-cone pines of insular California. Leaflets of Western Botany 3 (1): 1-8.

Howell, J. 1942. A short list of plants from Cedros Island, Lower California. Leaflets of Western Botany 3 (8): 180-185.

Irwin, M. 1945. San Nicolas Island expedition. Museum Leaflet 20 (5): 51-56. Santa Barbara Museum of Natural History, Santa Barbara, CA.

Jepson, W. 1908. Field notes for 11-13 July 1908. Unpublished manuscript on file at Jepson Herbarium, University of California, Berkeley, CA.

Jepson, W. 1916. Field notes for 14 November 1916. Unpublished manuscript on file at Jepson Herbarium, University of California, Berkeley, CA.

Johnson, D. 1980. Episodic vegetation stripping, soil erosion, and landscape modifications in prehistoric and recent historic time, San Miguel Island, California. Pp. 103-121 IN: D. Power (ed.) The California Islands: proceedings of a multi-disciplinary symposium. Santa Barbara Museum of Natural History, Santa Barbara, CA.

Jones, P. 1969. San Nicolas Island archaeology in 1901, edited by Robert F. Heizer. Masterkey 43 (3): 84-98.

Junak, S. and J. Vanderwier. 1990. An annotated checklist of the vascular plants of San Nicolas Island, California. Pp. 121-145 IN: Proceedings of the Fifth Biennial Mugu Lagoon/San Nicolas Island Ecological Research Symposium. Naval Air Station, Point Mugu, CA.

Junak, S., R. Philbrick, and C. Drost. 1993. A revised flora of Santa Barbara Island: an annotated catalog of the ferns and flowering plants and a brief history of botanical exploration. Santa Barbara Botanic Garden, Santa Barbara, CA. 59 pp.

Junak, S., T. Ayers, R. Scott, D. Wilken, and D. Young. 1995a. A flora of Santa Cruz Island. Santa Barbara Botanic Garden, Santa Barbara, CA and California Native Plant Society, Sacramento, CA. 397 pp.

Junak, S., W. Halvorson, C. Schwemm, and T. Keeney. 1995b. Sensitive plants of San Nicolas Island, California (Phase 1). Technical Report 51. National Biological Service, Cooperative Park Studies Unit, University of Arizona, Tucson, AZ. 91 pp.

Junak, S., W. Halvorson, C. Schwemm, and T. Keeney. 1996. Sensitive plants of San Nicolas Island, California (Phase 2). Technical Report 57. U.S. Geological Survey, Cooperative Park Studies Unit, University of Arizona, Tucson, AZ. 104 pp.

Junak, S., S. Chaney, R. Philbrick, and R. Clark. 1997. A checklist of vascular plants of Channel Islands National Park. 2nd edition. Southwest Parks and Monuments Association, Tucson, AZ. 43 pp.

Junak, S. 2003a. Distribution of native cacti (*Opuntia* spp.) and boxthorn *(Lycium californicum)* on San Nicolas Island, California. Technical Report 3. Santa Barbara Botanic Garden, Santa Barbara, CA. Report prepared for Southwest Division, Naval Facilities and Engineering Command, San Diego, CA. 30 pp.

Junak, S. 2003b. Exotic plant survey, Outlying Landing Field, San Nicolas Island, California. Technical Report 4. Santa Barbara Botanic Garden, Santa Barbara, CA. Report prepared for Southwest Division, Naval Facilities and Engineering Command, San Diego, CA. 151 pp.

Junak, S. 2003c. Sensitive plant survey, Outlying Landing Field, San Nicolas Island, California. Technical

Report 5. Santa Barbara Botanic Garden, Santa Barbara, CA. Report prepared for Southwest Division, Naval Facilities and Engineering Command, San Diego, CA. 98 pp.

Junak, S., D. Knapp, J. Haller, R. Philbrick, A. Schoenherr, and T. Keeler-Wolf. 2007. The California Channel Islands. Pp. 229-252 IN: M. Barbour, T. Keeler-Wolf, and A. Schoenherr (eds.) Terrestrial vegetation of California (3rd edition). University of California Press, Berkeley, CA.

Junger, A. and D. Johnson. 1980. Was there a Quaternary land bridge to the Northern Channel Islands? Pp. 33-39 IN: D. Power (ed.) The California Islands: proceedings of a multi-disciplinary symposium. Santa Barbara Museum of Natural History, Santa Barbara, CA.

Kerbavaz, J. 1985. Status reports on invasive weeds: pampas grass. Fremontia 12 (4): 18-19.

Kelley, J. 1923. Description of a trip to St. Nicholas Island in the year 1897. Unpublished manuscript on file at the San Diego Public Library, San Diego, CA. 25 pp.

Kemnitzer, L. 1933. Geology of San Nicolas and Santa Barbara islands, southern California. M.S. thesis, California Institute of Technology, Pasadena, CA. 45 pp., maps.

Kimberly, J. 1961. Account of my father's life. Unpublished manuscript on file at Santa Barbara Historical Society, Santa Barbara, CA. 6 pp.

Kimberly, J. (as told to M. Phillips). 1988. Fifty years and more in Santa Barbara. Noticias 34 (3): 50-62.

Kimberly, M. 1858. Pre-emption claim to 160 acres on San Nicolas Island. Santa Barbara County Official Records, Miscellaneous Book A, p. 105.

Kimberly, M. 1870. Indenture between M.M. Kimberly, William Hamilton, and Abraham Halsey, 15 September 1870. Ventura County Official Records, Deeds Book D, pp. 396-398.

Lamberth, R. and E. Lamberth. 1939. Log of San Nicolas Island 1938-1939. Unpublished manuscript on file at Santa Barbara Botanic Garden and Santa Cruz Island Foundation, Santa Barbara, CA.

Laughrin, L., M. Carroll, A. Bromfield, and J. Carroll. 1994. Trends in vegetation changes with removal of feral animal grazing pressures on Santa Catalina Island. Pp. 523-530 IN: W. Halvorson and G. Maender (eds.) The Fourth California Islands Symposium: update on the status of resources. Santa Barbara Museum of Natural History, Santa Barbara, CA.

Mabberley, D. 1997. The plant-book: a portable dictionary of the vascular plants (2nd edition). Cambridge University Press, Cambridge, U.K. 858 pp.

MacMullen, J. 1938. Expedition off for Channel; *Quaker* arrives. San Diego Union, 12 April 1938, p. II 7.

Manns, U. and A. Anderberg. 2005. Molecular phylogeny of *Anagallis* (Myrsinaceae) based on ITS, *trn*L-F, and *ndh*F sequence data. International Journal of Plant Sciences 166: 1019-1028.

Martz, P. 2005. Prehistoric settlement and subsistence on San Nicolas Island. Pp. 65-82 IN: D. Garcelon and C. Schwemm (eds.) Proceedings of the sixth California Islands symposium, Ventura, California, December 1-3, 2003. Institute for Wildlife Studies, Arcata, CA.

Mast, A. and J. Reveal. 2007. Transfer of *Dodecatheon* to *Primula* (Primulaceae). Brittonia 59 (1): 79-82.

Mathis, S. 1899. The lone woman of San Nicolas. Los Angeles Times, 8 January 1899, p. B11.

McCawley, W. 1997. Out where the wind blows and the breakers roll high: sheep ranching at the north shore ranch of San Nicolas Island. Technical Report 97-23. Statistical Research, Inc., Tucson, AZ and Redlands, CA. Report prepared for Naval Air Weapons Station, Point Mugu, CA. 131 pp.

McCoy, W. 1917. San Nicolas Island, time's hour glass. Outing 70: 775-787.

Meadows, D. 1939. Progress report of the Los Angeles Museum-Channel Islands Biological Survey: 4th expedition. Unpublished manuscript on file at Santa Barbara Botanic Garden, Santa Barbara, CA. 37 pp.

Miller, L. 1938. San Nicolas Island expedition. Unpublished field notes on file at the Santa Cruz Island Foundation, Santa Barbara, CA. 9 pp.

Miller, S. 1995. Terrestrial arthropod species considered endemic to the California Channel Islands. Unpublished manuscript on file at Santa Barbara Botanic Garden, Santa Barbara, CA. 11 pp.

Mills, H. and E. Tuttle. 1882. Indenture between Hiram W. Mills, E.L. Tuttle, E. Elliott, and J.V. Elliott, 7 August 1882. Ventura County Official Records, Deeds Book 24, pp. 60-62.

Millspaugh, C. and L. Nuttall. 1923. Flora of Santa Catalina Island, California. Botany Series, Publication

212. Field Museum of Natural History, Chicago, IL. 413 pp.

Moody, A. 2000. Analysis of plant species diversity with respect to island characteristics on the Channel Islands, California. Journal of Biogeography 27 (3): 711-723.

Moran, R. 1995. The subspecies of *Dudleya virens* (Crassulaceae). Haseltonia 3: 1-9.

Moran, R. 1996. The flora of Guadalupe Island, Mexico. Memoirs of the California Academy of Sciences 19: 1-190.

Mosyakin, S.L. 1996. A taxonomic synopsis of the genus *Salsola* (Chenopodiaceae) in North America. Annals of the Missouri Botanical Garden 83 (3): 387-395.

Neill, W. 1985. Status reports on invasive weeds: tamarisk. Fremontia 12 (4): 22-23.

Oberbauer, T. 2002. Analysis of vascular plant species diversity of the Pacific Coast islands of Alta and Baja California. Pp. 201-211 IN: D. Browne, K. Mitchell, and H. Chaney (eds.) Proceedings of the fifth California Islands symposium. 2 volumes. Santa Barbara Museum of Natural History, Santa Barbara, CA.

Olmstead, R., K-J. Kim, R. Jansen, and S. Wagstaff. 2000. The phylogeny of the Asteridae *sensu lato* based on chloroplast *ndh*F gene sequences. Molecular Pylogenetics and Evolution 16: 96–112.

Olmstead, R. 2002. Whatever happened to the Scrophulariaceae? Fremontia 30 (2): 13-22.

Orr, P. 1945. Return to San Nicolas. Museum Leaflet 20 (7): 75-79. Santa Barbara Museum of Natural History, Santa Barbara, CA.

Orr, P. 1968. Prehistory of Santa Rosa Island. Santa Barbara Museum of Natural History, Santa Barbara, CA. 253 pp.

Philbrick, R. 1963. Biosystematic studies of two Pacific Coast opuntias. Unpublished Ph.D. dissertation, Cornell University, Ithaca, NY. 177 pp.

Philbrick, R. 1964. *Opuntia oricola,* a new Pacific Coast species. Cactus and Succulent Journal of America 36 (6): 163-165.

Philbrick, R. 1972. The plants of Santa Barbara Island, California. Madroño 21 (5), part 2: 329-393.

Philbrick, R. and J. Haller. 1977. The Southern California Islands. Pp. 893-906 IN: M. Barbour and J. Major (eds.) Terrestrial vegetation of California. John Wiley and Sons, New York, NY.

Philbrick, R. 1980. Distribution and evolution of endemic plants of the California Islands. Pp. 173-187 IN: D. Power (ed.) The California Islands: proceedings of a multi-disciplinary symposium. Santa Barbara Museum of Natural History, Santa Barbara, CA.

Raven, P. 1963. A flora of San Clemente Island, California. Aliso 5 (3): 289-347.

Raven, P. 1967. The floristics of the California Islands. Pp. 57-67 IN: R. Philbrick. (ed.) Proceedings of the symposium on the biology of the California Islands. Santa Barbara Botanic Garden, Santa Barbara, CA.

Ray, M. 1998. New combinations in *Malva* (Malvaceae: Malveae). Novon 8: 288-295.

Rett, E. 1947. A report on the birds of San Nicolas Island. Condor 49 (4): 165-168.

Reveal, J. 1989. The eriogonoid flora of California (Polygonaceae: Eriogonoideae). Phytologia 66 (4): 295-414.

Rhodes, H. 1924. Memo to Commissioner of Lighthouses, Washington, D.C., 8 October 1924. U.S. National Archives, Record Group 26, Lighthouse Board Correspondence (1911-1939), File 253 (San Nicolas Island).

Roberts, L. 1991. San Miguel Island: Santa Barbara's fourth island west. Cal Rim Books, Carmel, CA. 214 pp.

Rodman, J., R. Price, K. Karol, E. Conti, K. Sytsma, and J. Palmer. 1993. Nucleotide sequences of the *rbcL* gene indicate monophyly of mustard oil plants. Annals of Missouri Botanical Garden 80: 686-699.

Rollins, R. 1993. The Cruciferae of continental North America: systematics of the mustard family from the Arctic to Panama. Stanford University Press, Stanford, CA. 976 pp.

Ross, T., S. Boyd, and S. Junak. 1997. Additions to the vascular flora of San Clemente Island, Los Angeles County, California, with notes on clarifications and deletions. Aliso 15 (1): 27-40.

Schumacher, P. 1877. Researches in the *kjokkenmoddings* and graves of a former population of the Santa

Barbara Islands and adjacent mainland. Bulletin of the U.S. Geological and Geographical Surveys of the Territories 3 (1): 37-56.

Schwartz, S. and P. Martz. 1992. An overview of the archaeology of San Nicolas Island, southern California. Pacific Coast Archaeological Society Quarterly 28 (4): 46-75.

Schwartz, S. 1994. Ecological ramifications of historic occupation on San Nicolas Island. Pp. 171-180 IN: W. Halvorson and G. Maender (eds.) The fourth California Islands symposium: update on the status of resources. Santa Barbara Museum of Natural History, Santa Barbara, CA.

Sebree, U. and E. Davis. 1901. Letter to Lighthouse Board, Washington, D.C., 25 September 1901. U.S. National Archives, Record Group 26, Lighthouse Board Correspondence (1901-1910), File 2998 (San Nicolas Island).

Smith, C. 1976. A flora of the Santa Barbara region, California. Santa Barbara Museum of Natural History, Santa Barbara, CA. 331 pp.

Strike, S. 1994. Ethnobotany of the California Indians. Volume 2. Aboriginal uses of California's indigenous plants. Koeltz Scientific Books, Champaign, IL. 210 pp.

Swanson, M. 1993. Historic sheep ranching on San Nicolas Island. Technical Series 41. Statistical Research, Inc., Tucson, AZ. Report prepared for Naval Air Weapons Station, Point Mugu, CA. 82 pp.

Sweet Van Hook, S. 1985. Status reports on invasive weeds: European beachgrass. Fremontia 12 (4): 20-21.

Tanaka, T. 1976. Tanaka's cyclopedia of edible plants of the world. Keigaku Publishing Company, Tokyo, Japan. 924 pp.

Thomsen, C., G. Barbe, W. Williams, and M. George. 1986. 'Escaped' artichokes are troublesome pests. California Agriculture 40 (3&4): 7-9.

Thorne, R. 1967. A flora of Santa Catalina Island, California. Aliso 6 (3): 1-77.

Thorne, R. 1969a. A supplement to the floras of Santa Catalina and San Clemente islands, Los Angeles County, California. Aliso 7 (1): 73-83.

Thorne, R. 1969b. The California Islands. Annals of Missouri Botanical Garden 56: 391-408.

Timbrook, J. 1984. Chumash ethnobotany: a preliminary report. Journal of Ethnobiology 4: 141-169.

Trask, B. 1899. Field notes from Santa Catalina Island. Erythea 7: 135-146.

Trask, B. 1900. Dying San Nicolas. Land of Sunshine 13 (2): 97-100.

Trask, B. 1904. Flora of San Clemente Island. Bulletin of the Southern California Academy of Sciences 3: 76-78, 90-95.

U.S. Department of Agriculture. 1985. Soil survey of Channel Islands area, California: San Nicolas Island part. Interim Report. U.S. Department of Agriculture, Soil Conservation Service, National Cooperative Soil Survey, Washington, D.C. 248 pp.

U.S. Navy. 1934. Revocable permit for sheep grazing on San Nicolas Island, granted to R.E. Agee and L.P. Elliott by Secretary of the Navy, 11 June 1934. Copy on file in Brooks binder, Santa Cruz Island Foundation, Santa Barbara, CA. 5 pp.

Uphof, J. 1968. Dictionary of economic plants. 2nd Edition. J. Cramer, Lehre, Germany. 599 pp.

Vedder, J. and R. Norris. 1963. Geology of San Nicolas Island, California. U.S. Geological Survey Professional Paper 369. U.S. Government Printing Office, Washington, D.C. 65 pp., maps.

Vedder, J. and D. Howell. 1980. Topographic evolution of the southern California borderland during late Cenozoic time. Pp. 7-31 IN: D. Power (ed.) The California Islands: proceedings of a multi-disciplinary symposium. Santa Barbara Museum of Natural History, Santa Barbara, CA.

Ventura Signal. 1876. Supplement to the regular newspaper on 16 Feb 1876 (Delinquent tax list for year 1875-1876).

Visel, C. 1923. Letter to the Secretary of Commerce, Washington, D.C., 22 November 1923. U.S. National Archives, Record Group 26, Lighthouse Board Correspondence (1911-1939), File 253 (San Nicolas Island).

Visel, C. 1926. Letter to H.W. Rhodes, Superintendent of Lighthouses, San Francisco, California, 31 March 1926. U.S. National Archives, Record Group 26, Lighthouse Board Correspondence (1911-1939), File 253 (San Nicolas Island).

Visel, C. 1928. Letter to H.W. Rhodes, Superintendent of Lighthouses, San Francisco, California, 7 June 1928. U.S. National Archives, Record Group 26, Lighthouse Board Correspondence (1911-1939), File 253 (San Nicolas Island).

Wallace, G. 1985. Vascular plants of the Channel Islands of southern California and Guadalupe Island, Baja California, Mexico. Natural History Museum of Los Angeles County, Contributions in Science 365: 1-136.

Weissman, D. and D. Rentz. 1976. Zoogeography of the grasshoppers and their relatives (Orthoptera) on the California Channel Islands. Journal of Biogeography 3: 105-114.

Wenner, A. and D. Johnson. 1980. Land vertebrates of the California Channel Islands: sweepstakes or bridges? Pp. 497-530 IN: D. Power (ed.) The California Islands: proceedings of a multi-disciplinary symposium. Santa Barbara Museum of Natural History, Santa Barbara, CA.

Westec Services, Inc. 1978. Survey of archaeological and biological resources of San Nicolas Island. Report prepared for Pacific Missile Test Center. 51 pp., maps & appendices.

Wheeler, S. 1944. California's little known Channel Islands. U.S. Naval Institute Proceedings 70 (3): 257-267.

Wilvert, C. 1980. Kikuyu grass, an African invader. Pacific Horticulture 41 (3): 45-47.

Windle, E. 1940. Windle's history of Santa Catalina Island. The Catalina Islander, Avalon, CA. 160 pp.

Woodward, A. 1939. Archaeological survey of San Nicolas Island, July 22-28, 1939. Unpublished field notes. Original document on file at Arizona Historical Society, Tucson, AZ. Copy transcribed and edited by S. Schwartz in 1993 on file at Naval Base Ventura, Point Mugu.

GLOSSARY OF TERMS

Compiled by Julie Broughton

abundant. Distribution on San Nicolas Island for which a taxon is exceptionally widespread and known from more than 30 localities.

acaulescent. Apparently without a stem; leaves and inflorescences usually basal; sometimes with a subterranean stem or a stem protruding only slightly above-ground.

achene. A dry, indehiscent fruit with 1 locule and 1 seed.

acuminate. Tapered (often abruptly) to a point, the margins concave.

acute. Tapered to an often sharp point; margins straight or somewhat convex but converging at less than a right angle.

adnate. Fusion of unlike parts, such as stamens fused to the corolla.

alternate. Either referring to leaf arrangement in which each node bears 1 leaf or to arrangement of flower parts, such as stamens located between petals rather than in front of petals.

annual. Completing a life cycle (from germination to reproduction and death) in one year or growing season.

anther. Pollen-producing part of a stamen.

anthesis. Flowering period; the time during which a flower is reproductively functional.

apex. The tip or distal end of a structure.

apiculate. Terminated abruptly with a minute point.

appressed. Pressed against or closely applied to the surface of a structure.

attenuate. Tapered gradually to a point, the margins straight.

ascending. Curving or angling upward from base.

awn. A bristle-like appendage, either as an appendage of a larger structure, as in the glumes and lemmas of some Poaceae, or as a separate part, as in the pappus of some Asteraceae.

axil. Angle formed at a node by a leaf or branch of the stem (adjective: axillary).

beak. With a prolonged, usually narrow, and thick apex.

biennial. Completing a life cycle (from germination to reproduction and death) in two years or growing seasons.

bilabiate. Bilateral flower with 2 (upper and lower) lips.

bilateral. A form of symmetry in which the structure, usually a flower, is divisible into 2 halves (mirror images).

bisexual. Flower with both fertile stamens and fertile pistils.

blade. The flattened, expanded part of a leaf or petal.

bract. A reduced, often modified, leaf-like structure. (adjective: bracteate).

callus. A thick, sharp to blunt extension of the lemma base in Poaceae.

calyx. Collective term for the outer series of the perianth or sepals, which are usually green and enclose the flower in bud.

campanulate. Bell-shaped.

canescent. Covered with fine, usually grayish to white trichomes.

capitate. Head-shaped or subglobose.

capsule. A dry fruit developing from a compound ovary (of one or more carpels), dehiscent from the apex by one or more lines of dehiscence.

caryopsis. The fruit of Poaceae, characterized by the fusion of the seed coat to the ovary wall.

catkin. A spike composed of unisexual flowers with inconspicuous perianths, sometimes pendent, and often with conspicuous bracts.

caulescent. With an obvious, leafy stem.

cauline. Borne on a stem; not basal.

ciliate. Leaves, petals, or sepals with marginal trichomes (diminutive: ciliolate).

common. Distribution on San Nicolas Island for which a taxon is widespread and conspicuous within one or more habitats and is known from 21-30 localities.

cordate. Heart-shaped.

corolla. Collective term for the inner series of the perianth or petals, which are usually colorful.

cyme. A branched inflorescence in which the terminal flower(s) open(s) before the lateral ones on any axis (adjective: cymose).

deciduous. Falling at the end of one season of growth or life; not evergreen.

decumbent. Stems whose bases lie or rest on the ground, but whose distal portions ascend or curve upward.

decussate. Usually applied to leaf arrangement in which pairs of opposite leaves alternate at right angles with those above and below.

deflexed. Abruptly bent downward.

dehiscent. A fruit or anther that opens by means of

sutures, lids, pores, or teeth.

deltate. Usually equilaterally triangular.

dentate. Margins bearing teeth (often sharp) that project at right angles to the leaf axis (diminutive: denticulate).

dichotomous. Branching repeatedly in pairs or forks, as in some stems or leaf venation.

didynamous. With 4 stamens, 2 of which are shorter than the other pair.

dioecious. Plants unisexual, the staminate and pistillate flowers on separate plants of the same species.

disjunct. A geographical distribution in which a significantly wide gap separates the distribution of plants within a species, as in some California taxa that also occur in Chile.

dorsal. The back or outer side (facing away from the axis) of a structure, such as leaves, bracts, petals, or fruits (opposite of ventral).

ellipsoid. A three-dimensional structure with the shape of an ellipse.

elliptic. A two-dimensional outline, shaped like an ellipse, with the length about twice as long as the width.

evergreen. Remaining green during the dormant season; usually applied to plants that retain their leaves throughout the year.

exserted. Extending beyond or out of, as with stamens that protrude beyond the throat of the corolla.

farinose. Covered with a meal-like powder.

fascicle. A dense cluster of flowers or leaves.

filiform. Thread-like.

flexuous. Having a more or less zigzag form; often applied to stems or stem-like structures.

foliaceous. Leaf-like; applied to sepals or bracts that resemble leaves in texture or appearance.

funnelform. Funnel-shaped, usually referring to a corolla, with the tube gradually widening toward the throat.

glabrate. Becoming glabrous with maturity or almost glabrous.

glabrous. Surfaces without hairs.

glandular. With gland-like trichomes or secretions.

glaucous. Covered with whitish to grayish, waxy or powdery film that is often rubbed off easily.

globose. A three-dimensional structure that is spherical or round in outline.

hastate. With the shape of an arrowhead, but with the basal lobes spreading almost at right angles.

hirsute. Covered with coarse, stiff trichomes (diminutive: hirsutulous).

hispid. Covered with stiff or bristly trichomes (diminutive: hispidulous).

hyaline. Translucent to almost transparent.

hypanthium. Structure derived from the fusion of sepals, petals, and stamens, often forming a tube.

imbricate. Overlapping, as in the scales of a cone or the bracts of some involucres.

incurved. Bending inwards.

indehiscent. A fruit or anther that does not open by means of sutures, lids, pores, or teeth.

indusium. In ferns, the leaf appendage that covers or encloses the sorus or sori.

inflorescence. The arrangement of flowers on the rachis or in a cluster.

intergrade. To merge gradually from one form to another through a more or less continuous series of intermediates.

internode. The portion of the stem between two adjacent nodes.

involucre. One or more whorls of bracts, usually subtending 1 or more flowers.

involute. Margins that are rolled toward the upper surface.

keeled. With a ridge or crease, usually along the dorsal side of a folded structure, as in the glumes or lemmas of certain Poaceae.

laciniate. Margins that are deeply and often irregularly cut.

lanate. Covered with loosely entangled, long, woolly trichomes.

lanceolate. Shaped like a lance; narrow, tapered at both ends and widest at a point between the middle and the petiole.

lenticular. Shaped like a lens or discus, with both sides convex.

ligule. In Asteraceae, the strap-shaped part of a ray corolla; in Poaceae, the appendage located at the junction of the blade and the sheath.

limb. The expanded, often flat, apical portion of a structure (e.g., corolla).

linear. Shape that is long, narrow, and with a uniform width.

lobe. A segment or division of a structure, usually with sinuses that extend up to one-half the distance to the midrib or center of a leaf blade, petal, or ovary.

locule. The chamber(s) within an ovary, anther or fruit.

margin. The edge of a leaf blade or perianth part.

membranous. With the consistency of a membrane; usually thin and soft.

mericarp. One segment of a schizocarp (indehiscent fruit) that falls from the fruit when the fruit splits lengthwise along the septum; the seeds remain enclosed in the original locule.

monoecious. Plants bisexual, but with both kinds of unisexual flowers (staminate and pistillate) on the same plant.

mucronate. With a sharp terminal point or spiny tip, as at the apex of a leaf or bract.

naturalized. Populations of alien (non-native) plant taxa that reproduce (usually by seed) from one generation to the next without human intervention.

nectary. Gland-like structure that secretes nectar.

node. The joint of a stem at which leaves are attached and axillary buds are produced.

nutlet. Small, dry, indehiscent fruit with hard walls.

ob-. Prefix used to describe a shape with attachment at other end (e.g., obconic: cone-shaped, but with point of attachment at narrow end; oblanceolate: a lanceolate-shaped structure, but with point of attachment at narrowest end; obovate: egg-shaped, but with point of attachment at narrow end).

oblong. Shape that is longer than broad, but with parallel sides.

occasional. Distribution on San Nicolas Island for which a taxon is fairly widespread within one or more habitats but with a scattered distribution and known from 11-20 localities.

opposite. Either referring to leaf arrangement in which each node bears 2 leaves or to arrangement of flower parts, such as stamens located in front of petals.

ovary. The part of the pistil that encloses the ovules and that develops into a fruit after pollination.

ovate. Outline in the shape of an egg, with attachment at widest end.

ovoid. A three dimensional structure with the shape of an egg.

ovule. Structure inside ovary that develops into a seed.

palea. A chaff-like or scale-like bract. In Asteraceae, a scale-like pappus; in Poaceae, the upper and smaller of two bracts subtending a flower.

paleaceous. Chaff- or scale-like.

palmate. Radiating from one common point, like the fingers of an open hand; usually used to describe the venation or lobing of a leaf or other structure.

panicle. A compound inflorescence in which the branches are racemose (adjective: paniculate).

papilionaceous. A bilateral corolla in the Fabaceae, composed of a banner, 2 wings, and 2 additional petals usually fused into a keel.

papilla. Swollen or rounded projections.

papillate. Covered with papillae (diminutive: papillose).

pappus. The highly modified calyx of Asteraceae, composed of awns, bristles, capillary (hair-like) segments, or scales.

pedicel. The stalk that subtends a single flower (adjective: pedicellate).

peduncle. The stalk that subtends an inflorescence (adjective: pedunculate).

peltate. A flat structure subtended by a stalk that is attached to its lower surface instead of the margin.

perennial. Living for more that two years or growing seasons and flowering and fruiting repeatedly throughout the life of the plant.

perianth. Collective term for the calyx and corolla.

petal. One segment or lobe of the corolla.

petaloid. Petal-like; usually used to describe a structure (often a sepal) that resembles a petal in color, shape, or position.

petiole. A stalk that subtends a leaf blade and connects it to the stem.

pilose. Covered with soft, straight trichomes.

pinnate. Radiating in 2 rows from a linear axis, like the parts of a feather; used to describe the veins or lobes of a leaf or to describe the arrangement of leaflets in a pinnately compound leaf.

pinnatifid. A deeply lobed or dissected leaf with pinnate venation.

pistil. Female reproductive structure of a flower, composed of an ovary, style, and stigma.

pistillate. A unisexual flower bearing only pistils.

placentation. The arrangement or orientation of the placentas within the ovary, and to which the ovules are attached.

plumose. Plume-like; usually used to describe a structure with lateral, hair-like processes or appendages, such as the pappus of Asteraceae.

pollinium. The pollen mass derived from anthers in either the Asclepiadaceae or Orchidaceae.

polygamous. With both unisexual and bisexual flowers on the same plant.

prostrate. Usually used to describe stems that are

lying flat on the ground.

puberulent. The diminutive form for pubescent, in which trichomes are scarcely visible to the unaided eye.

pubescent. Usually used to describe a surface with short, soft trichomes.

punctate. A surface characterized by translucent to opaque, often glandular dots, pits, or depressions.

pustulate. Trichomes with swollen bases, like a blister (diminutive: pustulose).

raceme. A simple inflorescence (without branches) composed of a single rachis bearing pedicellate flowers (adjective: racemose).

rachis. The central axis of an inflorescence or of a pinnately compound leaf.

radial. A form of floral symmetry in which the perianth parts radiate from the center, like the spokes of a wheel.

rare. Distribution on San Nicolas Island for which a taxon is known from only one to three, often very restricted localities.

receptacle. The part of the flower on which the calyx, corolla, stamens, and pistils are attached; in Asteraceae the flat surface of the head, on which flowers are attached.

recurved. Curved gradually, either downward or backward.

reflexed. Bent downward.

reniform. Kidney-shaped.

reticulate. Net-like; with a network of veins or minute ridges.

retrorse. Usually used to describe trichomes that are bent abruptly downward or backward.

revolute. Margins that are rolled toward the lower surface.

rhizome. A horizontal, underground stem, with scale-like leaves and often producing leafy shoots (adjective: rhizomatous).

rhombic. Outline in the shape of a diamond, angular, often with 4 straight edges, and with the widest point midway between the base and the apex.

rosette. An arrangement of leaves in which leaves are crowded and basal.

rotate. A corolla or calyx with a very short tube and an abruptly spreading, flat, often circular limb.

salverform. A corolla with a long, slender tube and an abruptly spreading, flat limb.

scaberulous. Diminutive form of scabrous.

scabrous. Surface that feels rough to the touch, like sand-paper; covered with short, stiff trichomes.

scape. An inflorescence and peduncle usually arising directly from a basal rosette of leaves (adjective: scapose).

scarce. Distribution on San Nicolas Island for which a taxon is locally common but known only from 4-10 localities.

scarious. Usually used to describe leaf or bract margins that are thin, dry, tan to brownish, and often translucent.

schizocarp. A dry indehiscent fruit that splits at maturity into 1-few-seeded, closed segments (mericarps).

scurfy. A surface covered with minute, flat scale-like trichomes.

sepaloid. A sepal-like structure (often a bract) that resembles a sepal in color, shape, or position.

septum. Wall that separates locules in an ovary or fruit.

sericeous. Usually used to describe long, slender, very soft (silky) trichomes that are often appressed.

serrate. Margins bearing teeth (often sharp) that project forward or toward the leaf apex (diminutive: serrulate).

sepal. One segment or lobe of the calyx.

sessile. Attached directly by the base, such as leaves without petioles and flowers without pedicels.

setaceous. With a bristle-like structure or bristle-like apex.

setose. A surface covered with bristle-like trichomes.

sinuate. A margin that appears wavy or undulate, with waves that are parallel to the plane of the blade.

sinus. The indentation or cleft between the lobes of a margin.

sorus. The part of a fern leaf or leaflet that bears sporangia (plural: sori).

spatulate. Shaped like a spatula, with a rounded apex and tapered gradually to the base.

spike. A simple inflorescence (without branches) composed of a single rachis bearing sessile flowers (adjective: spikate).

spine. Stiff, sharp-pointed structure; usually arising from the epidermis (adjective: spinose; diminutive: spinulose).

spinescent. Spine-like

sporangium. The sac or case-like structure enclosing spores in plants not producing seeds.

spore. A simple, reproductive, single-celled structure, usually haploid and capable of developing into a new individual.

squarrose. With tips spreading rigidly at right

angles or reflexed; usually used to describe phyllaries in the Asteraceae.

stamen. Male reproductive structure of a flower, usually composed of a stalk-like filament and a terminal, pollen-producing anther.

staminate. A unisexual flower bearing only stamens.

staminode. A sterile stamen, often highly modified in appearance, and sometimes petal-like

stellate. Star-shaped; often referring to a trichome with three or more branches radiating from a common point.

stigma. The part of the pistil on which pollen is normally deposited and germinates (adjective: stigmatic).

stipe. Usually used to describe the stalk subtending an ovary in the genus *Astragalus* and the family Euphorbiaceae (adjective: stipitate).

stipules. A pair of appendages at the base of the petiole, variable in form but often leaf-like or bract-like (e.g., Rubiaceae).

stolon. A horizontal, aerial stem, often with long internodes producing leafy shoots (adjective: stoloniferous).

strigose. A surface covered with stiff, appressed, straight hairs (diminutive: strigulose).

style. The narrow segment of a pistil that separates the ovary from the stigma.

sub-. A prefix, usually meaning "almost", "slightly", or "somewhat".

subulate. Triangular in shape and sharp-pointed, like an awl.

subtend. Located below or beneath a structure, often in close proximity.

succulent. Fleshy (with a high water content) and often thickened.

taproot. The primary root of an annual or biennial plant, usually with lateral secondary roots.

tendril. A slender, coiled structure by which a climbing plant becomes attached to its support and derived either from leaves, leaflets, stipules, or stems.

terete. Cylindrical and usually round in cross section.

terminal. At the apex or top of a structure.

ternate. Usually used to describe a compound leaf divisible into 3 leaflets.

tomentose. Covered with densely interwoven, usually matted trichomes (diminutive: tomentulose).

trichome. A hair-like, epidermal structure, composed of 1 or more cells.

trifoliate. With 3 leaves.

trifoliolate. A compound leaf with 3 leaflets.

truncate. Usually used to describe an apex or base that appears cut off at right angles to the axis.

tube. A hollow structure, such as the fused portion of a corolla.

tuberculate. Covered with tubercles, which are swollen, wart-like or tuber-like projections on a surface.

turbinate. The opposite of conic or obconic; top-shaped.

umbel. A simple inflorescence with pedicels of equal length and arising from a common point (adjective: umbellate).

undulate. A margin that appears wavy with waves perpendicular to the plane of the blade.

unisexual. Flower with either fertile stamens or fertile pistils, but not both.

utricle. An indehiscent, bladder-like fruit, usually with 1 seed.

vascular. With conductive tissue (e.g., xylem and phloem).

ventral. The front or inner side (facing the axis) of a structure, such as leaves, bracts, petals, or fruits (opposite of dorsal).

villous. With long, soft, unmatted trichomes (diminutive: villosulous).

whorled. Either referring to leaf arrangement in which each node bears 3 or more leaves or to the arrangement of flower parts, in which many parts radiate from the center.

APPENDIX I. Representative Herbarium Specimens.

For most taxa, no more than six specimens are cited. Voucher specimens are listed in essentially the same geographic order used in the taxonomic treatments and are deposited at the Santa Barbara Botanic Garden (SBBG) unless otherwise indicated. Standard herbarium abbreviations (Holmgren *et al.*1990) are used for other depositories (i.e., CAS for California Academy of Sciences, DS for Dudley Herbarium of Stanford University, GH for Harvard University Herbaria, MO for Missouri Botanical Garden, ND-G for Notre Dame-Greene Herbarium of Notre Dame University, NY for New York Botanical Garden, OSC for Oregon State University at Corvallis, POM for Pomona College, RSA for Rancho Santa Ana Botanic Garden, SD for San Diego Natural History Museum, UC for University of California at Berkeley, and US for United States National Herbarium).

No attempt has been made to cite every herbarium where a specific collection may be housed. Multiple specimens from the same location are listed chronologically. Herbarium accession numbers are cited only if there is no collector number. Label data on some specimens have been edited for consistency (e.g., punctuation). Most place names used by collectors can be found on the map in this publication. Locations of benchmarks and triangulation stations can be found on the U.S. Geological Survey topographic map of the island.

Pteridophytes
POLYPODIACEAE

Polypodium californicum Kaulf.: lower portion of Live-forever Canyon, 25 Apr 1989, *Junak SN-381;* small side gully in lower portion of Celery Canyon, 25 Apr 1985, *Junak et al. SN-131;* n of airfield, in e fork of first large canyon e of Airfield Grade on Beach Road, 10 Mar 1992, *Junak SN-687.*

PTERIDACEAE

Adiantum jordanii Müller Halle: ne coastal flats, in second canyon w of "L" Canyon, 23 May 1995, *Junak SN-1219;* w fork of Jetty Canyon, 16 May 1995, *Junak SN-1184.*

Pellaea andromedifolia (Kaulf.) Fée: along Beach Road at Airfield Grade, in first large canyon e of upper hairpin turn, 10 Mar 1992, *Junak SN-682;* ne coastal flats, in second canyon w of "L" Canyon, 23 May 1995, *Junak SN-1218;* ne corner of island, on s side of S Spur Canyon, 24 Mar 1993, *Junak SN-1068.*

Pentragramma triangularis (Kaulf.) Yatsk., Windham & E. Wollenw. subsp. **triangularis:** n escarpment, between forks of E Mesa Canyon, ca. 0.3 mi nw of Benchmark 396, 10 Mar 1993, *Junak SN-982;* w fork of Jetty Canyon, 16 May 1995, *Junak SN-1182;* ne portion of mesa, in upper reaches of E Mesa Canyon, 11 Apr 1989, *Junak SN-333.*

Dicotyledonous Angiosperms
AIZOACEAE

Carpobrotus chilensis (Molina) N.E.Br.: along road to Red Eye Beach, 0.3 mi from main road, 10 Jun 1992, *Junak SN-946;* e portion of Red Eye Beach, N of road, 31 Mar 1992, *Junak SN-750;* E Sea Lion Beach, s of Jackson Hill, 10 Jun 1969, *Philbrick and Benedict B69-174.*

Carpobrotus edulis (L.) N.E.Br.: canyon running ne from camp area to ocean, 23 Apr 1961, *Blakley 4169;* living compound, next to tennis courts, 29 Nov 1988, *Junak SN-274.*

Delosperma litorale (Kensit) L.Bolus: lower end of Tule Creek and dunes toward coast, ese of Thousand Springs, 3-4 Apr 1979, *Thorne et al. 52424;* sand dunes on w side of Tule Creek, just downstream from pumping station at Humphrey Sump, 24 May 1995, *Junak SN-1234.*

Malephora crocea (Jacq.) Schwantes var. **crocea:** s edge of mesa, just s of Building #112, 20 May 1993, *Junak SN-1118.*

Mesembryanthemum crystallinum L.: San Nicolas Island, Apr 1901, *Trask 46a* (GH,US); nw end of island, Seal Beach, on sandy slopes just above ocean, 23-24 Apr 1966, *Raven & Thompson 20704;* clay flat s of camp area, 22 Apr 1961, *Blakley 4148.*

Mesembryathemum nodiflorum L.: sea beach, Apr 1897, *Trask s.n.* (US 340164); San Nicolas Island,

Apr 1901, *Trask 46* (GH); nw end of island, Seal Beach, on sandy slopes just above ocean, 23-24 Apr 1966, *Raven & Thompson 20692* (SD); Corral Harbor, on sandy back beach, 22 Jul 1939, *Dunkle 8302* (RSA,SBBG); small canyon e of living quarters, 28 Jul 1965, Foreman et al. 72 (UC); open field near airport terminal, 27 Jul 1965, *Foreman et al.* 4; sand dune area near where escarpment road starts up escarpment, 7 Apr 1966, *Foreman & Smith 183* (UC).

Tetragonia tetragonioides (Pallas) Kuntze: w end of Red Eye Beach, 25 Apr 1985, *Junak & Laughrin SN-132.*

ANACARDIACEAE

Schinus molle L.: living compound, along road to Public Works building, 29 Nov 1988, *Junak SN-272.*
Toxicodendron diversilobum (Torr. & A.Gray) Greene: n coastal flats on w side of lower portion of Celery Canyon, at base of n escarpment, 22 May 1990, *Junak SN-508.*

APIACEAE

Apiastrum angustifolium Nutt.: among *Opuntia,* Apr 1901, *Trask 48* (GH); amid cacti, Apr 1901, *Trask 49* (RSA); ne end of island, in N Spur Canyon, 23 Mar 1993, *Junak SN-1035;* ne coastal flats, e of Beach Road, just s of first canyon s of S Spur Canyon, 24 Mar 1993, *Junak SN-1059.*
Apium graveolens L.: Corral Harbor, in seepage of sea cliffs, 22 Jul 1939, *Dunkle 8305* (RSA,SBBG); mouth of cove near old landing pier, 16 Mar 1945, *Rett & Orr s.n.* (SBBG 78824); near pump house near Windmill Springs, 29 Jun 1983, *Vanderwier s.n.* (SBBG 94205); nw end of mesa, along s side of Tufts Road in Wells area, 15 Mar 1994, *Junak SN-1136.*
Berula erecta (Hudson) Coville: spring just e of Thousand Springs, 11 Jun 1969, *Philbrick & Benedict B69-196;* western Thousand Springs, 23 Jun 1977, *Philbrick & Kritzman B77-22;* seepage behind foredunes at Thousand Springs, 30 Jun 1978, *Wier & Beauchamp s.n.* (RSA); Thousand Springs, 29 Jul 1979, *Daily & Bromfield SNI-172.*
Conium maculatum L.: Tule Canyon, 23 Jul 1939, *Dunkle 8333;* near lower end of Tule Creek, 22 Apr 1961, *Blakley 4113* (CAS,SBBG), Tule Canyon, at Thousand Springs pump area, 7 Apr 1966, *Foreman & Smith 184;* spring in Thousand Springs area, 23-24 Apr 1966, *Raven & Thompson 20762* (DS,SBBG); Tule Creek, 29 Jul 1979, *Daily SNI-163.*
Daucus carota L.: Windmill Spring area, 18 Jul 1981, *Newman s.n.* (SBBG 94204); along Jackson Highway, 100 yds e of Shannon and Jackson roads, on s side of intersection, 12 Oct 1983, *Vanderwier s.n.* (SBBG 94146).
Daucus pusillus Michx.: ne escarpment sw of Rock Jetty, in e fork of Jetty Canyon, 16 May 1995, *Junak SN-1177;* mesa s of housing area, 22 Apr 1961, *Blakley 4098;* s escarpment, along Theodolite Road, 21 May 1990, *Junak SN-481.*
Foeniculum vulgare L.: canyon running ne from camp area to ocean, 23 Apr 1961, *Blakley 4171;* in mouth of small canyon e of living quarters, 28 Jul 1965, *Foreman et al. 74* (UC); near coastline in n-central portion of island, 3 Jun 1987, *Junak SN-224.*
Lomatium insulare (Eastw.) Munz: sand cliffs and arroyos, Apr 1901, *Trask 81* (GH); n coastal flats, along Beach Road n of airfield runways, 20 Mar 1983, *Junak et al. SN-43;* mesa, in e fork of second large canyon w of airfield runways, 1 Apr 1992, *Junak SN-784;* sw coast, 11 Feb 1949, *Moran 3169* (SD); se end of island, 18 May 1967, *Boutin & Gonderman 1601.*
Sanicula arguta J.M.Coult. & Rose: dry rocky heights, Apr 1897, *Trask s.n.* (US 340257); one locality, Apr 1901, *Trask 94* (RSA,US); slopes and arroyos above pier near se end of island, 23-24 Apr 1966, *Raven & Thompson 20780;* n edge of mesa, in upper w fork of W Mesa Canyon, 4 Apr 1989, *Junak SN-295;* s escarpment, in upper e fork of Twin Rivers drainage, 28 Mar 1991, *Junak SN-602.*
Torilis nodosa (L.) Gaertn.: in upper reaches of Mineral Canyon, 21 May 2001, *Junak SN-1703.*

ASTERACEAE

Achillea millefolium L.: ne coast, just w of Rock Jetty, 10 Mar 1992, *Junak SN-676;* ne portion of mesa, just s of w end of airfield runways, 7 Mar 1991, *Junak SN-538;* s escarpment, in upper e fork of Twin

Rivers drainage, 28 Mar 1991, *Junak SN-603;* se end of island, near Beach Road, 6 Jun 1977, *Hesseldenz SNI-13;* s side of island, just sw of Sand Spit, 23 May 1990, *Junak SN-514.*

Amblyopappus pusillus Hook. & Arn.: covering the uplands, Apr 1897, *Trask s.n.* (RSA 416889); upland ridges, Apr 1901, *Trask s.n.* (GH); n side, canyon bottom, 23 Jul 1939, *Dunkle 8336* (DS,NY,SBBG); ne coastal flats between N and S Spur canyons, 16 Mar 1994, *Junak SN-1141;* mesa s of housing area, 22 Apr 1961, *Blakley 4097;* se escarpment, on slope overlooking Sand Spit, 1 May 1991, *Junak SN-629.*

Ambrosia chamissonis (Less.) Greene: sand dunes, Apr 1901, *Trask 77* (GH,RSA); flats above Vizcaino Point, 21 Aug 1990, *Junak SN-530;* near Gardens, on sandy bench near sea, 23 Jul 1939, *Dunkle 8315* (NY,RSA,SBBG); along Shannon Road, near center of island, 18 May 1967, *Boutin & Gonderman 1561;* mesa s of housing area, 22 Apr 1961, *Blakley 4104.*

Ambrosia psilostachya DC.: ne side of island, on gravel pile along Beach Road, 0.2 mi e of "L" Canyon, 19 Dec 1995, *Junak SN-1263;* mesa, w of airfield terminal along s edge of runways, 21 Mar 1983, *Junak et al. SN-57;* ne portion of mesa, on sw side of airfield runways, 19 Nov 1991, *Junak SN-666.*

Artemisia californica Less.: n coastal flats, at first canyon w of Celery Canyon, 22 Feb 1990, *Junak SN-439;* se escarpment, in Towers Canyon, 11 Mar 1992, *Junak SN-711;* extreme e end of island, at head of second canyon s of Divide Ridge, 18 May 1993, *Junak SN-1103.*

Artemisia nesiotica Raven: canyon mouth near ranch, 25 Jul 1939, *Dunkle 8346;* upper portion of canyon ne of barracks, ca. 0.5 mi e of Celery Creek, 12 Jun 1969, *Philbrick & Benedict B69-204;* slopes and arroyos above pier near se end of island, 23-24 Apr 1966, *Raven & Thompson 20787;* s escarpment, just sw of Twin Towers, 13 May 1992, *Junak SN-906.*

Baccharis pilularis DC.: nw end of island, ca. 1 mi ese of Seal Beach, 23-24 Apr 1966, *Raven & Thompson 20716;* n side, deep canyon, 28 Mar 1945, *Rett & Orr s.n.* (SBBG 79596); mesa s of housing area, 22 Apr 1961, *Blakley 4095;* Big Creek Canyon, 25 Jul 1939, *Dunkle 8345.*

Baccharis salicifolia (Ruiz & Pav.) Pers.: wells area on n side of Tufts Road, 7 Mar 1991, *Junak SN-555;* mesa just ne of living compound, in upper portion of w fork of W Mesa Canyon, 21 May 1990, *Junak SN-479;* borrow pit along Monroe Drive, sw of w end of airfield runways, 22 Mar 1984, *Junak SN-95;* artificial pond, s of w end of runway, 12 Jun 1969, *Philbrick & Benedict B69-210;* Daytona Beach, at barge landing area, 24 Apr 1989, *Junak SN-358.*

Calendula officinalis L.: w of distillation plant at Coast Guard Beach, 22 Jul 1977, *Philbrick & Kritzman B77-11;* e side of Beach Road near desalination plant, 4 Aug 1982, *Vanderwier s.n.* (SBBG 94197); along Beach Road ca. 0.5 mi w of abandoned desalination plant at Rock Jetty, 22 Mar 1984, *Junak SN-90.*

Centaurea melitensis L.: one locality, Apr 1897, *Trask 69* (MO); living compound, 21 May 1990, *Junak SN-475;* mesa, 0.5 mi se of barracks, between Celery Creek and artificial pond, 10 Jun 1969, *Philbrick & Benedict B69-137;* flats at southern edge of mesa, e of Desert Fan Canyon, 26 May 1992, *Junak SN-925;* summit of island, 23-24 Apr 1966, *Raven & Thompson 20769;* sw end of island, 9 Jun 1992, *Junak SN-945.*

Centaurea solstitialis L.: living compound, along w rim of Celery Canyon, n of Public Works building, 14 Sep 1999, *Junak SN-1435;* living compound, in disturbed areas along n side of Public Works building, 15 Sep 1999, *Junak SN-1442.*

Chrysanthemum coronarium L.: well area, Shannon Road near headwaters of Tule Creek, 22 Jun 1977, *Philbrick & Kritzman B77-18;* wells area of mesa, 26 Apr 1983, *Vanderwier s.n.* (SBBG 94193); nw portion of mesa, near Building #120 on s side of Tufts Road, 22 May 1985, *Junak et al. SN-136;* along Tufts Road at wells area, 28 Mar 1991, *Junak SN-594.*

Cirsium occidentale (Nutt.) Jeps. var. **occidentale:** n escarpment, in e fork of Keyhole Canyon, 10 Jun 1993, *Junak SN-1130;* ne side of island, in large canyon just e of hairpin turn in Beach Road, 23 May 1990, *Junak SN-519;* ne escarpment sw of Rock Jetty, in w fork of Jetty Canyon, 16 May 1995, *Junak SN-1183;* slopes and arroyos above pier near se end of island, 23-24 Apr 1966, *Raven & Thompson 20778;* ne edge of mesa, just n of Benchmark 396, 23 Mar 1993, *Junak SN-1020.*

Conyza bonariensis (L.) Cronquist: spring seepage at northwestern end of mesa, 23-24 Apr 1966, *Raven & Thompson 20735* (DS,RSA); about living quarters on mesa, 27 Jul 1979, *Daily SNI-105;* Public Works yard, 12 May 1983, *Vanderwier s.n.* (SBBG 94195).

Conyza canadensis (L.) Cronquist: Thousand Springs Road near the mouth of Tule Canyon, 19 Dec 1965,

Foreman & Lloyd 149 (UC); ne coastal flats, on gravel pile w of Rock Jetty, 19 Nov 1991, *Junak SN-662;* small canyon due e of living quarters, 28 Jul 1965, Foreman et al. 77 (RSA,SBBG); on mesa, at living compound, 17 Oct 1985, *Junak et al. SN-169.*

Coreopsis gigantea (Kellogg) H.M.Hall: San Nicolas Island, Apr 1901, Trask s.n. (POM 356046); three localities, at foot of *Opuntia,* Apr 1901, *Trask 76* (GH); along old marine terrace 0.2 mi nw of Coast Guard pier, 28 Jul 1965, *Foreman et al. 64;* midway along sw coast, 18 May 1967, *Boutin & Gonderman 1603;* s escarpment, just e of Theodolite Road, 8 Mar 1993, *Junak SN-963.*

Cotula coronopifolia L.: San Nicolas Island, 22 Sep 1945, *Rett 667;* Celery Canyon, 7 Apr 1966, *Foreman & Smith 198;* wet arroyo about 1 mi e of Celery Canyon, on road to Navy dock, 7 Aug 1969, *Benedict s.n.* (SBBG 33778); small canyon se of Triangulation Point "Port", 21 Apr 1961, *Blakley 4089;* Building #120, 29 Jul 1979, *Daily SNI-157.*

Cynara cardunculus L. subsp. **flavescens** Wiklund: nw portion of mesa, just w of Building #120 on Tufts Road, 23 May 1985, *Junak & Vanderwier SN-147;* along s side of Tufts Road, just w of water treatment plant, 23 May 2000, *Junak SN-1547.*

Deinandra clementina (Brandegee) B.G. Baldwin: one locality, sea cliffs, Apr 1901, *Trask 75* (NY); ne coast, 25 Jul 1939, *Dunkle 8350;* mesa, s of western portion of airfield runways, 22 Mar 1984, *Junak SN-103;* sw escarpment, in main fork of lower Cattail Canyon, 19 May 1993, *Junak SN-1109;* s coastal flats, just e of Daytona Beach, 27 May 1992, Junak *SN-928.*

Encelia californica Nutt.: along road to Thousand Springs, ca. 0.25 mi from coastline, 19 Mar 1983, *Junak SN-23;* well area, Shannon Road near headwaters of Tule Creek, 22 Jun 1977, *Philbrick & Kritzman B77-19;* wells area on Tufts Road, 7 Mar 1991, *Junak SN-554.*

Gnaphalium palustre Nutt.: upland ridge, Apr 1901, *Trask 81* (GH); mesa s of airfield runways, w of passenger terminal building, just s of access road from Beach Road, 18 Oct 2001, *Junak SN-1722.*

Grindelia hirsutula Hook. & Arn.: along Shannon Road just n of Tufts Road, 2 Jun 1987, *Junak SN-196.*

Helianthus annuus L.: n coastal flats, on gravel piles along Beach Road, ca. 0.2 mi w of Rock Jetty, 11 Jun 1996, *Junak SN-1269.*

Heterotheca grandiflora Nutt.: w end, large dune area, 20 Jul 1981, *Newman s.n.* (SBBG 94304); mesa at living quarters, 27 Jul 1979, *Daily SNI-98;* mesa, at living compound, 17 Oct 1985, *Junak et al. SN-170;* roadside near intersection of Monroe and Owens roads, 9 Nov 1984, *Vanderwier s.n.* (SBBG 94302); along Monroe Road near air terminal, 29 Mar 1983, *Vanderwier s.n.* (SBBG 94305).

Hypochaeris radicata L.: mesa, outside weight room at compound, 27 Jul 1979, *Daily SNI-125.*

Isocoma menziesii (Hook. & Arn.) G.L.Nesom var. **menziesii:** marine terrace e of living quarters, 17 Dec 1965, *Foreman & Lloyd 128* (SD); in open dunes ca.1 mi ese of Seal Beach, 23-24 Apr 1966, *Raven & Thompson 20715* (SD); near ocean e of Dutch Harbor, 23-24 Apr 1966, *Raven & Thompson 20756* (SD).

Isocoma menziesii var. **vernonioides** (Nutt.) G.L.Nesom: on beach at mouth of canyon running sw from Jackson Hill, 23 Apr 1961, *Blakley 4161* (SBBG,SD).

Lactuca serriola L.: n coastal flats, at foot of road from living compound, 31 May 1984, *Junak SN-114;* nw end of mesa, along Tufts Road near water well buildings, 6 Sep 1989, *Junak SN-421;* along airfield runways and roads, 28 Jul 1965, *Foreman 112* (UC); along road to artificial pond, s of airstrip, 7 Aug 1969, *Benedict s.n.* (SBBG 33777); se end of airstrip, 16-17 Nov 1971, *Benedict s.n.* (SBBG 43410).

Lasthenia gracilis (DC.) Greene: common on ridges, Apr 1901, *Trask 74* (GH); ne coastal flats, just e of E Mesa Canyon, 10 Mar 1993, *Junak SN-985;* near n edge of mesa, on w side of first canyon e of living compound, 19 Mar 1983, *Junak & Kuizenga SN-31;* n edge of mesa, nw of airfield runways, 27 Mar 1991, *Junak SN-565;* se end of mesa, just s of Building #121, 7 Mar 1991, *Junak SN-545;* se escarpment, just uphill from barge landing at Daytona Beach, 24 Apr 1989, *Junak SN-360.*

Malacothrix foliosa subsp. **polycephala** W.Davis: covering large areas on the ridge, Apr 1897, *Trask s.n.* (MO 1892070); just e of Tule Creek, near Triangulation Point "Corral", 11 Jun 1969, *Philbrick & Benedict B69-124;* near mouth of Celery Canyon, 28 Jul 1965, *Foreman et al. 80;* n edge of mesa, nw of airfield runways, 27 Mar 1991, *Junak SN-568;* sw escarpment, w of Benchmark 688, 12 May 1992, *Junak SN-868;* s escarpment, in upper e fork of Twin Rivers drainage, 2 May 1991, *Junak SN-644.*

Malacothrix incana (Nutt.) Torr. & A.Gray: dunes e of Corral Harbor, 29 Apr 1978, *Daily 73;* Celery

Canyon, 3 Apr 1979, *Wallace et al. 1627* (NY); stabilized dunes between Beach Road and NavFac, 4 May 1982, *Vanderwier 71;* nw end of island, near Tufts Road, 15 Apr 1982, *Vanderwier 52;* mesa, along Jackson Highway, 20 Jul 1981, *Newman s.n.* (SBBG 94293).

Malacothrix saxatilis var. **implicata** (Eastw.) H.M.Hall: sand and wind-carved cliffs, Apr 1897, *Trask s.n.* (MO 1892072); Celery Creek canyon, 28 Jul 1965, *Foreman et al. 87;* ne coast, at canyon head, 26 Jul 1939, *Dunkle 8357* (MO,SBBG); s escarpment ca. 0.5 mi sw of Benchmark 688, 22 May 1985, *Junak & Vanderwier SN-141;* midway along sw coast, 18 May 1967, *Boutin & Gonderman 1603;* e of Dutch Harbor, on sandy slopes near ocean, 23-24 Apr 1966, *Raven & Thompson 20760* (DS,SBBG).

Malacothrix saxatilis var. **tenuifolia** (Nutt.) A.Gray: mesa, at ne side of living compound, 12 May 1983, *Junak & Vanderwier SN-78;* mesa, just e of recreation center at living compound, 4 Jun 1987, *Junak SN-231.*

Microseris douglasii (DC.) Sch.Bip. subsp. **douglasii:** se end of mesa, ca. 0.2 mi n of Peak 606, 0.1 mi w of Building #137, 9 Apr 1992, *Junak SN-860;* se end of mesa, ca. 0.1 mi w of buildings #121 and #137, 7 May 1998, *Junak SN-1310* (CAS,OSC,SBBG).

Microseris douglasii subsp. **tenella** (A.Gray) K.Chambers: se end of mesa, ca. 0.1 mi w of buildings #121 and #137, 7 May 1998, *Junak SN-1311* (CAS,OSC,SBBG).

Microseris elegans Greene: se end of mesa, ca. 0.1 mi w of buildings #121 and #137, 7 May 1998, *Junak SN-1309* (CAS,OSC,SBBG).

Picris echioides L.: wells area along Tufts Road, 17 Mar 1999, *Junak SN-1327;* living compound, across road from Public Works building, 13 Jul 1989, *Junak SN-398;* w side of Owens Road, 0.6 mi s of Monroe Drive, 4 Jun 1987, *Junak SN-230;* along Jackson Highway, just n of Jackson Hill, 19 Mar 1983, *Junak SN-25;* along Jackson Highway at driveway to Building #112 and Tower #68, 10 May 1988, *Junak SN-253;* along Jackson Highway ca. 0.1 mi e of intersection with Shannon Road, 23 Jul 1992, *Junak SN-954.*

Pseudognaphalium biolettii Anderberg: San Nicolas Island, 26 Mar 1945, *Orr & Rett s.n.* (SBBG 79684); ne bench rim, 26 Jul 1939, *Dunkle 8358* (RSA,SBBG); ne coastal flats, in first canyon w of "L" Canyon, 23 May 1995, *Junak SN-1217;* ne portion of mesa, in upper reaches of E Mesa Canyon, 11 Apr 1989, *Junak SN-332;* ne edge of mesa, just n of Benchmark 396, 23 Mar 1993, *Junak SN-1019.*

Pseudognaphalium californicum (DC.) Anderberg: ne escarpment s of Sissy Cove, 22 Mar 2000, *Junak SN-1464;* ne escarpment at top of Airfield Grade on Beach Road, in upper e fork of Keyhole Canyon, 10 Jun 1993, *Junak et al. SN-1133.*

Pseudognaphalium luteo-album (L.) Hilliard & B.L.Burtt: near lower end of Tule Creek, 22 Apr 1961, *Blakley 4112;* Tule Canyon, just s of pump station, 11 Jun 1969, *Philbrick & Benedict B69-191;* along road to Thousand Springs, at top of grade, 19 Mar 1983, *Junak & Timbrook* SN-24; around barracks buildings of main compound, 27 Jul 1979, *Daily & Bromfield SNI-123;* airfield, on median strip between runway and taxiway, 3 Jun 1987, *Junak SN-210.*

Pseudognaphalium stramineum (Kunth) Anderberg: Coral [sic] Harbor, 23 Jul 1939, *Dunkle 8312;* n coastal flats, just w of Celery Canyon, 24 May 2000, *Junak SN-1552;* mesa s of housing area, 22 Apr 1961, *Blakley 4102;* mesa, just w of end of airfield runways, 7 Apr 1992, *Junak SN-823;* sw end of mesa, ca. 0.3 mi s of Army Springs, 19 Mar 1983, *Junak et al. SN-7;* sw portion of island, on s escarpment ca. midway between Cormorant Rock and Benchmark 688, 22 May 1985, *Junak & Vanderwier SN-144.*

Senecio vulgaris L.: n coastal flats, at base of road from living compound, 18 Mar 1983, *Junak et al. SN-1;* living compound, across road from Transportation Yard, 1 Mar 1989, *Junak SN-279;* n edge of mesa, along Beach Road on nw side of airfield runways, 20 Mar 1983, *Junak et al. SN-36;* grassland surrounding Building #121, 8 Apr 1982, *Vanderwier 14;* se escarpment s of Peak 606, on ridge overlooking barge landing, 11 Mar 1992, *Junak SN-692.*

Silybum marianum (L.) Gaertn.: mesa s of airfield runways, s of road between airfield terminal and Building #121, 8 Jun 1993, *Junak SN-1119;* mesa s of airfield runways, s of access road to Building #121, ca. 0.5 mi e of Monroe Road, 23 May 2000, *Junak SN-1545.*

Sonchus asper (L.) Hill: moist slopes, Apr 1897, *Trask s.n.* (MO); fertile flats, Apr 1901, *Trask s.n.* (RSA 474093); San Nicolas Island, 13 Mar 1932, *Howell 8221* (CAS); s escarpment, se of Jackson Peak, 13 May 1992, *Junak SN-893.*

Sonchus oleraceus L.: moist slope, Apr 1897, *Trask s.n.* (MO); bluff overlooking Tule Creek, 28 Jul 1965, *Foreman et al. 95* (SBBG,UC); Corral Harbor, in seepage of sea cliffs, 22 Jul 1939, *Dunkle 8306* (MO,SBBG); Big Creek Canyon, 23 Jul 1939, *Dunkle 8330;* mesa s of housing area, 22 Apr 1961, *Blakley 4099;* mesa, along Beach Road between runway and barracks, 10 Jun 1969, *Philbrick & Benedict B69-141.*

Sonchus tenerrimus L.: moist slopes, Apr 1897, *Trask s.n.* (MO 2526501, US 340276); cliffs, Apr 1897, *Trask 23* (US); moist slopes, Apr 1900, *Trask s.n.* (RSA 474098); San Nicolas Island, Apr 1901, *Trask 68* (GH,NY); n escarpment, on ridgetop ne of barge landing at Daytona Beach, 24 Apr 1989, *Junak SN-365;* s escarpment, on ridge along e side of Grand Canyon, 19 May 1993, *Junak SN-1117.*

Stebbinsoseris heterocarpa (Nutt.) K.L.Chambers: mesa, just e of Beach Road, s of sw end of airfield runway, 11 Apr 1989, *Junak SN-329.*

Taraxacum officinale F.H.Wigg.: mesa, in lawn of Navy barracks, 28 Jul 1979, *Daily SNI-146.*

Tragopogon porrifolius L.: ne coastal flats, along Beach Road near Rock Jetty, 24 Apr 1989, *Junak SN-351;* mesa, just n of fire station at intersection of Owens Road and Monroe Drive, 4 Jun 1987, *Junak SN-229;* ne part of mesa, at sw end of airfield runways, 11 Apr 1989, *Junak SN-325;* near Telemetry Buildings, 29 Jun 1983, *Vanderwier s.n.* (SBBG 94288).

Uropappus lindleyi (DC.) Nutt.: one locality, fertile lowland by the sea, Apr 1897, *Trask s.n.* (MO); one locality, Apr 1897, *Trask 20* (US); one locality, Apr 1897, *Trask 66* (NY); fertile flat, Apr 1901, *Trask 66* (GH).

Xanthium strumarium L.: barge landing area at Daytona Beach, 1 Sep 1988, *Junak SN-267;* barge landing area at Daytona Beach, 21 Aug 1990, *Junak SN-521.*

BORAGINACEAE

Amsinckia menziesii var. **intermedia** (Fisch. & C.A.Mey.) Ganders: n coastal flats, along Beach Road at Celery Canyon, 23 May 2001, *Junak SN-1713;* n edge of mesa, sse of Corral Harbor, 12 Mar 1992, *Junak SN-732;* past fire station on Owens Road, Mar 1980, *Newman 89;* on Owens Road, 100 yards n of Building #265, 15 Apr 1982, *Vanderwier & Dow 59;* mesa, on w side of Owens Road, 0.6 mi s of intersection with Monroe Drive, 7 Mar 1991, *Junak SN-547,* mesa, se of air terminal building, w of Building #121, 22 Feb 1990, *Junak SN-453.*

Amsinckia spectabilis Fisch. & C.A.Mey. var. **spectabilis:** sandhills, 18 Jul 1938, *Cockerell s.n.* (GH); ridgetop just e of Vizcaino Point, 7 Apr 1992, *Junak SN-826;* e end of Red Eye Beach, 19 Mar 1983, *Junak SN-17;* n coastal flats, behind first small beach w of Tender Beach, 2 Apr 1992, *Junak SN-810;* Coral [sic] Harbor, on sandy beach, 23 Jul 1939, *Dunkle 8339;* n coastal flats, along Beach Road between Corral Harbor and w end of Tranquility Beach, 2 May 1991, *Junak SN-658.*

Cryptantha maritima (Greene) Greene var. **maritima:** ne escarpment, on ridgetop ne of barge landing at Daytona Beach, 24 Apr 1989, *Junak SN-366;* se edge of mesa, at top of s escarpment, se of Peak 606, 7 Apr 1992, *Junak SN-829;* sw escarpment, on ridgetop along e side of Cattail Canyon, 12 May 1992, *Junak SN-886;* s escarpment, se of Jackson Peak, 13 May 1992, *Junak SN-892;* se coastal flats above Daytona Beach, just w of barge landing on n side of road, 10 Mar 1992, *Junak SN-671.*

Cryptantha traskiae I.M.Johnst.: n escarpment, at head of small valley overlooking Tender Beach, n of Peak 616, 28 Mar 1991, *Junak SN-597;* n coastal flats just w of W Mesa Canyon, 23 Mar 1993, *Junak SN-1043;* nw end of mesa, along n side of Tufts Road, just e of Water Tank #130, 8 Apr 1992, *Junak SN-839;* mesa, ca. 0.2 mi n of intersection of Beach Road and Monroe Drive, 1 Apr 1992, *Junak SN-771;* s escarpment, s of Jackson Peak, in upper e fork of Twin Rivers drainage, 2 May 1991, *Junak SN-645.*

Heliotropium curassavicum L.: beach, two localities, Apr 1897, *Trask 55* (MO); near Juana Maria Point, at extreme w end of Red Eye Beach, 10 May 1988, *Junak SN-246;* Gardens, 23 Jul 1939, *Dunkle 8319* (RSA,SBBG); Thousand Springs area, 29 Jul 1965, *Foreman 119;* in first canyon w of Live-forever Canyon, 18 May 1993, *Junak SN-1097;* along road to Vizcaino Point, 0.25 mi n of Rock Crusher buildings, 21 Aug 1990, *Junak SN-526.*

Pectocarya linearis subsp. **ferocula** (I.M.Johnst.) Thorne: on "the ridge" at 1000 ft above sea, Apr 1897, *Trask s.n.* (US 340190); ne portion of mesa, along e side of E Mesa Canyon, 11 Apr 1989, *Junak SN-330* (RSA,SBBG); mesa, ca. 0.2 mi n of intersection of Beach Road and Monroe Drive, 1 Apr 1992, *Junak SN-*

766 (RSA,SBBG); ne portion of mesa, on n side of Beach Road, just e of n runway access road, 23 Mar 1993, *Junak SN-1023* (RSA,SBBG); se escarpment, on ridgetop ne of barge landing at Daytona Beach, 24 Apr 1989, *Junak SN-364;* se coastal flats, above w end of Daytona Beach, 11 Mar 1993, *Junak SN-1004.*

BRASSICACEAE

Brassica nigra (L.) Koch: about a spring ca. 2 mi ese of Seal Beach, 23-24 Apr 1966, *Raven & Thompson 20730* (DS,SBBG).

Brassica tournefortii Gouan: mesa on e side of Building #176, 17 Mar 1999, *Junak & Smith SN-1324;* mesa n of Jackson Highway near Building #273, 23 Mar 2000, *Junak SN-1469.*

Cakile edentula (Bigelow) Hook. subsp. **edentula:** big sand dune, 13 Mar 1945, *Rett & Orr s.n.* (SBBG 77989); dune area near nw end of island, 23 Apr 1961, *Blakley 4150.*

Cakile edentula (Bigelow) Hook. X **C. maritima** Scop.: dune area near nw end of island, 23 Apr 1961, *Blakley 4151;* along old road on s coastal flats, sw of Benchmark 688, 22 May 1985, *Junak & Vanderwier SN-138.*

Cakile maritima Scop.: ridge above Vizcaino Point, at Triangulation Point "Hyd", 7 Mar 1991, *Junak SN-552;* near nw end of island, 18 May 1967, *Boutin & Gonderman 1607;* ne side of island near tank farm and Navy pier, 21 Apr 1961, *Blakley 4057;* sand dunes near Dutch Harbor, where escarpment road starts up escarpment, 7 Apr 1966, *Foreman & Smith 182;* near beaches along s shore of island, below runway, 10 Apr 1966, *Foreman 236.*

Dithyrea maritima Davidson: one locality on kitchen middens, Apr 1897, *Trask s.n.* (US 340377); one locality in kitchen middens, Apr 1901, *Trask 29* (NY,US); beach e of nw tip of island, 22 Apr 1961, *Blakley 4142;* in gully at extreme e end of Red Eye Beach, 31 Mar 1992, *Junak SN-753;* ca. 0.3 mi e of Vizcaino Point, on s side of ridge, 7 Apr 1992, *Junak SN-827;* se of Cormorant Rock, 22 May 1985, *Vanderwier s.n.* (SBBG 94467).

Draba cuneifolia var. **integrifolia** S.Watson: n coastal flats, just w of W Mesa Canyon, 27 Mar 1991, *Junak SN-584;* n coastal flats, on e side of W Mesa Canyon, 4 Apr 1989, *Junak SN-303;* ne coastal flats, at e edge of small branch of E Mesa Canyon, 10 Mar 1993, *Junak SN-988.*

Hirschfeldia incana (L.) Lagr.-Foss: 1000 ft w of Tule Creek Mouth, 4 Jul 1978, *Wier & Beauchamp s.n.* (RSA 289102); just se of Benchmark 43, nw of Coast Guard Beach, 22 Jun 1977, *Philbrick & Kritzman B77-15;* nw portion of mesa, s of Tufts Road, ca. 0.4 mi w of intersection with Shannon Road, 23 May 1985, *Junak & Vanderwier SN-152;* canyon running ne from camp area to ocean, 23 Apr 1961, *Blakley 4170;* field near motor pool area, 27 Jul 1965, *Foreman et al. 9.*

Hornungia procumbens (L.) Hayek: by a brackish spring, Apr 1901, *Trask 98* (US); one locality, spring, Apr 1901, *Trask s.n.* (US 960444); just s of tower at Vizcaino Point, 7 Apr 1992, *Junak SN-825;* at e end of western half of Red Eye Beach, 31 Mar 1992, *Junak SN-747;* e end of Tender Beach, 8 Apr 1992, *Junak SN-836;* n coastal flats, near Beach Road between Thousand Springs and Corral Harbor, just w of Balloon Inflation Building, 11 May 1983, *Junak & Vanderwier SN-63;* mesa, ca. 0.2 mi n of intersection of Beach Road and Monroe Drive, 1 Apr 1992, *Junak SN-772.*

Lepidium lasiocarpum Torr. & A.Gray var. **lasiocarpum:** San Nicolas Island, 18 Apr 1940, *Kanakoff s.n.* (RSA 429616); ridge above Vizcaino Point, at Triangulation Point "Hyd", 7 Mar 1991, *Junak SN-550;* n side of island, inland from cove just w of N Range Marker poles, 5 May 1993, *Junak SN-1080;* se edge of mesa, ca. 0.5 mi w of Peak 606, 11 Mar 1993, *Junak & McEachern SN-995* (RSA,SBBG); sw portion of island, ca. 0.6 mi se of Cormorant Rock, 22 May 1985, *Junak & Vanderwier SN-137;* near beach in vicinity of Dutch Harbor, 23-24 Apr 1966, *Raven & Thompson 20796* (RSA,SBBG).

Lepidium nitidum Torr. & A.Gray var. **nitidum:** two or three localities, Apr 1897, *Trask 29* (MO); n coastal flats, just w of second large canyon e of W Mesa Canyon, 27 Mar 1991, *Junak SN-579;* ne corner of island, on coastal flats between N and S Spur canyons, 24 Mar 1993, *Junak SN-1064;* swales on top of island, 9 Mar 1945, *Rett & Orr s.n.* (SBBG 78014); ne edge of mesa, near Benchmark 396, 18 Apr 1995, *Junak SN-1163;* eastern portion of mesa, near Building #121, 8 Apr 1982, *Vanderwier 12.*

Lepidium oblongum var. **insulare** C.L.Hitchc.: n coastal flats, near mouth of W Mesa Canyon, 4 Apr 1989, *Junak SN-299;* n edge of mesa, e of Shannon Road near Building #115, 31 Mar 1992, *Junak SN-742;*

se edge of mesa, se of Peak 606, 7 Apr 1992, *Junak SN-828;* sw coastal flats, between base of Drop-off Road and Cattail Canyon, ca. 0.5 mi s of Benchmark 688, 19 May 1993, *Junak SN-1107;* se coastal flats, above w end of Daytona Beach, 11 Mar 1993, *Junak SN-1003.*

Lobularia maritima (L.) Desv.: well area, Building #120, Shannon Road near headwaters of Tule Creek, 22 Jun 1977, *Philbrick & Kritzman B77-17;* building toward w end of mesa, 28 Apr 1978, *Daily SN-82;* main compound, in disturbed garden area outside mess hall, 27 Jul 1979, *Daily & Bromfield SNI-113;* s edge of mesa and top of s escarpment at Jackson Peak, w of Building #112, 21 Feb 1990, *Junak SN-434;* s escarpment, in upper e fork of Twin Rivers drainage, sw of Building #112, 2 May 1991, *Junak SN-647.*

Nasturtium officinale R.Br.: lower portion of Tule Creek, 22 Mar 1984, *Junak SN-91;* Tule Creek, 23 Jul 1992, *Junak SN-955* (RSA,SBBG); Tule Creek, at Humphrey Sump, 24 May 1995, *Junak SN-1232* (RSA,SBBG).

Raphanus sativus L.: probably from coastal bluff area, Red Eye Beach, 30 Apr 1978, *Daily 45;* Daytona Beach, 1 Jul 1978, *Wier & Beauchamp s.n.* (RSA 289043).

Sisymbrium irio L.: sw coastal flats, on w side of canyon ca. half-way between conspicuous offshore shipwreck and Grenedier Point, 11 May 1992, *Junak SN-863.*

Sisymbrium orientale L.: near Vizcaino Point, at Missile Site A, 22 Jun 1977, *Philbrick & Kritzman B77-16;* n coastal flats, at se end of island, ca. 0.25 mi w of old desalination plant at Rock Jetty, 24 Apr 1985, *Junak et al. SN-120;* sw end of mesa, along spur road to Building #165, ca. 0.3 mi s of Army Springs, 19 Mar 1983, *Junak et al. SN-6;* mesa, just n of barracks, ca. 0.5 mi e of Celery Creek, 11 June 1969, *Philbrick & Benedict B69-177;* se end of island, just w of barge landing at Daytona Beach, 23 May 1990, *Junak SN-509.*

CACTACEAE

Opuntia ficus-indica (L.) J.Mill.: mesa, at living compound, on flats just e of Building #20, 27 Oct 1988, *Junak SN-270.*

Opuntia littoralis (Engelm.) Cockerell: canyon running ne from camp area to ocean, 23 Apr 1961, *Blukley 4167;* ne escarpment, in large gully just e of "L" Canyon, near top of escarpment, 13 Jun 1995, *Junak & Stone SN-1248;* sw end of island, in main fork of lower Cattail Canyon, 19 May 1993, *Junak SN-1108;* road from Jackson Hill down w side of island to beach, 23 Apr 1961, *Blakley 4156.*

Opuntia littoralis (Engelm.) Cockerell X **O. oricola** Philbrick: n coastal flats, just e of bottom of NavFac Grade, 18 May 1993, *Junak SN-1098;* slopes and arroyos above pier near se end of island, 23-24 Apr 1966, *Raven & Thompson 20790;* ne escarpment, in short amphitheater-like canyon between "L" and Jetty canyons, 13 Jun 1995, *Junak & Stone SN-1246.*

Opuntia oricola Philbrick: road from Jackson Hill down w side of island to beach, 23 Apr 1961, *Blakley 4155.*

Opuntia prolifera Engelm.: on old beach terrace near Navy pier, ne of airstrip, 23 Apr 1961, *Blakley 4165;* flats at southern edge of mesa, ca. 0.25 mi wsw of Peak 606, 26 May 1992, *Junak SN-924.*

CARYOPHYLLACEAE

Herniaria hirsuta var. **cinerea** (DC.) Loret & Barrandon: ne coast of island, near mouth of W Mesa Canyon, 25 Apr 1989, *Junak SN-378.*

Loeflingia squarrosa Nutt.: ne edge of mesa, at Benchmark 396, 11 Apr 1989, *Junak SN-338;* ne edge of mesa, at Benchmark 396, 24 Apr 1989, *Junak SN-340.*

Sagina apetala Ard.: s central portion of mesa, just e of Shannon Road, ca. 0.2 mi n of intersection with Jackson Highway, 9 Mar 1993, *Junak SN-970;* mesa, on caliche flats just w of upper end of Mineral Canyon, 21 May 2001, *Junak SN-1697;* near n rim of mesa, n of airfield, ese of Benchmark 396, 10 Apr 2001, *Junak SN-1657.*

Silene gallica L.: one locality, cliff overhanging the brackish stream, Apr 1897, *Trask 25* (MO); seldom seen, Apr 1901, *Trask 24* (NY); ne coastal flats, in first canyon w of "L" Canyon, 23 May 1995, *Junak SN-1215;* ne escarpment, in upper portion of "L" Canyon, 17 May 1995, *Junak SN-1202;* mesa, on flats just w of upper reaches of Mineral Canyon, 21 May 2001, *Junak SN-1699;* mesa, in e fork of E Mesa

Canyon, 23 May 2001, *Junak SN-1718;* se coastal flats, just inland from e end of Daytona Beach, 9 Apr 1992, *Junak SN-848.*

Spergularia bocconi (Scheele) Graebner: ne end of mesa, along n side of airfield runways, 24 Apr 1989, *Junak SN-347;* near e edge of mesa, at se end of airfield runways, just e of Building #121, 27 May 1986, *Junak SN-183.*

Spergularia macrotheca (Cham. & Schlecht.) Heyn. var. **macrotheca:** San Nicolas Island, 21 Apr 1938, *Bilderback s.n.* (SD 21255); nw end of island, about spring ca. 2 mi ese of Seal Beach, 23-24 Apr 1966, *Raven & Thompson 20739;* small valley 50 yards ne of Benchmark 616, overlooking Tender Beach, 6 Apr 1966, *Foreman & Smith 158;* ne side of island, near barracks, 6 Jun 1977, *Hesseldenz SNI-5;* mesa, w of airfield terminal, along s edge of runways, 21 Mar 1983, *Junak et al. SN-56.*

Spergularia salina J.Presl & C.Presl: n side of island, in lower portion of W Mesa Canyon, 4 Apr 1989, *Junak SN-297;* ne end of island, at e end of Coast Guard Beach, w of Rock Jetty, 3 Jun 1987, *Junak SN-221;* mesa, at living compound, 12 May 1983, *Junak & Vanderwier SN-81;* mesa, just s of terminal building, 4 Jun 1987, *Junak SN-234;* excavation, along southeastern edge of dry lake, 0.5 mi s of air terminal, 17 Nov 1971, *Benedict s.n.* (SBBG 43435).

CHENOPODIACEAE

Aphanisma blitoides Nutt.: one locality, Apr 1897, *Trask 21* (MO); among cacti, Apr 1901, *Trask 20* (NY); at foot of cacti, May 1901, *Trask 20* (GH); ne escarpment, in large gully just e of "L" Canyon, 13 Jun 1995, *Junak & Stone SN-1249;* ne end of island, just s of Beach Road in N Spur Canyon, 23 Mar 1993, *Junak SN-1036;* se end of island, on s escarpment ca. 0.5 mi wsw of Triangulation Station "Cliff", 13 May 1992, *Junak SN-909.*

Atriplex argentea var. **mohavensis** (M.E.Jones) S.L.Welsh: extreme ne end of mesa, just e of ne end of airfield runway, 5 Sep 2000, *Junak SN-1586.*

Atriplex californica Moq. in DC.: sea shore sands, Apr 1897, *Trask s.n.* (US 339987); sea beach, Apr 1897, *Trask s.n.* (US 339984); on beach, Apr 1901, *Trask 17* (GH); Thousand Springs, 29 Jul 1979, *Daily & Bromfield SNI-170;* n side of island, at first canyon w of Live-forever Canyon, 18 May 1993, *Junak SN-1095;* sw escarpment, in lower w fork of Cattail Canyon, 12 May 1992, *Junak SN-876;* s escarpment, at bottom of Theodolite Road, 9 Jun 1992, *Junak SN-943;* sandy areas behind Jehemy Beach, 28 Jul 1965, *Foreman et al. 53.*

Atriplex canescens (Pursh) Nutt. var. **canescens:** mesa, at e end of borrow pit along Monroe Drive, sw of airfield runways, 23 Jan 1991, *Junak & Halvorson SN-533.*

Atriplex coulteri (Moq.) D.Dietr.: one locality, moist upland, Apr 1901, *Trask s.n.* (GH).

Atriplex lentiformis (Torr.) S.Watson: mesa, at ne side of living compound, 12 May 1983, *Junak & Wanderwier SN-79;* main Navy compound, just e of recreation center, 4 Jun 1987, *Junak SN-232;* mesa, along e side of Owens Road, ca. 0.1 mi n of intersection with Monroe Drive, 3 Jun 1987, *Junak SN-226.*

Atriplex leucophylla (Moq.) D.Dietr.: on beaches, Apr 1897, *Trask 18* (MO); seashore, Apr 1901, *Trask s.n.* (GH, RSA 478333); seashore, Apr 1901, *Trask 93* (US); Coral [sic] Harbor, sandy beach, 23 Jul 1939, *Dunkle 8340* (RSA,SBBG); ne side of island near tank farm and Navy pier, 21 Apr 1961, *Blakley 4056;* nw end of Sand Spit, 11 May 1988, *Junak SN-255;* near mouth of canyon running sw from Jackson Hill, 23 Apr 1961, *Blakley 4162.*

Atriplex pacifica A.Nelson: ne escarpment, in lower part of "L" Canyon, 17 May 1995, *Junak SN-1191;* ne coastal flats, just e of N Spur Canyon, 24 Mar 1993, *Junak SN-1046;* ne coastal flats, between N and S Spur canyons, 15 Mar 1994, *Junak SN-1137.*

Atriplex prostrata DC.: small canyon se of Triangulation Point "Port", 21 Apr 1961, *Blakley 4084;* e end of Coast Guard Beach, w of Rock Jetty, 3 Jun 1987, *Junak SN-222;* near motor pool, 27 Jul 1965, *Foreman et al. 15* (US).

Atriplex semibaccata R.Br.: nw end of island, Seal Beach, sandy slopes just above ocean, 23-24 Apr 1966, *Raven & Thompson 20702;* Coral Harbor [sic], sea bluffs, 22 Jul 1939, *Dunkle 8304;* small canyon e of living quarters, 28 Jul 1965, *Foreman et al. 73;* vernal pool at top of bluff nw of Sand Spit, 21 Apr 1961, *Blakley 4074.*

Atriplex watsonii A.Nelson: seashore sands, Apr 1897, *Trask s.n.* (US 339980); flats above Vizcaino Point, 21 Aug 1990, *Junak SN-528;* nw end of island, Seal Beach, sandy slopes just above ocean, 23-24 Apr 1966, *Raven & Thompson 20695;* E Thousand Springs, crest of bluff above main spring, 11 Jun 1969, *Philbrick & Benedict B69-197;* w end of Dutch Harbor, 3 Jun 1987, *Junak SN-219;* e end of Daytona Beach, 20 Mar 1983, *Junak SN-48.*

Bassia hyssopifolia (Pallas) Kuntze: n coastal flats, ca. 0.1 mi w of N Range Marker Poles, 23 Jul 1992, *Junak SN-957;* mouth of gully se of Benchmark 43, nw of Coast Guard Beach, 22 Jun 1977, *Philbrick & Kritzman B77-14;* ne coastal flats, along Beach Road just n of turn at Sand Spit, near light station, 27 May 1992, *Junak SN-934;* mesa, along Harrington Road, 29 Jul 1979, *Daily SNI-174.*

Chenopodium ambrosioides L.: living compound, 22 Feb 1990, *Junak SN-448;* n side of water tanks just sw of intersection of Owens Road and Monroe Drive, 1 Sep 1988, *Junak SN-266;* e end of mesa, on w side of Building #121, 24 May 1995, *Junak SN-1235.*

Chenopodium berlandieri Moq.: n coastal flats, just w of road from airfield, 27 May 1986, *Junak et al. SN-178.*

Chenopodium californicum (S.Watson) S.Watson: moist flat, Apr 1901, *Trask 15* (GH); ne escarpment, in swale on w side of first large canyon e of living compound, 1 Apr 1992, *Junak SN-791;* slopes and arroyos above pier near se end of island, 23-24 Apr 1966, *Raven & Thompson 20788* ne escarpment, in swale on n side of second canyon n of "L" Canyon, 27 Jul 2000, *Junak SN-1573.*

Chenopodium murale L.: about *Opuntia,* Apr 1897, *Trask s.n.* (US 340029); San Nicolas Island, Apr 1901, *Trask 16* (GH); Gardens, 23 Jul 1939, *Dunkle 8322;* Thousand Springs, at base of cliff, 22 Apr 1961, *Blakley 4136;* around airport terminal, 27 Jul 1965, *Foreman et al. 1* (UC); se end of island, on s escarpment ca. 0.5 mi wsw of Triangulation Station "Cliff", 13 May 1992, *Junak SN-910.*

Salicornia depressa Standl.: Sand Spit, Apr 1901, *Trask 89* (MO); sandy beach at base of Sand Spit, on salty playa, 21 Apr 1961, *Blakley 4062;* in and surrounding salt water pool behind Jehemy Beach, 27 Jul 1965, *Foreman & Rainey 47;* sandy slopes near ocean e of Dutch Harbor, in salt marsh, 23-24 Apr 1966, *Raven & Thompson 20750.*

Salsola kali subsp. **pontica** (Pall.) Мозуakin. near road above Sand Spit, n.d., *Foreman et al. 42* (US).

Salsola tragus L.: back beach area of Sand Spit, se end of island, 7 Apr 1966, *Foreman & Smith 208.*

Suaeda taxifolia (Standl.) Standl.: w end of Red Eye Beach, 22 Jul 1992, *Junak SN-949;* Coral [sic] Harbor, on arroyo bank, 23 Jul 1939, *Dunkle 8313;* sea bluffs near Ranch landing, 25 Jul 1939, *Dunkle 8341;* base of bluff near end of Sand Spit, 21 Apr 1961, *Blakley 4066;* se escarpment, on slopes above e end of Daytona Beach, 24 Apr 1989, *Junak SN-357.*

CONVOLVULACEAE

Calystegia macrostegia subsp. **amplissima** Brummitt: San Nicolas Island, Apr 1897, *Trask 54* (MO); moist flats, Apr 1901, *Trask 53* (GH); n of n end of Shannon Road, 18 May 1967, *Boutin & Gonderman 1574;* n coastal flats, ca. 0.5 mi w of bottom of grade from airfield, 20 Mar 1983, *Junak SN-50;* mesa, just n of barracks, ca. 0.5 mi e of Celery Creek, 11 Jun 1969, *Philbrick & Benedict B69-176;* s side of island, just e of base of Theodolite Road, 21 May 1990, *Junak SN-485;* se escarpment, sw of Building #121, 8 Apr 1992, *Junak SN-831.*

Calystegia malacophylla subsp. **pedicellata** (Jepson) Munz: ne coastal flats, on large gravel pile, 0.2 mi w of Building #199 and Rock Jetty, 28 Mar 1991, *Junak SN-611;* ne coastal flats, on large gravel pile, 0.2 mi w of Building #199 and Rock Jetty, 1 May 1991, *Junak SN-621.*

Calystegia soldanella (L.) R.Br.: w end of Red Eye Beach, 25 Apr 1985, *Junak & Laughrin SN-133;* peninsula at w end of Red Eye Beach, 4 Jun 1987, *Junak SN-233;* at extreme e end of Red Eye Beach, 22 May 1990, *Junak SN-492;* behind w end of Tender Beach, 19 Mar 1983, *Junak SN-22;* Tender Beach, 2 Apr 1992, *Junak SN-805.*

Convolvulus arvensis L.: living compound, 1 Sep 1988, *Junak SN-265;* living compound, 1 Mar 1989, *Junak SN-284;* living compound, 6 Sep 1989, *Junak SN-419.*

Cressa truxillensis Kunth: sandy flats at se end of island, 0.54 mi ssw of Sand Spit, 19 Oct 2000, *Junak SN-1607.*

CRASSULACEAE

Crassula connata (Ruiz & Pav.) A.Berger: Tender Beach, 27 Mar 1991, *Junak SN-561;* mesa, just e of Fire Station at intersection of Owens Road and Monroe Drive, 4 Apr 1989, *Junak SN-291;* ne edge of mesa, at Benchmark 396, 11 Apr 1989, *Junak SN-337;* sw portion of s escarpment, sw of Benchmark 688, 9 Mar 1993, *Junak SN-965;* s side of island, at base of Theodolite Road, 8 Mar 1993, *Junak SN-961.*

Dudleya virens subsp. **insularis** (Rose) Moran: n escarpment s of Corral Harbor, 10 Jun 1993, *Junak SN-1128;* n coastal flats, at Live-forever Canyon, 8 Jun 1993, *Junak SN-1121;* mesa, just e of upper Shannon Road, 9 Jun 1993, *Junak SN-1122;* w Jehemy Beach, 10 Jun 1969, *Philbrick & Benedict B69-170;* se end of island, in n fork of second canyon e of Towers Canyon, 27 May 1992, *Junak SN-933.*

CUCURBITACEAE

Marah macrocarpus var. **major** (Dunn) Stocking: ne bench rim, at canyon head w of E Fence Line, 26 Jul 1939, *Dunkle 8359;* ne escarpment, in gully along Beach Road at Airfield Grade, 1 Mar 1989, *Junak SN-282;* ne side of island, in large canyon just e of hairpin turn in Beach Road, 23 May 1990, *Junak SN-518.*

EUPHORBIACEAE

Ricinus communis L.: on canyon side below Telemetry Building #182, 30 Jul 1979, *Daily & Bromfield SNI-181.*

FABACEAE

Astragalus didymocarpus Hook. & Arn. var. **didymocarpus:** on cliffs of a briny water-way, Apr 1897, *Trask 41* (MO); on cliffs of a briny stream, one locale, Apr 1897, *Trask s.n.* (US).

Astragalus traskiae Eastw.: two localities, high dry gulches over briny streams, Apr 1897, *Trask 42* (MO); two localities, high dry gulches, Apr 1897, *Trask s.n.* (US 339976); on ridge forming base of nw tip of island, 22 Apr 1961, *Blakley 4143;* Tender Beach, 2 Apr 1992, *Junak SN-808;* at base of bluff near end of Sand Spit, 21 Apr 1961, *Blakley 4064;* sw end of mesa, along spur road to Building #165, 19 Mar 1983, *Junak et al. SN-5;* old marine terrace above Jehemy Beach, 7 Apr 1966, *Foreman & Smith 171.*

Lotus argophyllus var. **argenteus** Dunkle: beach dunes at mouth of Tule Creek, 22 Apr 1961, *Blakley 4121;* along Shannon Road just n of Tufts Road, 18 May 1967, *Boutin & Gonderman 1565;* mesa top above Sand Dune Canyon, 27 Jul 1965, *Foreman et al. 20;* side of road ca. 0.5 mi se of Rock Crusher Point, 6 Jun 1977, *Hesseldenz SN-12;* s side of island, just w of base of Theodolite Road, 21 May 1990, *Junak SN-488.*

Lotus salsuginosus Greene var. **salsuginosus:** n coastal flats, on gravel pile along Beach Road, ca. 0.2 mi w of Rock Jetty, 11 Jun 1996, *Junak SN-1270;* mesa, on gravel pile along road leading to Building #121, s side of airfield, 0.25 mi e of Monroe Drive, 12 Jun 1996, *Junak SN-1283.*

Lupinus albifrons var. **douglasii** (J.Agardh) C.P.Sm.: n coast canyon above the back of Corral Harbor, 23 Jul 1939, *Dunkle 8334* (MO); beach dunes at mouth of Tule Creek, 22 Apr 1961, *Blakley 4127;* nw end of mesa, along Tufts Road 0.3 mi n of intersection with Jackson Highway, 2 Jun 1987, *Junak et al. SN-205;* mesa s of housing area, 22 Apr 1961, *Blakley 4101;* s side of island, just w of base of Theodolite Road, 21 May 1990, *Junak SN-487;* sandy slopes near ocean e of Dutch Harbor, 23-24 Apr 1966, *Raven & Thompson 20759.*

Lupinus bicolor Lindl.: w of airfield terminal along s edge of runways, 21 Mar 1983, *Junak et al. SN-55;* ne part of mesa, just s of airfield runways, w of passenger terminal building, 11 Apr 1989, *Junak SN-319;* mesa just n of airfield, e of n runway access road, across airfield from control tower, 10 Apr 2001, *Junak SN-1670.*

Lupinus succulentus Koch: on gravel pile along Monroe Road near cement plant, 29 Mar 1983, *Vanderwier s.n.* (SBBG 94175); Daytona Beach, on gravel pile at barge landing area, 24 Apr 1989, *Junak SN-359;* e end of Daytona Beach, on gravel pile at barge landing, 1 May 1991, *Junak SN-631.*

Medicago polymorpha L.: two localities, fertile flats, Apr 1897, *Trask 40* (MO); two slopes, Apr 1897, *Trask s.n.* (US 340155); moist flat, Apr 1901, *Trask 39* (GH); clay flats, 13 Mar 1932, *Howell 8215* (CAS); nw end of island, on open dunes ca. 1 mi ese of Seal Beach, 23-24 Apr 1966, *Raven & Thompson 20709*

(DS,SBBG); Gardens, 23 Jul 1939, *Dunkle 8323;* Thousand Springs pump area in Tule Canyon, 7 Apr 1966, *Foreman 238* (UC); mesa s of housing area, 22 Apr 1961, *Blakley 4109;* grassland outside Building #121, 8 Apr 1982, *Vanderwier 10.*

Medicago sativa L.: one locality, flat above the brackish stream, Apr 1897, *Trask s.n.* (US 340158); one locality, above the brackish stream, Apr 1897, *Trask 39* (MO); in old corral, Apr 1901, *Trask s.n.* (POM 348077); near an old corral, Apr 1901, *Trask 38* (GH); mesa, ca. 70 m ne of Building #113, nw of Jackson Hill, 29 Sep 1991, *Smith s.n.* (SBBG 97909).

Melilotus albus Medik.: nw end of island, about a spring ca. 2 mi ese of Seal Beach, 23-24 Apr 1966, *Raven & Thompson 20738* (DS,SBBG); near lower end of Tule Creek, in moist bottom of canyon, 22 Apr 1961, *Blakley 4116;* dry bluffs above Tule Creek, 28 Jul 1965, *Foreman et al. 103* (UC); ne edge of mesa, at top of grade along Beach Road, 19 Nov 1991, *Junak SN-664;* sw portion of island, sw of Benchmark 688, 22 May 1985, *Junak & Vanderwier SN-140.*

Melilotus indicus (L.) All.: nw end of island, in open dunes ca. 1 mi ese of Seal Beach, 23-24 Apr 1966, *Raven & Thompson 20710* (DS,SBBG); n side, canyon bottom, 23 Jul 1939, *Dunkle 8337;* small canyon se of Triangulation Point "Port", 21 Apr 1961, *Blakley 4092;* open dirt area surrounding airport terminal, 27 Jul 1965, *Foreman et al. 8;* s side of island, just e of base of Theodolite Road, 21 May 1990, *Junak SN-483.*

Robinia pseudoacacia L.: Mineral Canyon, a few trees along streambed, 29 Jul 1979, *Daily & Bromfield SNI-180.*

Spartium junceum L.: wells area, Shannon Road near headwaters of Tule Creek, 22 Jun 1977, *Philbrick & Kritzman B77-20;* intersection of Tufts and Shannon roads, 29 Jun 1983, *Vanderwier s.n.* (SBBG 94495); along Tufts Road, just w of intersection with Shannon Road, 2 Jun 1987, *Junak et al. SN-200.*

Trifolium albopurpureum Torr. & A.Gray: upland flats, Apr 1901, *Trask 37* (GH); ne coastal flats, just w of first canyon w of "L" Canyon, s of Beach Road, 26 Apr 2001, *Junak SN-1695;* mesa n of Monroe Drive, on w side of second large canyon w of airfield runways, 9 Apr 1992, *Junak SN-855;* mesa top, 100 yds sw of air terminal, 10 Apr 1966, *Foreman 233;* se end of mesa, ca. 0.1mi w of buildings #121 & #137, 8 Apr 1992, *Junak SN-846;* se end of mesa, ca. 0.1mi w of buildings #121 & #137, 7 May 1998, *Junak SN-1308.*

Trifolium depauperatum var. **truncatum** (Greene) Isely: mesa n of Monroe Drive, on w side of second large canyon w of airfield runways, 9 Apr 1992, *Junak SN-856;* mesa, on n side of large canyon sw of airfield runways, w of Beach Road, 28 Mar 1991, *Junak SN-613;* n edge of mesa, along Beach Road on nw side of airfield runways, 20 Mar 1983, *Junak et al. SN-35;* ne end of mesa, just w of Beach Road, near sw end of airfield runways, 23 Mar 1993, *Junak SN-1016;* se end of mesa, ca. 0.1mi w of buildings #121 & #137, 8 Apr 1992, *Junak SN-847.*

Trifolium gracilentum Torr. & A.Gray: slope, Apr 1897, *Trask 35* (MO); one locality, Apr 1897, *Trask s.n.* (US 340318); ne end of mesa, just e of top of n runway access road, on n side of airfield runways, 6 May 1993, *Junak SN-1083;* ne end of mesa, just e of top of n runway access road, on n side of airfield runways, 6 May 1993, *Junak SN-1086.*

Trifolium microcephalum Pursh: N coastal flats, just n of Beach Road, just w of road to Sissy Beach, 28 Mar 1991, *Junak SN-587;* near base of n escarpment, just e of "L" Canyon, 23 Mar 1993, *Junak SN-1028;* ne coastal flats, n of Beach Road, just e of S Spur Canyon, 24 Mar 1993, *Junak SN-1051;* near n edge of mesa, on w side of first canyon e of living compound, 19 Mar 1983, *Junak SN-34;* ne edge of mesa, just s of Benchmark 396, 7 May 1998, *Junak SN-1315.*

Trifolium microdon Hook. & Arn.: n coastal flats, at base of n escarpment just w of second large canyon e of living compound, 1 Apr 1992, *Junak SN-797;* mesa, on w side of second large canyon w of airfield runways, 9 Apr 1992, *Junak SN-853;* mesa, on e side of second large canyon w of airfield runways, 1 Apr 1992, *Junak SN-776;* mesa, ca. 0.2 mi n of intersection of Beach Road and Monroe Drive, along e edge of canyon, 1 Apr 1992, *Junak SN-761.*

Trifolium palmeri S.Watson: one locality, Apr 1897, *Trask s.n.* (US 340296); dunes e of Corral Harbor, 29 Apr 1978, *Daily SNI-71;* ne coastal flats, ca. 0.4 mi w of base of Sand Spit, 18 Apr 1995, *Junak SN-1161;* mesa, just n of Beach Road, just e of intersection with road to Benchmark 396, 10 Apr 2001, *Junak SN-1655;* ne end of mesa, just e of top of n runway access road, on n side of airfield runways, 6 May 1993,

Junak SN-1084; se end of mesa, ca. 0.3 mi w of Building #137, 9 Apr 1992, *Junak SN-858.*

Trifolium willdenovii Spreng.: hillside e of Corral Harbor, 29 Apr 1978, *Daily SNI-75;* ne end of island, near base of n escarpment, just e of "L" Canyon, 23 Mar 1993, *Junak SN-1030;* steep slopes and arroyos above pier near se end of island, 23-24 Apr 1966, *Raven & Thompson 20783;* mesa edge near Beach Road and end of runway, 26 Apr 1983, *Vanderwier s.n.* (SBBG 94497); mesa, ca. 0.2 mi n of intersection of Beach Road and Monroe Drive, along e edge of canyon, 1 Apr 1992, *Junak SN-762.*

Vicia hassei S.Watson: about cacti, Apr 1901, *Trask 85* (MO,US); ne coastal flats just n of S Spur Canyon, 20 Mar 2003, *Junak SN-1743.*

Vicia sativa L. subsp. **sativa:** near n edge of mesa, just n of water tanks n of w end of airfield, 18 Apr 1995, *Junak SN-1162.*

Vicia villosa subsp. **varia** (Host) Corbiere: along dry stream bed of Tule Canyon, 28 Jul 1965, *Foreman et al. 93;* near spring in Thousand Springs area, 23-24 Apr 1966, *Raven & Thompson 20766;* n central portion of island, at n end of Shannon Road, 18 May 1967, *Boutin & Gonderman 1569.*

FRANKENIACEAE

Frankenia salina (Molina) I.M.Johnst.: Coral [sic] Harbor, on bank with seepage, 22 Jul 1939, *Dunkle 8300;* n shore, just e of Sissy Cove, 13 Jul 1989, *Junak SN-408;* sw end of island, near Rock Crusher building, 2 Jun 1987, *Junak et al. SN-207;* e of Dutch Harbor, on sandy slopes near ocean, 23-24 Apr 1966, *Raven & Thompson 20751;* back of Jehemy Beach, 27 Jul 1965, *Foreman et al. 48.*

GENTIANACEAE

Zeltnera venusta (A.Gray) Mansion.: nw of living compound, in lower portion of Live-forever Canyon, 28 May 1986, *Junak SN-190;* n escarpment, in side gully which drains into Celery Canyon, 25 Apr 1985, *Junak et al. SN-129;* ne coastal flats, in first canyon w of "L" Canyon, 23 May 1995, *Junak SN-1212;* ne escarpment, in e fork of Jetty Canyon, 16 May 1995, *Junak SN-1173;* ne escarpment, on e side of first large canyon w of Sand Spit, 27 May 1992, *Junak SN-936.*

GERANIACEAE

Erodium botrys (Cav.) Bertol.: n rim of mesa, near Benchmark 396, 4 Apr 1989, *Junak SN-316;* ne end of mesa, near Benchmark 396, 7 May 1998, *Junak SN-1314;* ne end of mesa, just w of Beach Road, near sw end of airfield runways, 23 Mar 1993, *Junak SN-1014;* mesa, behind fire station at airfield, 22 Mar 1984, *Junak SN-101;* mesa, flats just se of air terminal building, just s of road to Building #121, 22 Feb 1990, *Junak SN-451.*

Erodium cicutarium (L.) Aiton.: one moist slope, Apr 1897, *Trask s.n.* (US 340066); moist flats, Apr 1901, *Trask 43* (GH); ne coast, along road between living quarters and Pirate's Cove, 9 Apr 1966, *Foreman 228;* near Triangulation Point "Corral", just e of Tule Creek, 11 Jun 1969, *Philbrick & Benedict B69-186;* terrace, nne of living quarters, 22 Jun 1977, *Philbrick & Kritzman B77-5;* along n rim of canyon on s side of Jackson Highway, ca. 0.2 mi sw of Army Springs, 6 May 1998, *Junak SN-1299.*

Erodium moschatum (L.) Aiton: San Nicolas Island, 13 Mar 1932, *Howell 8216* (CAS); steep slopes and arroyos above pier near se end of island, 23-24 Apr 1966, *Raven & Thompson 20781* (DS,SBBG); mesa, at living compound, 1 Mar 1989, *Junak SN-278;* small canyon se of Triangulation Point "Port", 21 Apr 1961, *Blakley 4087* (CAS,SBBG); se escarpment, on ridgetop between forks of Towers Canyon, 11 Mar 1992, *Junak SN-709.*

Pelargonium peltatum (L.) L'Her.: n end of mesa top, on road to rock crusher, 28 Jul 1965, *Foreman et al. 106* (UC): behind water tank along road to w end, 30 Apr 1978, *Daily 65;* around watertank, n end of island, 28 Jul 1979, *Daily & Bromfield SNI-141.*

HYDROPHYLLACEAE

Nemophila pedunculata Benth.: n coastal flats, on e side of W Mesa Canyon, 4 Apr 1989, *Junak SN-305;* n escarpment, n of airfield, 27 Mar 1991, *Junak SN-571;* ne escarpment, along Beach Road on Airfield Grade, 10 Mar 1992, *Junak SN-677;* mesa, in e fork of second large canyon w of airfield runways, 1 Apr

1992, *Junak SN-782;* n rim of mesa, just w of Benchmark 396, 4 Apr 1989, *Junak SN-313.*
Phacelia distans Benth.: two localities, moist flats, Apr 1901, *Trask s.n.* (ND-G 042098).

LAMIACEAE
Marrubium vulgare L.: wash bottom, Jun 1980, Newman 114; n side of island, in first canyon w of
Celery Canyon, 22 May 1990, *Junak SN-507;* n coastal flats, in lower portion of Celery Canyon, 14 Jun
1995, *Junak & Stone SN-1256;* e fork of Cattail Canyon, 19 May 1993, *Junak SN-1112;* s escarpment, in
upper e drainage of Twin Rivers, 28 Mar 1991, *Junak SN-605.*

LOASACEAE
Mentzelia affinis Greene: sw escarpment, on ridge on e side of Grand Canyon, 19 May 1993, *Junak SN-
1116;* s central portion of island, in wash draining into e end of Elephant Seal Beach, 22 May 1985, *Junak
SN-139;* s escarpment, se of Jackson Peak, along first ridge w of Range Marker Poles at w end of Dutch
Harbor, 13 May 1992, *Junak SN-897;* s escarpment, on e side of ridge just sw of Twin Towers (Building
#186), 13 May 1992, *Junak SN-911.*

LYTHRACEAE
Lythrum hyssopifolia L.: n escarpment, in drainage ditch above head of major canyon near e end of airfield runways, 21 Mar 1984, *Junak SN-87;* mesa, at wells area along Tufts Road, 17 Mar 1999, *Junak SN-
1328;* mesa, at Nicktown, 18 Mar 1999, *Junak SN-1330.*

MALVACEAE
Lavatera assurgentiflora Kellogg subsp. **assurgentiflora:** n escarpment, 0.1 mi w of Dump Canyon [W
Mesa Canyon], 8 Apr 1992, *Junak & Clark SN-842;* base of n escarpment, just w of W Mesa Canyon, 12
Jun 1996, *Junak SN-1288;* wash behind Telemetry Building on Harrington Road, 27 Jul 1979, *Daily SNI-
94;* mesa, near buildings, Jackson Hill, 30 Apr 1978, *Daily 86.*
Malva parviflora L.: one locality, flat above the brackish stream, Apr 1897, *Trask s.n.* (US 340150):
Gardens, on moist ditch bank, 23 Jul 1939, *Dunkle 8326;* mesa, along Tufts Road at wells area, 28 Mar
1991, *Junak SN-592;* small canyon se of Triangulation Point "Port", 2 Apr 1961, *Blakley 4091;* weedy
area near airport terminal, 27 Jul 1965, *Foreman et al. 11* (SBBG,UC).

MYOPORACEAE
Myoporum laetum Forster f.: n escarpment, in lower portion of Celery Canyon, 25 Apr 1985, *Junak et al.
SN-130;* near coastline in n central portion of island, ca. 1.3 mi w of base of Beach Road, 3 Jun 1987, *Junak
SN-225;* ornamental around fire station and mess hall, 28 Jul 1965, *Foreman et al. 104.*

NYCTAGINACEAE
Abronia maritima S.Watson: nw end of island, just inland from w end of Tender Beach, 17 May 1995,
Junak SN-1188; e end of island, at base of Sand Spit, 21 Apr 1961, *Blakley 4061;* sandy slopes above Seal
Beach, 1 mi s of Vizcaino Point, 27 Jul 1965, *Foreman et al. 36;* Dutch Harbor, 24 Nov 1940, *Dunkle s.n.*
(SBBG 88377); e end of Daytona Beach, just e of Barge Landing, 19 Apr 1995, *Junak SN-1172.*
Abronia umbellata Lam. var. **umbellata:** flats above Vizcaino Point, 21 Aug 1990, *Junak SN-529;* n
side near end of Shannon Road, 17 May 1967, *Boutin 1568;* e end of island, at base of Sand Spit, 21 Apr
1961, *Blakley 4060;* e of Dutch Harbor, on sandy slopes near ocean, 23-24 Apr 1966, *Raven & Thompson
20748* (DS,SBBG); marine terrace above Jehemy Beach, 7 Apr 1966, *Foreman & Smith 177.*

ONAGRACEAE
Camissonia cheiranthifolia (Spreng.) Raim. subsp. **cheiranthifolia:** beach dunes at mouth of Tule Creek,
22 Apr 1961, *Blakley 4125;* e end of Tranquility Beach, near Anchor Point, 11 May 1988, *Junak SN-260;*
nw portion of mesa, just e of Building #120, 23 May 1985, *Junak & Vanderwier SN-151;* sw end of island,
ca. 0.5 mi se of Rock Crusher, above Bachelor Beach, 10 May 1988, *Junak SN-241;* s escarpment, in upper

e fork of Twin Rivers drainage, 2 May 1991, *Junak SN-646*.

Camissonia cheiranthifolia subsp. **suffruticosa** (S.Watson) Raven: nw end of island, open dunes ca. 1 mi ese of Seal Beach, 23-24 Apr 1966, *Raven & Thompson 20761* (RSA,SD); nw end of mesa, on w side of canyon w of Tufts Road, 20 Apr 1999, *Junak SN-1361*.

OROBANCHACEAE

Orobanche fasciculata Nutt.: s side of dirt road to Dutch Harbor, 29 Apr 1978, *Daily 36*.

Orobanche parishii subsp. **brachyloba** Heckard: n escarpment, just w of Live-forever Canyon, 10 Jun 1993, *Junak SN-1124;* ne end of island, near coastline ca. 0.5 mi w of base of Beach Road, 3 Jun 1987, *Junak SN-223;* ne coastal flats, ne of Benchmark 396, 13 Jun 1995, *Junak SN-1251;* sw escarpment, on ridgetop along e side of Cattail Canyon, 12 May 1992, *Junak SN-887;* se portion of island, above e end of Daytona Beach, 12 May 1983, *Junak & Vanderwier SN-72*.

OXALIDACEAE

Oxalis corniculata L.: living compound, in disturbed area near mess hall, 27 Jul 1979, *Daily SNI-114;* airfield, in planter in front of air control tower, 8 Apr 1999, *Junak SN-1349*.

Oxalis pes-caprae L.: mesa, at Nicktown, in front of chapel (Building #109), 16 Mar 1999, *Junak & Smith SN-1318;* mesa, on e side of Owens Road, 0.6 mi s of intersection with Monroe Drive, just s of Building #265, 7 Mar 1991, *Junak SN-549;* mesa, along e side of Owens Road, ca. 0.1 mi s of Building #265, 23 Mar 2000, *Junak SN-1468*.

PAPAVERACEAE

Eschscholzia californica Cham. subsp. **californica:** mesa, in living compound, on nw side of Barracks Building #150, 1 Sep 1988, *Junak SN-264*.

Eschscholzia ramosa Greene: s escarpment, along first ridge w of Range Marker Poles, 13 May 1992, *Junak SN-898;* s escarpment, above w end of Dutch Harbor, on ridge e of first canyon e of Range Marker Poles, 13 May 1992, *Junak SN-899;* se escarpment, on ridge on w side of Desert Fan Canyon, 1 May 1991, *Junak SN-634;* s escarpment above Daytona Beach, 18 Apr 1995, *Junak SN-1153;* steep slopes just w of Spit at e end of island, 1 Jul 1978, *Weir & Beauchamp s.n.* (RSA 289095, UC 1443177).

Platystemon californicus Benth.: nw portion of island, near e end of Red Eye Beach, 19 Mar 1983, *Junak SN-14;* n escarpment, at head of small valley overlooking Tender Beach, n of Peak 616, 28 Mar 1991, *Junak SN-598;* ne coast of island, just w of Northeast Point, near coastal navigation light n of Benchmark 396, 25 May 1995, *Junak SN-1243;* se side of island, near beach in vicinity of Dutch Harbor, 23-24 Apr 1966, *Raven & Thompson 20793;* coastal strand along Daytona Beach, 29 Apr 1978, *Daily 50*.

PLANTAGINACEAE

Plantago coronopus L.: ne coastal flats, along n side of Beach Road, 0.2 mi s of S Spur Canyon, 22 May 2001, *Junak SN-1710;* mesa, adjacent to Building #148, Jackson Hill, 16 Mar 1999, *Junak SN-1317;* s edge of mesa near Buildings #173 and #266, at s end of Radar Row complex, 17 Mar 1999, *Junak SN-1322;* s edge of mesa, just e of Building #176, 23 Mar 2000, *Junak SN-1474;* living compound, on w side of Public Works Building #147, 18 Mar 1999, *Junak SN-1332;* mesa, along Monroe Road ca. 0.4 mi w of airfield, 5 May 1999, *Junak SN-1374*.

Plantago major L.: San Nicolas Island, 1985, *Vanderwier s.n.* (SBBG 94732); mesa, at living compound, along road to Public Works Building, 6 Sep 1989, *Junak SN-418*.

Plantago ovata Forssk.: riven sand flats by the sea, Apr 1897, *Trask 62* (MO); seashore sand flats, Apr 1901, *Trask 63* (NY); ne coastal flats, near mouth of Keyhole Canyon, 4 Apr 1989, *Junak SN-308;* mesa s of housing area, 22 Apr 1961, *Blakley 4096;* sw portion of island, in lower w fork of Cattail Canyon, 12 May 1992, *Junak SN-878;* se coastal flats, above w end of Daytona Beach, 11 Mar 1993, *Junak SN-1006*.

POLEMONIACEAE

Gilia nevinii A.Gray: one locality, sand cliffs above a brackish stream, Apr 1897, *Trask s.n.* (US 340269);

one slope, Apr 1901, *Trask 52* (GH,NY).

POLYGONACEAE

Eriogonum cinereum Benth.: s-facing slope behind Building #112, 23 Aug 1984, *Vanderwier s.n.* (SBBG 94604); top of s escarpment, on s side of Jackson Peak, downslope from Building #112, 23 May 1985, *Junak & Vanderwier SN-156;* s edge of mesa at Jackson Peak, on slopes and flats just s and w of Building #112, 21 Feb 1990, *Junak & Murphey SN-432;* Jackson Peak, at top of s escarpment behind Building #112, 23 Jan 1991, *Junak SN-534.*

Eriogonum fasciculatum var. **polifolium** (A.DC.) Torr. & A.Gray: slope behind Building #112, 23 Aug 1984, *Vanderwier s.n.* (SBBG 94603); top of s escarpment, on s side of Jackson Peak, downslope from Building #112, 23 May 1985, *Junak & Vanderwier SN-155;* s edge of mesa at Jackson Peak, on slopes and flats just s and w of Building #112, 21 Feb 1990, *Junak & Murphey SN-433.*

Eriogonum grande var. **timorum** Reveal: along W NavFac Road, ca. 0.1 mi e of Balloon Launch Building, 22 May 1990, *Junak SN-505;* n coastal flats, along Beach Road n of airfield runways, 20 Mar 1983, *Junak & Timbrook SN-44;* ne coastal flats near S Spur Canyon, 16 Mar 1994, *Junak SN-1149;* cliffs at e end of island, above base of Sand Spit, 21 Apr 1961, *Blakley 4059;* se end of mesa, se of Building #121, 21 Mar 1984, *Junak SN-83;* s side of island, just w of base of Theodolite Road, 21 May 1990, *Junak SN-486;* e of Dutch Harbor, on sandy slopes near ocean, 23-24 Apr 1966, *Raven & Thompson 20744;* se escarpment, se of Peak 606, 24 Apr 1989, *Junak SN-367.*

Polygonum arenastrum Boreau: mesa, at airfield, just se of air terminal building, 2 Jun 1987, *Junak SN-195.*

Polygonum argyrocoleon Kunze: Thousand Springs, 7 Aug 1969, *Benedict s.n.* (SBBG 33751); dry lake 0.5 mi sw of air terminal, 17 Nov 1971, *Benedict s.n.* (SBBG 43405); on mesa, at Borrow Pit sw of airfield runways, s of Monroe Road, 3 Jun 1987, *Junak SN-227;* e end of mesa, at Borrow Pit just se of intersection of Monroe and Beach roads, 23 Jul 1992, *Junak SN-953.*

Pterostegia drymarioides Fisch. & C.A.Mey.: ne coastal flats, just s of E NavFac Road, ca. 0.7 mi w of intersection with Beach Road, 10 Mar 1993, *Junak SN-976;* ne escarpment, on flats just nw of Benchmark 396, n of airfield, 4 Apr 1989, *Junak SN-309;* ne coastal flats, in lower part of "L" Canyon, 17 May 1995, *Junak SN-1200;* ne end of island, in N Spur Canyon, 23 Mar 1993, *Junak SN-1033;* ne portion of mesa, in upper reaches of E Mesa Canyon, w of nw end of airfield runways, 11 Apr 1989, *Junak SN-335;* ne edge of mesa, at Benchmark 396, near water reclamation tanks at nw end of airfield, 24 Apr 1989, *Junak SN-341.*

Rumex crispus L. subsp. **crispus:** northern slope, Jun 1980, *Newman 123;* one location, poor clayey grassland at summit of island, 23-24 Apr 1966, *Raven & Thompson 20768* (CAS); across from Mess Hall, 4 Jul 1978, *Wier & Beauchamp s.n.* (UC 1443179); ne part of mesa, at airfield, just s of Passenger Terminal Building, 26 Apr 1989, *Junak SN-396.*

Rumex obtusifolius L.: mesa, on n side of airfield, in drainage ditch opposite Air Traffic Control Tower, 3 Jun 1987, *Junak SN-213.*

Rumex salicifolius Weinm. var. **salicifolius:** nw end of island, about a spring, ca. 2 mi ese of Seal Beach, 23-24 Apr 1966, *Raven & Thompson 20729;* spring, 2.25 mi se of Vizcaino Point, s of Carrier Reef, 11 Jun 1969, *Philbrick & Benedict B69-201;* n side of island, just inland from w end of Tender Beach, in swale behind dunes, 22 May 1990, *Junak SN-501;* Tule Canyon, 23 Jul 1939, *Dunkle 8331;* ne coastal flats, in lower portion of W Mesa Canyon, 13 Jul 1989, *Junak SN-412.*

PORTULACACEAE

Calandrinia ciliata (Ruiz & Pav.) DC.: mesa, ca. 0.6 mi n of intersection of Beach and Monroe roads, on e side of second large canyon w of airfield runways, 1 Apr 1992, *Junak SN-780;* mesa, ca. 0.2 mi n of intersection of Beach and Monroe roads, along e edge of canyon, 1 Apr 1992, *Junak SN-764;* ne end of mesa, just s of Benchmark 396, along road to water reclamation tanks at nw end of airfield, 24 Apr 1989, *Junak SN-346;* near n rim of mesa, n of airfield, ese of Benchmark 396, 10 Apr 2001, *Junak SN-1659.*

Claytonia parviflora Hook. subsp. **parviflora:** n escarpment s of Corral Harbor, in first large canyon e of Corral Harbor, 12 Mar 1992, *Junak SN-724;* Fern Canyon, 6 Mar 1985, *Vanderwier s.n.* (SBBG 94696); ne escarpment, in upper e fork of "L" Canyon, on w side of ridge between "L" and Jetty canyons, 11 Mar

1992, *Junak SN-697;* mesa, in e fork of second large canyon w of airfield runways, 1 Apr 1992, *Junak SN-783;* n edge of mesa, n of airfield, 10 Apr 2001, *Junak SN-1663.*

Claytonia perfoliata subsp. **mexicana** (Rydb.) J.M.Miller & K.L.Chambers: Tender Beach, inland from center of beach, at base of hill behind dunes, 27 Mar 1991, *Junak SN-559;* n coastal flats at base of n escarpment, just e of W Mesa Canyon, 17 Apr 1995, *Junak SN-1151;* ne coastal terrace, in lower portion of "L" Canyon, 11 Mar 1992, *Junak SN-706;* ne escarpment, on flats just nw of Benchmark 396, n of airfield, 4 Apr 1989, *Junak SN-310;* near n edge of mesa, on w side of first canyon e of living compound, 19 Mar 1983, *Junak SN-27.*

PRIMULACEAE

Anagallis arvensis L.: living compound, in planted area at Building #109, 16 Mar 1999, *Junak & Smith SN-1319;* mesa, at water tank #4 along Owens Road, just s of generator building, 17 Mar 1999, *Junak SN-1321;* mesa, at airfield, in planter in front of Air Traffic Control Tower, 3 Jun 1987, *Junak & DiVittorio SN-211;* at top of s escarpment, s of Building #176, at E.A.T.S. antenna site, 17 Mar 1999, *Junak SN-1325.*

Primula clevelandii subsp. **insulare** (H.J.Thomps.) A.R.Mast & Reveal: n coastal flats, just n of Beach Road, ca. 0.2 mi e of bottom of NavFac Grade, just w of road to Sissy Beach, 28 Mar 1991, *Junak SN-586;* n coastal flats, ca. 75 ft n of E NavFac Road, 0.3 mi e of base of NavFac Grade, 23 Mar 1993, *Junak SN-1040;* open grassy area 0.5 mi nw of pier, nw of Sand Spit, 10 Jun 1969, *Philbrick & Benedict B69-164;* ne escarpment, in main branch of first large canyon w of Sand Spit, sse of Triangulation Point "Spur", 11 Mar 1992, *Junak SN-715.*

RESEDACEAE

Oligomeris linifolia (M.Vahl) J.F.Macbr.: beach dunes at mouth of Tule Creek, 22 Apr 1961, *Blakley 4120;* ne escarpment, on e side of first large canyon w of Sand Spit, 27 May 1992, *Junak SN-935;* eroded slopes near summit of grade from nw end of island, 23-24 Apr 1966, *Raven & Thompson 20741;* back of marine terrace above Seal Beach, against adjacent hills, 7 Apr 1966, *Foreman & Smith 205* (SBBG,UC); sw portion of island, in lower e fork of Cattail Canyon, 12 May 1992, *Junak SN-882;* se escarpment, on ridge on w side of Desert Fan Canyon, 1 May 1991, *Junak SN-635.*

RUBIACEAE

Galium aparine L.: San Nicolas Island, Apr 1901, *Trask s.n.* (GH); bottom of road which leads down to Tranquility Beach, n of main Navy compound, 18 Mar 1983, *Junak et al. SN-2;* se end of island, 12 Mar 1932, *Howell 8224* (CAS); slopes and arroyos above pier near se end of island, 23-24 Apr 1966, *Raven & Thompson 20779;* ne escarpment, in upper portion of "L" Canyon, 18 Apr 1990, *Junak SN-470;* ne escarpment, in w fork of Jetty Canyon, 16 May 1995, *Junak SN-1185;* edge of mesa, nw of runway, 17 Feb 1985, *Vanderwier s.n.*

SALICACEAE

Salix exigua Nutt.: sw portion of island, in lower part of Grand Canyon, 12 May 1992, *Junak SN-890.*

Salix lasiolepis Benth.: seepage area at northern end of mesa top, on road to Rock Crusher, 28 Jul 1965, *Foreman et al. 105;* nw end of island, about spring ca. 2 mi ese of Seal Beach, 23-24 Apr 1966, *Raven & Thompson 20733;* along Jackson Highway, at w end of island, 18 May 1967, *Boutin & Gonderman 1613;* below w end of mesa, at Army Springs, 19 Mar 1983, *Junak et al. SN-10;* low flat spot, head of Tule Canyon, 23 Apr 1961, *Blakley 4163;* small canyon se of Triangulation Point "Port", at outlet from camp waste water, 21 Apr 1961, *Blakley 4090.*

SAURURACEAE

Anemopsis californica (Nutt.) Hook. & Arn.: at a spring, Apr 1901, *Trask 98* (US); seep on bluff, point on e end of Red Eye Beach, 30 Apr 1978, *Daily 54;* about seep on coastal bluff, Red Eye Beach, 28 Jul 1979, *Daily SNI-132.*

SAXIFRAGACEAE
Jepsonia malvifolia (Greene) Small: n escarpment, in lower portion of Live-forever Canyon, 29 Nov 1988, *Junak SN-276;* n escarpment, on e side of Mineral Canyon, in small side gully above waterfall, 2 Apr 1992, *Junak & Soiseth SN-815;* ne escarpment, along Beach Road, just w of upper hairpin turn, 30 Nov 1989, *Junak SN-430;* ne escarpment, in upper portion of "L" Canyon, 18 Apr 1990, *Junak SN-469;* ne end of island, near base of n escarpement, just e of "L" Canyon, 23 Mar 1993, *Junak SN-1026;* ne escarpment, in e fork of first large canyon w of Sand Spit, 11 Mar 1992, *Junak SN-713.*

SCROPHULARIACEAE
Castilleja densiflora (Benth.) T.I.Chuang & Heckard: n coastal flats, just e of first large canyon e of living compound, 1 Apr 1992, *Junak SN-802;* n coastal flats, on e side of E Mesa Canyon, 23 May 2001, *Junak SN-1720;* near n edge of mesa, on w side of first large canyon e of living compound, 19 Mar 1983, *Junak et al. SN-30;* mesa, n of Monroe Drive, on w side of second large canyon w of airfield runways, 9 Apr 1992, *Junak SN-851;* n edge of mesa, nw of airfield runways, just n of water project tanks, 27 Mar 1991, *Junak SN-566.*

SOLANACEAE
Lycium brevipes Benth.: Celery Canyon, 12 Jun 1969, *Philbrick & Benedict B69-205;* Celery Canyon, along road, 12 Jun 1969, *Philbrick B69-223.*
Lycium californicum Nutt.: sea cliffs, Coral [sic] Harbor, 25 Jul 1939, *Dunkle 8307* (RSA,SBBG); 0.5 mi nw of old sheep dock, 22 Apr 1961, *Blakley 4132;* slopes and arroyos above pier near se end of island, 23-24 Apr 1966, *Raven & Thompson 20789;* top of bluff above Sand Spit, 21 Apr 1961, *Blakley 4067;* mesa, w of airfield terminal, along s edge of runways, 21 Mar 1983, *Junak et al. SN-53;* back of Jehemy Beach and adjacent escarpment, 28 Jul 1965, *Foreman et al. 52.*
Lycium verrucosum Eastw.: San Nicolas Island, Apr 1901, *Trask s.n.* (RSA 350721).
Nicotiana glauca Graham: w end of island, at S.L.A.M. missile target site, 19 Oct 1999, *Junak SN-1451;* ne coastal flats, along Beach Road ca. 0.25 mi w of old desalination plant at Rock Jetty, 24 Apr 1985, *Junak et al. SN-122;* se end of island, along n side of Beach Road, directly across from road that leads down to Sand Spit, 14 Mar 2000, *Junak SN-1452;* launch site ca. 2 mi sw of Thousand Springs, 1.5 mi se of Carrier Reef, 16 Nov 1971, *Benedict s.n.* (SBBG 43586).
Solanum americanum Mill.: nw end of island, about spring ca. 2 mi ese of Seal Beach, 23-24 Apr 1966, *Raven & Thompson 20734;* seep on coastal bluffs, Red Eye Beach, 29 Apr 1978, Daily SNI-57; above seepage area at end of Thousand Springs Road, 7 Apr 1966, *Foreman & Smith 194;* Thousand Springs, 7 Aug 1969, *Benedict s.n.* (SBBG 33782); near lower end of Tule Creek, 22 Apr 1961, *Blakley 4117;* canyon running ne from camp area to ocean, 23 Apr 1961, *Blakley 4168;* below cliffs, base of Sand Spit, 7 Aug 1969, *Benedict s.n.* (SBBG 33787); ne part of mesa, at airfield, just w of Building #67, 13 Jul 1989, *Junak SN-417.*
Solanum douglasii Dunal: nw portion of island, near Tufts Road, 15 Apr 1982, *Vanderwier 53.*
Solanum lycopersicum L. var. **lycopersicum:** mesa at living compound, just n of Building #19, 22 Feb 1990, *Junak SN-449.*

TAMARICACEAE
Tamarix ramosissima Lebed.: n of Tufts and Shannon, 12 Oct 1983, *Vanderwier s.n.* (SBBG 94684); nw end of mesa, just e of Building #110, w or upper reaches of Tule Creek, 29 Nov 1989, *Junak SN-428;* e end of mesa, at ne corner of Borrow Pit along Monroe Drive, 24 May 1995, *Junak SN-1236.*

URTICACEAE
Parietaria hespera var. **californica** B.D.Hinton: ne escarpment, in lower portion of e fork of E Mesa Canyon, ca. 0.25 mi nw of Benchmark 396, 10 Mar 1993, *Junak SN-990;* ne side of island, near base of n escarpment, just e of "L" Canyon, 23 Mar 1993, *Junak SN-1031;* se end of island, on s escarpment ca. 0.5 mi wsw of Triangulation Station "Cliff", 13 May 1992, *Junak SN-907;* se side of island, in n fork of

second canyon e of Towers Canyon, 27 May 1992, *Junak SN-931.*
Soleirolia soleirolii (Req.) Dandy: Thousand Springs, 22 Apr 1961, *Blakley 4138;*   cliff face above Thousand Springs, 29 Jul 1965, *Foreman 120;* moist cliff e of Thousand Springs, 11 Jun 1969, *Philbrick & Benedict B69-195;* above catchment basin at Thousand Springs, 22 May 1990, *Junak SN-502.*

VERBENACEAE
Verbena lasiostachys Link var. **lasiostachys:** nw portion of mesa, s of Tufts Road, just e of Building #120, ca. 0.4 mi w of intersection with Shannon Road, 23 May 1985, *Junak & Vanderwier SN-150.*

Monocotyledonous Angiosperms
ALLIACEAE
Dichelostemma capitatum (Benth.) A.W.Wood subsp. **capitatum:** at foot of *Opuntia,* Apr 1901, *Trask 87* (NY); ne escarpment sw of Rock Jetty, in e fork of Jetty Canyon, 16 May 1995, *Junak SN-1176;* e edge of mesa near Building #121, 29 Mar 1983, *Vanderwier s.n.* (SBBG 95037); s escarpment,in upper e fork of Twin Rivers drainage, 28 Mar 1991, *Junak SN-604.*

CYPERACEAE
Eleocharis macrostachya Britton: small pond, 26 Mar 1945, *Rett & Orr s.n.* (SBBG 77213); mesa, on n side of airfield, in drainage ditch opposite Air Traffic Control Tower, 3 Jun 1987, *Junak SN-214;* mesa, at sw end of airfield runways, w of Beach Road, 10 Mar 1992, *Junak SN-670;* artificial pond, s of w end of runway, 12 Jun 1969, *Philbrick & Benedict B69-209;* Borrow Pit along Monroe Drive, sw of w end of airfield runways, 22 Mar 1984, *Junak SN-93.*
Schoenoplectus americanus (Pers.) Schinz & R.Keller: spring in Thousand Springs area, 23-24 Apr 1966, *Raven & Thompson 20765;* springy canyon bank, Big Creek Canyon, 23 Jul 1939, *Dunkle 8329* (RSA, SBBG); near lower end of Tule Creek, 22 Apr 1961, *Blakley 4118;* Tule Canyon bottom, near pumping station, 28 Jul 1965, *Foreman et al. 92;* lower portion of Tule Creek, 22 Mar 1984, *Junak SN-92;* Tule Creek, 23 Jul 1992, *Junak SN-956.*

JUNCACEAE
Juncus bufonius L.: near lower end of Tule Creek, 22 Apr 1961, *Blakley 4115;* mesa, ca. 0.6 mi n of intersection of Beach Road and Monroe Drive, on e side of second large canyon w of airfield runways, 1 Apr 1992, *Junak SN-778;* ne edge of mesa, at Benchmark 396, 24 Apr 1989, *Junak SN-343;* ne end of mesa, just w of Beach Road, near sw end of airfield runways, 23 Mar 1993, *Junak SN-1013;* artificial pond s of w end of airstrip, 7 Aug 1969, *Benedict s.n.* (SBBG 33785); along Building #120, 28 Jul 1979, *Daily & Bromfield SNI-158.*

POACEAE
Achnatherum diegoense (Swallen) Barkworth: one locality, about *Opuntia,* Apr 1897, *Trask 6* (MO); n escarpment, on e side of Mineral Canyon, 2 Apr 1992, *Junak & Soiseth SN-814;* ne coastal flats, just w of second canyon w of "L" Canyon, 23 May 1995, *Junak SN-1221;* ne escarpment, just above major forks in first significant canyon w of Sand Spit, 27 May 1992, *Junak SN-937;* sw portion of island, in lower Grand Canyon, 12 May 1992, *Junak SN-889;* se end of island, on s escarpment at head of canyon just s of Twin Towers (Building #186), 13 May 1992, *Junak SN-905;* s escarpment above Daytona Beach, at head of w fork of first canyon e of Desert Fan Canyon, 26 May 1992, *Junak SN-922.*
Ammophila arenaria (L.) Link: below w end of mesa, across road from Army Springs, 19 Mar 1983, *Junak et al. SN-8;* sand dune at head of Tule Creek, 22 Apr 1961, *Blakley 4110;* upland dune area along Thousand Springs Road, 28 Jul 1965, *Foreman et al. 101;* dune, Thousand Springs Road, 11 Jun 1969, *Philbrick & Benedict B69-180;* along Shannon Road, ca. 1.5 mi n of intersection with Tufts Road, e of Tule Creek, 2 Jun 1987, *Junak et al. SN-199.*
Arundo donax L.: nw side of island, in Tule Creek at Humphrey Sump, 24 May 1995, *Junak SN-1233;*

near road, e of Dutch Harbor light, 27 Jul 1965, *Foreman et al. 61;* between Jehemy Beach and Dutch Harbor, 10 Jun 1969, *Philbrick & Benedict B69-173.*

Avena barbata Link: San Nicolas Island, 19 Apr 1940, *Kanakoff s.n.* (SBBG 37877, US 3110940); old beach terrace below Triangulation Point "Hot", n of Navy boat landing, 21 Apr 1961, *Blakley 4080;* mesa, just e of Fire Station at intersection of Owens Road and Monroe Drive, 4 Apr 1989, *Junak SN-293;* e of Dutch Harbor, on sandy slopes near ocean, 23-24 Apr 1966, *Raven & Thompson 20749;* below se end of runway, on heavily dissected terrain above Jehemy Beach, 7 Apr 1966, *Foreman & Smith 166* (UC); near salt marsh pool at Jehemy Beach, 28 Jul 1965, *Foreman et al. 54.*

Avena fatua L.: one locality, about *Opuntia,* Apr 1897, *Trask 17* (US); San Nicolas Island, 16 Mar 1945, *Rett & Orr s.n.* (SBBG 77139); off NavFac Road, 6 Mar 1985, *Vanderwier s.n.* (SBBG 94623).

Brachypodium distachyon (L.) P.Beauv.: mesa, on flats between Buildings #273 and #176, 23 Mar 2000, *Junak SN-1474;* mesa, on disturbed flats between Buildings #273 and #176, 10 Apr 2000, *Junak SN-1478.*

Bromus arizonicus (Shear) Stebbins: San Nicolas Island, 13 Mar 1932, *Howell 8226* (CAS,US); e end of island, ne of airport terminal, 3-4 Apr 1979, *Thorne et al. 52361* (MO).

Bromus berteroanus Colla: one locality, about *Opuntia,* Apr 1897, *Trask 1* (MO); one locality, about *Opuntia,* Apr 1897, *Trask 15* (US).

Bromus carinatus Hook. & Arn. var. **carinatus:** on a flat, about *Opuntia,* Apr 1897, *Trask 2* (MO); one locality, *Opuntia* roots, Apr 1897, Trask 3 (MO); one locality, about *Opuntia,* Apr 1897, *Trask 13* (US); San Nicolas Island, Apr 1897, *Trask 14* (US); se side of island, near beach in vicinity of Dutch Harbor, *Raven & Thompson 20791* (DS).

Bromus catharticus Vahl var. **catharticus:** mesa, at Radar Row, 9 Nov 1984, *Vanderwier s.n.* (SBBG 94700); sw end of mesa, at Radar Row, between Building #167 and Building #169, 14 May 1992, *Junak SN-913;* outside weight room, barracks area of main compound, 27 Jul 1979, *Daily & Bromfield SNI-124*; airport parking lot, 30 Jun 1978, *Wier & Beauchamp s.n.* (UC 1443168).

Bromus diandrus Roth: San Nicolas Island, 16 Apr 1940, *Kanakoff s.n.* (POM 363377, US 3110924); nw end of island, in open dunes ca. 1 mi ese of Seal Beach, 23-24 Apr 1966, *Raven & Thompson 20724* (DS,SBBG); Thousand Springs pump area in Tule Canyon, 7 Apr 1966, *Foreman & Smith 188;* old beach terrace below Triangulation Point "Hot", n of Navy boat landing, 21 Apr 1961, *Blakley 4078;* n edge of mesa, e of Shannon Road near Building #115, 31 Mar 1992, *Junak SN-738;* s side of island, just e of base of Theodolite Road, 21 May 1990, *Junak SN-484.*

Bromus hordeaceus L. subsp. **hordeaceus:** Thousand Springs pump area in Tule Canyon, 7 Apr 1966, *Foreman & Smith 189* (UC); sea bluffs, Coral [sic] Harbor, 22 Jul 1939, *Dunkle 8308* (RSA,SBBG); ne coastal flats, ca. 0.1 mi w of light station near Sand Spit, 19 Apr 1995, *Junak SN-1169;* mesa, in small valley 50 yards ne of Benchmark 616, overlooking Tender Beach, 6 Apr 1966, *Foreman & Smith 153;* mesa, w of airfield terminal along s edge of runways, 21 Mar 1983, *Junak et al. SN-54;* se end of mesa, just s of Building #121, 7 Mar 1991, *Junak SN-543;* coastal bluffs between Daytona Beach and Flatrock, 29 Apr 1978, *Daily SNI-28.*

Bromus madritensis L.: n escarpment, in e fork of canyon above NavFac buildings, 26 Apr 1989, *Junak SN-390.*

Bromus maritimus (Piper) C.Hitchc.: mesa, along Tufts Road at wells area, 28 Mar 1991, *Junak SN-593.*

Bromus rubens L.: nw end of island, in open dunes ca. 1 mi ese of Seal Beach, 23-24 Apr 1966, *Raven & Thompson 20720;* Thousand Springs pump area in Tule Canyon, 7 Apr 1966, *Foreman & Smith 190;* n edge of mesa, at top of NavFac Grade, n of living compound, 31 Mar 1992, *Junak SN-737;* se end of mesa, just s of Building #121, 7 Mar 1991, *Junak SN-544;* se coastal flats above Daytona Beach, just w of barge landing, 10 Mar 1992, *Junak SN-673.*

Cortaderia selloana (Schult. & Schult.f.) Asch. & Graebn.: mesa, just w or Fire Station at airfield, 18 Apr 1990, *Junak SN-473.*

Cynodon dactylon (L.) Pers. var. **dactylon:** above a brackish stream, 1000 ft elev, Apr 1897, *Trask 4* (MO); sandy garden, 23 Jul 1939, *Dunkle 8325;* n side of island, along W NavFac Road just e of Building #279, 22 May 1990, *Junak SN-504;* ne coastal flats, in "L" Canyon, 17 May 1995, *Junak SN-1194;* main compound on mesa, between mess hall and neighboring buildings, 27 Jul 1979, *Daily & Bromfield SNI-*

112; mesa s of housing area, 22 Apr 1961, Blakley 4094; bare areas near airport terminal, 27 Jul 1965, *Foreman et al. 6.*

Distichlis spicata (L.) Greene: one locality, Sand Spit region, Apr 1897, Trask 10 (MO); on bluffs overlooking Tule Creek, 28 Jul 1965, *Foreman et al. 97;* Tule Canyon, near pumping station, 11 Jun 1969, *Philbrick & Benedict B69-192;* ne coast of island, at Rock Jetty, just e of Pump Building #36, 9 May 1988, *Junak SN-237;* one locality, Sand Spit region, Apr 1897, *Trask 10* (MO); seep spring near mouth of canyon running sw from Jackson Hill, 23 Apr 1961, *Blakley 4159;* se end of island, on sandy slopes near ocean e of Dutch Harbor, 23-24 Apr 1966, *Raven & Thompson 20757.*

Festuca arundinacea Schreb.: ne coastal flats, in "L" Canyon, 17 May 1995, *Junak SN-1192;* ne escarpment, in upper portion of "L" Canyon, 17 May 1995, *Junak SN-1201;* ne coastal flats, in Jetty Canyon, 29 Jun 1999, *Junak SN-1426;* nw portion of mesa, along Tufts Road at Building #120, 23 May 1985, *Junak & Vanderwier SN-148;* nw end of mesa, along Tufts Road, opposite Water Tank #130, 8 Apr 1992, *Junak SN-840;* living compound, on n side of recreation building, 21 May 1990, *Junak SN-474;* on mesa, across road from Fire Station #3 at airfield, 22 Mar 1984, *Junak SN-96.*

Hordeum brachyantherum subsp. **californicum** (Covas & Stebbins) Bothmer, N.Jacobsen & Seberg: San Nicolas Island, Apr 1901, *Trask s.n.* (POM 352493); along n rim of canyon on s side of Jackson Highway, ca. 0.2 mi sw of Army Springs, 6 May 1998, *Junak SN-1302;* nw end of island, by spring in Thousand Springs area, 23-24 Apr 1966, *Raven & Thompson 20764;* ne escarpment, in first canyon s of S Spur Canyon, 18 May 1993, *Junak SN-1100;* central portion of mesa, along Shannon Road 0.4 mi s of intersection with Tufts Road, 5 May 1993, *Junak SN-1073;* mesa s of housing area, 22 Apr 1961, *Blakley 4100;* sw portion of island, in lower e fork of Cattail Canyon, 12 May 1992, *Junak SN-880.*

Hordeum intercedens Nevski: s end of mesa, just ne of intersection of Jackson Highway and Owens Road, 8 Apr 1992, *Junak SN-845;* near cliffs behind Sand Spit, 7 Apr 1966, *Foreman 209* (SBBG,UC,US).

Hordeum marinum subsp. **gussoneanum** (Parl.) Thell.: mesa, e of Jackson Hill, along buried cable route between Owens Road and Jackson Highway, 4 May 1999, *Junak SN-1366;* mesa, just s of parking area at airfield terminal building, 5 May 1999, *Junak SN-1375.*

Hordeum murinum subsp. **glaucum** (Steud.) Tzvelev: in fertile spots and in sand, Apr 1897, *Trask 7* (MO); fertile spots in sand reaches, Apr 1897, *Trask s.n.* (US 340265); San Nicolas Island, 13 Mar 1932, *Howell 8211* (CAS); nw end of island, ca. 1 mi ese of Seal Beach, 23-24 Apr 1966, *Raven & Thompson 20725* (DS,SBBG); vernal pool at top of bluff nw of Sand Spit, 21 Apr 1961, *Blakley 4070* (CAS, SBBG,US); below se end of runway, on heavily dissected terrain above Jehemy Beach, 7 Apr 1966, *Foreman & Smith 168* (SBBG,UC).

Hordeum murinum subsp. **leporinum** (Link) Arcang.: San Nicolas Island, 17 Mar 1932, *Howell 8218* (CAS); along road between Army Camp Beach and living quarters, at bottom of grade, 17 Dec 1965, *Foreman & Lloyd 132* (UC,UCLA); on hillside, 100 yards above Sand Spit, 27 Jul 1965, *Foreman et al. 40* (SBBG,UC).

Hordeum vulgare L. var. **vulgare:** n edge of mesa, at e end of old landfill site, e of living compound, 1 Apr 1992, Junak SN-789.

Lamarckia aurea (L.) Moench: San Nicolas Island, 16 Apr 1940, *Kanakoff s.n.* (POM 352668, SBBG 37867); slopes and arroyos above pier near se end of island, 23-24 Apr 1966, *Raven & Thompson 20782;* dry hills and terraces near Coreopsis stand above Coast Guard pier, 28 Jul 1966, *Foreman et al. 66* (SBBG,UC); n edge of mesa, e of Shannon Road near Building #115, 31 Mar 1992, *Junak SN-739;* near se edge of mesa, in swale se of Building #121, 22 Feb 1990, *Junak SN-456;* s escarpment, in upper e fork of Twin Rivers drainage, 2 May 1991, *Junak SN-650;* se coastal flats, near Sand Spit, 24 Apr 1989, *Junak SN-352.*

Lolium multiflorum Lam.: nw end of island, about spring ca. 2 mi ese of Seal Beach, 23-24 Apr 1966, *Raven & Thompson 20737;* ne of living quarters near mouth of small canyon, 22 Jun 1977, Philbrick & Kritzman B77-6; slopes and arroyos above pier near se end of island, 23-24 Apr 1966, *Raven & Thompson 20800;* small canyon se of Triangulation Point "Port", 21 Apr 1961, *Blakley 4086;* midway along ne edge of airport runway, 18 May 1967, *Boutin & Gonderman 1597.*

Lolium perenne L.: around barracks, main compound, 27 Jul 1979, *Daily SNI-115;* 200 yards ssw of air terminal, 28 Jul 1965, *Foreman 111* (UC).

Melica imperfecta Trin.: ne escarpment, in swale on w side of first large canyon e of living compound, n of old landfill site, 1 Apr 1992, *Junak SN-792;* near n edge of mesa, at ne end of airfield runways, across from Building #121, 27 May 1986, *Junak SN-182.*

Nassella cernua (Stebbins & Love) Barkworth: Beach Road, 0.5 mi nw of pier, n of se end of runway, 10 Jun 1969, *Philbrick & Benedict B69-153;* mesa s of housing area, 22 Apr 1961, *Blakley 4107;* sw portion of island, in gully on s escarpment ca. 0.5 mi w of Benchmark 688, 22 May 1985, *Junak & Vanderwier SN-142;* sw portion of island, in lower e fork of Cattail Canyon, 12 May 1992, *Junak SN-884;* se escarpment, on slopes above Sand Spit, 24 Apr 1989, *Junak SN-355.*

Nassella lepida (Hitchc.) Barkworth: ne escarpment, in first canyon e of living compound, 1 Apr 1992, *Junak SN-786;* ne coastal flats, in second canyon w of "L" Canyon, 23 May 1995, *Junak SN-1220;* ne corner of island, on n side of N Spur Canyon, between Beach Road and coastal navigation light, 24 Mar 1993, *Junak SN-1065;* ne coastal flats, ne of Beach Road, between N and S Spur canyons, 24 Mar 1993, *Junak SN-1056;* se edge of mesa, on w side of canyon just sw of Twin Towers (Building #186), 13 May 1992, *Junak SN-912;* se escarpment, near base of ridge above w end of Daytona Beach, 11 Mar 1993, *Junak SN-999;* s coastal flats, just e of Daytona Beach, on e side of first canyon e of Towers Canyon, 27 May 1992, *Junak SN-927.*

Nassella pulchra (Hitchc.) Barkworth: ne coastal flats, in first canyon w of "L" Canyon, just sw of Beach Road, 18 May 1995, *Junak SN-1208;* ne coastal flats, e of Beach Road, just s of first canyon s of S Spur Canyon, 24 May 1993, *Junak SN-1058;* s edge of mesa, just e of Jackson Hill, s of Jackson Highway, 8 Apr 1992, *Junak SN-838;* mesa, ca. 0.5 mi n of intersection of Beach Road and Monroe Drive, on e side of second large canyon w of airfield runways, 1 Apr 1992, *Junak SN-775;* se end of mesa, ca. 0.2 mi n of Peak 606, 9 Apr 1992, *Junak SN-861;* ne end of mesa, on n side of airfield runways, near Triangulation Station "Spur", just n of S Spur Canyon, 27 May 1992, *Junak SN-938.*

Parapholis incurva (L.) C.E.Hubb.: San Nicolas Island, 14 Apr 1940, *Kanakoff s.n.* (RSA 213457, SBBG 37739); nw end of island, ca. 1 mi ese of Seal Beach, 23-24 Apr 1966, *Raven & Thompson 20719;* nw portion of island, at e end of Red Eye Beach, just e of Benchmark 21, 19 Mar 1983, *Junak SN-19;* canyon from camp area to old sheep dock, 23 Apr 1961, *Blakley 4174;* artificial pond s of w end of runway, 12 Jun 1969, *Philbrick & Benedict B69-213;* s side of island, just e of base of Theodolite Road, 21 May 1990, *Junak SN-482;* s escarpment, in upper e fork of Twin Rivers drainage, 2 May 1991, *Junak SN-652;* in stream bed cutting old marine terrace above Jehemy Beach, 7 Apr 1966, *Foreman & Smith 179* (SBBG,UC).

Pennisetum clandestinum Chiov.: small canyon just e of barracks, ca. 0.5 mi e of Celery Creek, 11 Jun 1969, *Philbrick & Benedict B69-179;* ne coastal flats, along Beach Road near Coast Guard Beach, just w of Building #199 at Rock Jetty, 22 Feb 1990, *Junak SN-447.*

Phalaris aquatica L.: n coastal flats, in Celery Canyon just s of NavFac Road, 14 Jun 1995, *Junak & Stone SN-1261;* n coastal flats, along Beach Road at base of grade below airfield, 28 Mar 1991, *Junak SN-610;* ne escarpment, in upper portion of "L" Canyon, 17 May 1995, *Junak SN-1203;* mesa, along Beach Road at w end of airfield runways, 1 May 1991, *Junak SN-620.*

Phalaris caroliniana Walter: one locality, fertile flat by the sea, Apr 1897, *Trask 9* (MO,US); one locality, fertile flat 100 ft above the sea, Apr 1897, *Trask s.n.* (US 340204).

Phalaris minor Retz: dirt road above Corral Beach, 30 Jun 1978, *Wier & Beauchamp s.n.* (RSA 289058); n coastal flats, along Beach Road between Corral Harbor and w end of Tranquility Beach, 2 May 1991, *Junak SN-657;* n escarpment, near head of e fork of major canyon near e end of airfield runways, 21 Mar 1984, Junak SN-85; n side of Jackson Highway, e of Hill 907, 30 Jun 1978, *Wier & Beauchamp s.n.* (UC 1443151); barracks, between Celery Creek and artificial pond, 10 Jun 1969, *Philbrick B69-134;* ne end of mesa, along s side of airfield runways, just w of road to Building #121, 27 May 1992, *Junak SN-940;* e end of mesa, off e end of airfield runways, just s of Triangulation Station "Harbor", 18 Apr 1990, *Junak SN-461;* se side of island, near beach in vicinity of Dutch Harbor, 23-24 Apr 1966, *Raven & Thompson 20792* (RSA,SBBG).

Phalaris paradoxa L.: mesa, on flats just n of airfield, 27 Jun 2000, *Junak SN-1563.*

Piptatherum miliaceum (L.) Cosson subsp. **miliaceum:** n escarpment, in e fork of Celery Canyon, 14 Jun 1995, *Junak & Stone SN-1258;* nw end of mesa, next to small shed on n side of Tufts Road, just e of wells area, 29 Nov 1989, *Junak SN-429;* ne edge of mesa, at top of grade along Beach Road, 19 Nov 1991, *Junak SN-663;* sw end of island, at end of small spur road on s side of main road, opposite Building #275, 0.2 mi w of turn-off to Red Eye Beach, 9 Jun 1992, *Junak SN-944;* sw escarpment, on w side of large canyon just e of Drop-off Road, ca. 0.25 mi w of Benchmark 688, 19 May 1993, *Junak SN-1104.*

Poa annua L.: ne end of mesa, along n side of airfield runways, near n runway access road, 24 Apr 1989, *Junak SN-348.*

Poa secunda J.Presl subsp. **secunda:** ne end of island, near base of n escarpment, just e of "L" Canyon, 23 Mar 1993, *Junak SN-1029;* ne end of island, on n escarpment just ssw of Rock Jetty, 10 Mar 1993, *Junak SN-973;* ne side of mesa, in upper reaches of E Mesa Canyon, 11 Apr 1989, *Junak SN-331.*

Polypogon monspeliensis (L.) Desf.: two localities, in sand-swept arroyos, Apr 1897, *Trask 11* (MO); along waterways, Apr 1901, *Trask s.n.* (RSA 484492); w side of nw point of island, 22 Apr 1961, *Blakley 4144;* seepage bluff near sea, Coral [sic] Harbor, 22 Jul 1939, *Dunkle 8301;* small canyon se of Triangulation Point "Port", 21 Apr 1961, *Blakley 4085;* se end of mesa, just s of Building #121, 28 May 1992, *Junak SN-942;* sw portion of island, in lower Cattail Canyon, 12 May 1992, *Junak SN-879.*

Schismus arabicus Nees: e end of island, at foot of Divide Ridge, just w of Beach Road, 27 Jun 2000, *Junak SN-1556;* s side of Divide Ridge, overlooking Sand Spit, 22 May 2001, *Junak et al. SN-1709.*

Stenotaphrum secundatum (Walter) Kuntze: mesa, near fire station at intersection of Monroe Drive and Owens Road, 6 Sep 1989, *Junak SN-420;* mesa, near unpaved access road along s side of airfield runways, w of airfield terminal building, 1 May 1991, *Junak SN-619;* ne portion of mesa, on sw side of airfield runways, along dirt access road, ca. 0.1 mi w of passenger terminal building, 19 Nov 1991, *Junak SN-665;* mesa, on e side of air terminal building, 27 May 1986, *Junak SN-172.*

Triticum aestivum L.: ne coastal flats, in "L" Canyon, ca. 125 yards s of Beach Road, 17 May 1995, *Junak SN-1193;* ne side of island, in N Spur Canyon, 6 Sep 2000, *Junak SN-1588.*

Vulpia myuros (L.) C.C.Gmel.: n coastal flats, along Beach Road ca. 0.25 mi w of NavFac facility, nw of living compound, 24 Apr 1985, *Junak et al. SN-124;* n coastal flats, flats just w of lower portion of Celery Creek, 24 Apr 1985, *Junak SN-126;* mesa, just e of barracks, ca. 0.5 mi e of Celery Creek, 12 Jun 1969, *Philbrick & Benedict B69-203;* mesa, w of airfield terminal along s edge of runways, 21 Mar 1983, *Junak et al. SN-51.*

Vulpia octoflora var. **hirtella** (Piper) Henrard: n coastal flats, just w of mouth of first large canyon e of living compound, just s of Beach Road, 1 Apr 1992, *Junak SN-794;* ne escarpment, in e fork of Jetty Canyon, 11 Mar 1992, *Junak SN-704;* ne escarpment just w of Sand Spit, on n side of first canyon n of Divide Ridge, 11 Mar 1992, *Junak SN-712;* mesa, at living compound, se of Public Works building, 4 Apr 1989, *Junak SN-286;* mesa, ca. 0.2 mi n of intersection of Beach Road and Monroe Drive, w of airfield buildings, 1 Apr 1992, *Junak SN-773;* se escarpment, s of Peak 606 on ridge leading down to barge landing, 11 Mar 1992, *Junak SN-694;* se coastal flats, above w end of Daytona Beach, 1 May 1991, *Junak SN-632.*

RUPPIACEAE

Ruppia maritima L.: Thousand Springs, in pond, 7 Aug 1969, *Benedict s.n.* (SBBG 33752).

TYPHACEAE

Typha domingensis Pers.: spring 2.5 mi se of Vizcaino Point, s of Carrier Reef, 11 Jun 1969, *Philbrick & Benedict B69-202b;* canyon from camp area to old sheep dock, 23 Apr 1961, *Blakley 4175;* e end of mesa, at borrow pit just se of intersection of Monroe Drive and Beach Road, 23 Jul 1992, *Junak SN-951;* sw portion of island, in lower e fork of Cattail Canyon, 12 May 1992, *Junak SN-881.*

Typha latifolia L.: nw end of island, about spring ca. 2 mi ese of Seal Beach, 23-24 Apr 1966, *Raven & Thompson 20731;* Tule Canyon, just above pumping station, 11 Jun 1969, *Philbrick & Benedict B69-189;* small canyon se of Triangulation Point "Port", 21 Apr 1961, *Blakley 4093;* mesa, on n side of airfield, in drainage ditch opposite air traffic control tower, 3 Jun 1987, *Junak SN-212.*

ZOSTERACEAE

Phyllospadix scouleri Hook.: e end of Tender Beach, 24 May 1995, *Junak SN-1230;* Seal Beach, 23-24 Apr 1966, *Raven & Thompson 20693.*

Phyllospadix torreyi S.Watson: Red Eye Beach, intertidal zone, 28 Jul 1979, *Daily & Bromfield SNI-126.*

Zostera pacifica S.Watson: e end of Tranquility Beach, 7 Nov 2000, *Junak SN-1611.*

APPENDIX II. Excluded Plants.

Plants listed below have either (1) been reported by previous authors on the basis of misidentified specimens, or (2) the present author has not seen specimens which verify the occurrence of these plants on San Nicolas Island and they are itemized here pending confirmation. This list is not intended to be an exhaustive treatment of undocumented plant records from San Nicolas Island; both Foreman (1967) and Wallace (1985) have additional information on this topic. Plant taxa are arranged alphabetically by genus.

Agoseris heterophylla (Nutt.) Greene: Reported by Foreman (1967), who collected only a sterile plant. No flowering specimen has been located.

Agropyron repens (L.) Beauv.: Tentatively reported by Foreman (1967) on the basis of a misidentified specimen: *Foreman 111* is *Lolium perenne* L. according to Wallace (1985).

Asparagus officinalis L.: Known from a single collection on ne coastal terrace near Rock Jetty (just east of Building 122) by S. Junak in 1989; has not naturalized.

Chenopodium multifidum L.: Reported by Wallace (1985) on the basis of a misidentified specimen; *Kanakoff s.n.,* 18 Apr 1940 (RSA 429616) is *Lepidium lasiocarpum* Torr. & A.Gray var. *lasiocarpum.*

Corethrogyne filaginifolia (Hook. & Arn.) Nutt.: Reported by Foreman (1967) on the basis of a misidentified specimen: *Foreman 162* is *Ambrosia chamissonis* (Less.) Greene.

Cryptantha clevelandii Greene var. **clevelandii:** Reported by Wallace (1985); no specimen has been seen.

Cryptantha clevelandii Greene var. **florosa** I.M.Johnst.: Reported by Wallace (1985); no specimen has been seen.

Cryptantha intermedia (A.Gray) Greene: Reported by Foreman (1967) on the basis of a misidentified specimen: *Kanakoff s.n.,* 12 April 1940 is *C. traskiae* I.M.Johnst.

Eucalyptus sp.: Reported by Foreman (1967); not naturalized.

Euphorbia aff. **pulcherrima** Willd.: Reported by Foreman (1967); not naturalized.

Gnaphalium beneolens Davids. Reported by Wallace (1985), who cited a specimen at SBBG. No specimen has been located.

Hordeum depressum (Scrib. & JG.Smith) Rybd.: Reported by Foreman (1967) on the basis of a misidentified specimen: *Foreman 209* is *H. intercedens* Nevski.

Lactuca saligna L.: Reported by Junak and Vanderwier (1990) on the basis of a misidentified sterile specimen that was actually an immature *L. serriola* L.

Hainardia cylindrica (Willd.) Greuter : Reported (as *Lepturus cylindricus*) by Dunkle (1950). No specimen has been seen.

Orobanche uniflora L. subsp. **occidentalis** Ferris: Reported by WesTec Services (1978) on the basis of a misidentified specimen: *Beauchamp s.n.,* 1 July 1978, (UC) is *O. fasciculata* Nutt.

Pinus sp.: Several species have been planted on San Nicolas Island, but they have not naturalized.

Populus x **parryi** Sarg. Reported (as *P. fremontii*) by Foreman (1967); not naturalized.

Potentilla pectinisecta Rydb.: Locality data for *Potentilla* labelled as *Dunkle 8316* (POM 366605) appears to be an error. This plant was probably not collected on San Nicolas Island. *Abronia umbellata* Lam. was also collected as *Dunkle 8316.*

Salvia mellifera Greene: Planted on San Nicolas Island, but not naturalized.

Sanicula hoffmannii (Munz) Shan & Constance: Reported by Foreman (1967) on the basis of a misidentified specimen: *Raven & Thompson 20780* is *S. arguta* J.M.Coult. & Rose.

Scandix pecten-veneris L.: Reported by Foreman (1967) on the basis of a misidentified sterile specimen: *Foreman 230* is *Daucus pusillus* L. according to Wallace (1985).

Scirpus sp.: Collected as *Junak SN-1723;* not processed in time for publication.

Trifolium dichotomum Hook. & Arn.: Reported by Eastwood (1898) and by Foreman (1967) on the basis of a misidentified specimen: *Trask 37* is *T. albopurpureum* Torr. & A.Gray according to Wallace (1985).

Typha angustifolia L.: Reported by Foreman (1967) on the basis of two misidentified specimens: *Blakley 4148* and *Blakley 4175* are *T. domingensis* Pers.

Vitis vinifera L.: Reported by WesTec Services (1978); no specimen has been seen.

APPENDIX III.
Monthly Precipitation (Inches) at San Nicolas Island from 1948-2005

SEASON	JUL	AUG	SEP	OCT	NOV	DEC	JAN	FEB	MAR	APR	MAY	JUN	TOTAL
1948-49	---	---	.00	T	T	1.22	.77	.30	.99	.03	.07	T	3.38
1949-50	T	.00	.00	T	.58	2.05	1.29	1.33	.40	.61	.05	T	6.31
1950-51	.13	T	T	T	1.42	0.10	1.73	.97	.30	1.65	.02	.16	6.48
1951-52	T	T	.01	.17	.23	4.20	4.61	1.46	2.55	.16	T	.00	13.39
1952-53	T	T	.00	T	1.02	1.69	.63	.01	.17	.40	T	.01	3.93
1953-54	T	.02	.08	.00	1.57	.00	3.04	.58	1.31	.15	.08	.02	6.85
1954-55	T	T	.00	.00	.44	.32	1.40	.97	.33	.39	.03	T	3.88
1955-56	T	T	T	.01	.42	1.30	1.20	.08	T	1.01	.25	T	4.27
1956-57	T	T	.00	.34	T	.24	1.71	.77	1.58	.81	T	T	5.45
1957-58	T	T	.01	1.60	.20	.65	.91	2.69	3.12	1.98	T	.00	11.17
1958-59	.00	T	.09	.02	.01	.04	1.40	2.28	.00	.59	T	T	4.43
1959-60	.01	T	.05	.03	T	.16	3.22	1.75	.09	.55	T	T	5.86
1960-61	.04	.00	.00	.00	1.57	.04	.80	T	.11	.02	T	.01	2.59
1961-62	T	T	.01	.00	1.38	1.31	.79	5.45	1.49	T	.00	.00	10.43
1962-63	T	.00	.00	.00	.02	.03	.37	2.20	1.05	.98	.04	.15	4.84
1963-64	.00	.00	.44	.43	1.29	.03	.92	.05	.32	.10	.15	.07	3.80
1964-65	.00	.00	.00	.44	.62	.72	.83	.45	1.28	2.68	T	.01	7.03
1965-6	T	T	.04	.00	5.63	2.15	.88	.73	.03	T	.06	T	9.52
1966-67	.00	T	.07	.05	1.16	2.59	1.16	.07	.58	1.78	T	.01	7.47
1967-68	.02	T	.04	.00	1.69	.54	.19	1.02	1.10	.47	T	.00	5.07
1968-69	.02	T	.04	.22	.83	1.16	4.35	2.26	.34	.43	.07	T	9.72
1969-70	.02	.00	T	.02	3.10	.23	1.63	.93	.64	.01	T	T	6.58
1970-71	.00	.00	.00	.01	3.09	1.80	.21	.27	.19	.61	.48	T	6.66
1971-72	T	T	.03	.07	.10	2.56	.03	.10	T	T	.12	.02	3.03
1972-73	T	.08	.03	.05	1.91	.45	1.93	2.54	1.31	.01	.02	.03	8.36
1973-74	.01	T	.02	.23	1.10	.47	3.44	.27	1.41	.18	.01	T	7.14
1974-75	.01	T	T	.44	.15	.70	.08	.50	2.22	1.61	T	T	5.71
1975-76	.01	.01	T	.11	.16	.01	.00	2.76	.25	.55	.10	.13	4.09
1976-77	T	.02	2.21	.09	.01	.23	.56	.11	.78	T	1.63	.09	5.73

SEASON	JUL	AUG	SEP	OCT	NOV	DEC	JAN	FEB	MAR	APR	MAY	JUN	TOTAL
1977-78	T	.84	.05	.01	.03	3.33	4.21	6.15	3.99	.91	.01	T	19.53
1978-79	.00	.01	.57	.01	.98	1.06	7.40	1.08	1.71	T	.06	.02	12.90
1979-80	T	.04	.03	.35	.79	.09	5.19	5.65	1.29	.22	.05	T	13.70
1980-81	.10	T	.09	.01	.03	.55	.67	1.59	3.08	.03	.01	.00	6.16
1981-82	.00	.00	.16	.04	.99	.59	1.75	.75	2.77	1.58	.06	T	8.69
1982-83	.02	.00	.21	.69	1.78	.87	2.73	5.51	4.62	1.47	T	T	17.90
1983-84	.00	1.01	.96	1.19	3.12	2.68	.06	.06	.05	.18	.00	T	9.31
1984-85	.00	.01	.38	.21	1.65	5.96	.59	.26	.38	.05	T	T	9.49
1985-86	T	T	T	1.40	2.30	1.54	2.41	2.48	2.96	.16	.01	T	13.26
1986-87	T	.00	.79	.00	.92	.59	1.66	1.19	2.02	.03	.02	.00	7.22
1987-88	.04	.00	.00	2.88	.15	3.06	1.62	1.03	.11	.79	T	T	9.68
1988-89	T	.00	T	T	.41	1.01	.97	2.72	.98	.05	.14	T	6.29
1989-90	.00	T	.18	.03	.18	.00	1.24	1.19	.06	.54	.50	.00	3.92
1990-91	.00	.00	.26	.00	.71	.55	1.59	2.04	4.01	.10	T	.14	9.40
1991-92	T	.20	.01	.07	.09	1.73	2.35	2.89	3.35	T	T	.01	10.70
1992-93	14	.00	.00	.54	T	2.64	6.41	3.72	1.33	.00	.15	.25	15.18
1993-94	.00	T	.00	.20	1.00	1.70	0.40	2.21	1.74	.22	.17	T	7.64
1994-95	T	.00	T	.15	.97	.20	11.53	.34	5.24	.33	.35	.17	19.28
1995-96	.00	T	.00	.06	.07	1.87	1.18	2.53	.55	.44	.07	.00	6.77
1996-97	T	T	T	.34	1.93	2.44	3.03	T	.00	.01	T	T	7.75
1997-98	T	.01	.10	T	1.68	7.66	0.98	7.20	2.69	1.09	.36	.00	21.77
1998-99	T	.00	.00	.01	.46	.25	.70	.10	2.42	1.04	.00	.04	5.02
1999-00	T	.00	.00	.00	.12	.00	.53	2.75	1.21	.92	T	.01	5.54
2000-01	.00	.00	.13	.30	.04	.24	2.92	2.32	.20	.36	T	.03	6.54
2001-02	T	T	.00	.09	1.22	.85	.13	.11	.32	.01	.13	T	2.86
2002-03	T	T	.04	.05	1.68*	.25	.00	5.08	1.48	.99	.33	.00	9.90
2003-04	.00	.00	.02	.15	.82	1.75	.13	1.44	.00	.01	.00	T	4.32
2004-05	.00	T	.00	2.49	.45	4.29	3.35	1.25	.98	.41	.71	T	13.93
AVERAGE	.01	.04	.13	.27	.95	1.31	1.86	1.69	1.29	.52	.11	.02	8.21

*=estimated amount

Data source: Charles Fisk, Naval Base Ventura County

INDEX

Only the taxonomic treatments are indexed. The index includes both common and scientific plant names. Plant family names (both common and scientific) are listed in **BOLD CAPITAL LETTERS.** Common names for genera and plant taxa at and below the species level are listed in CAPITAL LETTERS. Synonyms and misapplied names are listed in italics.

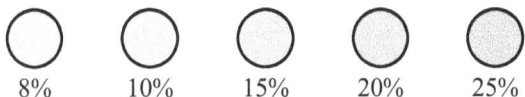

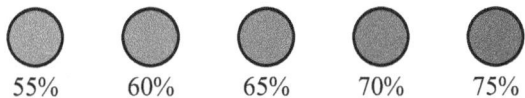

0%	2%	4%	5%	6%
8%	10%	15%	20%	25%
30%	35%	40%	45%	50%
55%	60%	65%	70%	75%
80%	85%	90%	95%	100%

www.ingramcontent.com/pod-product-compliance
Lightning Source LLC
Chambersburg PA
CBHW080608270326
41928CB00016B/2970